JASON BUSBY
MICHELE BOUSQUET

Mastering the Art of Production
with 3ds max™ 4

autodesk Press

THOMSON
LEARNING

Australia · Canada · Mexico · Singapore · Spain · United Kingdom · United States

Mastering the Art of Production with 3ds max™ 4
Jason Busby and Michele Bousquet

Business Unit Director:
Alar Elken

Executive Editor:
Sandy Clark

Acquisitions Editor:
James DeVoe

Development Editor:
John Fisher

Editorial Assistant:
Jasmine Hartman

Executive Marketing Manager:
Maura Theriault

Channel Manager:
Mary Johnson

Marketing Coordinator:
Karen Smith

Executive Production Manager:
Mary Ellen Black

Production Manager:
Larry Main

Production Editor:
Tom Stover

Art/Design Coordinator:
Mary Beth Vought

Cover art by Phillip Prahl,
Joe Clark, Joaquin Escondon,
and Jason Busby, Director.

COPYRIGHT © 2002 Thomson Learning.

Printed in Canada
1 2 3 4 5 XXX 05 04 03 02 01

For more information contact
Autodesk Press,
3 Columbia Circle, PO Box 15015,
Albany, NY 12212-5015.

Or find us on the World Wide Web at www.autodeskpress.com

All rights reserved. No part of this work covered by the copyright hereon may be reproduced or used in any form or by any means-graphic, electronic, or mechanical, including photocopying, recording, taping, Web distribution or information storage and retrieval systems-without written permission of the publisher.

For permission to use material from this text or product, contact us by
Tel   (800) 730-2214
Fax   (800) 730-2215
www.thomsonrights.com

Library of Congress Cataloging-in-Publication Data
Bousquet, Michele, 1962-
  Mastering the art of production with 3ds Max 4/Michele Bousquet, Jason Busby.
    p. cm.
  ISBN 0-7668-3470-0
  1. Computer animation.
  2. Computer graphics.
  3. 3ds Max 4 (Computer file)
  I. Busby, Jason. II. Title.

TR897.7 .B684 2001
006.6'96--dc21
                            2001047393

NOTICE TO THE READER
Publisher does not warrant or guarantee any of the products described herein or perform any independent analysis in connection with any of the product information contained herein. Publisher does not assume, and expressly disclaims, any obligation to obtain and include information other than that provided to it by the manufacturer.

The reader is expressly warned to consider and adopt all safety precautions that might be indicated by the activities herein and to avoid all potential hazards. By following the instructions contained herein, the reader willingly assumes all risks in connection with such instructions.

The Publisher makes no representation or warranties of any kind, including but not limited to, the warranties of fitness for particular purpose or merchantability, nor are any such representations implied with respect to the material set forth herein, and the publisher takes no responsibility with respect to such material. The publisher shall not be liable for any special, consequential, or exemplary damages resulting, in whole or part, from the readers' use of, or reliance upon, this material.

Trademarks: Autodesk, the Autodesk logo, and AutoCAD are registered trademarks of Autodesk, Inc., in the USA and other countries. Thomson Learning is a trademark used under license. Online Companion is a trademark and Autodesk Press is an imprint of Thomson Learning. Thomson Learning uses "Autodesk Press" with permission from Autodesk, Inc., for certain purposes. All other trademarks, and/or product names are used solely for identification and belong to their respective holders.

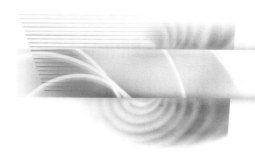

CONTENTS

CHAPTER 1 • THE JOB
The Job 2
    The Story 2
    Job Specifications 2
Paperwork 2
    Client Agreements 2
    The Schedule 5
Bidding the Job 7
    Good Workflow and the Importance Thereof 8
Preproduction 9
    Design Approval 9
    Sketches 9
    Scenery 12
    Storyboard 13
    Animatics 17
    Starting the Project 17

CHAPTER 2 • MODELING THE REMOTE CONTROL
Reference Images 20
    The Virtual Studio 20
Practice Makes Perfect 21
    Tutorial 2.1 Virtual Studio for the Remote Control ... 22
Box Modeling 33
    Sub-Objects 33
    The Anatomy of the Remote Control 35
    Tutorial 2.2 Remote Control Body 36
The Joystick and its Base 81
    Tutorial 2.3 Joystick Assembly 81
Finishing Touches 89
    Tutorial 2.4 Antenna and Hand Grip 90

CHAPTER 3 • MODELING THE ALIEN
Editable Poly Features 100
    NURMS Subdivision 100
    Mesh Control at Sub-Object Levels 101
    Tutorial 3.1 Alien Torso 102
    Tutorial 3.2 Arms and Hands 127
    Tutorial 3.3 Legs and Feet 144
    Tutorial 3.4 Finishing the Model 158

CHAPTER 4 • MODELING THE ROVER

Splines 162
  Creating Splines 162
  Vertex Types 162
The Surface Modifier 165
The Anatomy of the Rover 165
  Tutorial 4.1 Creating the Spline Network 166
  Tutorial 4.2 Surfacing the Rover 185
  Tutorial 4.3 Creating the Wheels 198
  Tutorial 4.4 Creating the Reactor 207
  Tutorial 4.5 Creating the Track Assembly 217
  Tutorial 4.6 Finishing the Rover 226

CHAPTER 5 • INTRODUCTION TO MAXSCRIPT

Understanding MAXScript 240
  Scripting Steps 240
  Accessing the MAXScript Options 241
  Writing a Script 242
  Tutorial 5.1 Basic Scripting Commands 243
The MAXScript Listener Window 244
  MAXScript Syntax 244
  Variables 245
  Tutorial 5.2 Variables 245
Arrays 247
The Script Editor 249
  Tutorial 5.3 A First Script 250
Functions 253
  Tutorial 5.4 Functions 254
Arguments 255
  Tutorial 5.5 Arguments 256
Program Flow 257
  The "If" Statement 258
  Loops 258
User Interface Design 260
  Tutorial 5.6 Custom User Interface 261
Tidy Programming Practices 270

CHAPTER 6 • TECHNICAL SETUP FOR THE ROVER

Scripting 272
The Rover Tracks 272
  Tutorial 6.1 Track Setup 273
Motion Capture 287
  Controllers 287
  MoCap/Controller Setup for the Rover 287
  Designing the Control System for the Rover 288
  Tutorial 6.2 Motion Capture for the Rover 289
MAXScripts to Control the Rover 299
  Improving Performance 299
  What Are We Scripting? 300
  Tutorial 6.3 Rover Control Scripts 300
Creating a User Interface 315
  Position Reset 315
  Motion Capture Playback 315
  Scripts to the Rescue 316
  Tutorial 6.4 Rover Control User Interface 317

CHAPTER 7 • SKELETAL SETUP FOR THE ALIEN

A Few Things You Should Know By Now 330
Character Animation 330
Character Rigs 331
   Tutorial 5.1 Alien Skeleton 332
Alien Control System 349
   IK Solvers 350
   Point Helpers 350
   Tutorial 5.2 IK Solvers and Point Helpers 350
   Tutorial 5.3 Hand Control System 364
   Tutorial 5.4 Foot Control System 373
   Tutorial 5.5 Mirroring and Completing the Skeleton 377

CHAPTER 8 • ALIEN SKINNING AND DEFORMATION SETUP

Skinning 390
   The Basics Of Skinning 390
   Skinning the Alien 392
   Tutorial 8.1 A Skinning Primer 393
   Tutorial 8.2 Skinning the Alien 399
Facial Expressions 420
   Morphing Basics 420
   Morphing with 3ds max 421
   Determining Facial Expressions 421
   Tutorial 8.3 Morph Target Setup 423
Eye Controls 429
   Tutorial 8.4 Eye Control System 429

CHAPTER 9 • ALIEN UI CONTROL SYSTEM

The User Interface 434
   Main Console 434
   Morph Targets 434
   Hand Controls 435
   Foot Controls 435
   Tutorial 9.1 Icons for User Interface 436
   Tutorial 9.2 Writing the User Interface 443

CHAPTER 10 • CAMERA TRACKING

Camera Matching 466
   Choosing a Location 466
   Shooting the Location 466
   Matching the 3D Camera to the Live Footage 467
   Tracking the Camera 467
Camera Matching Utilities 467
Reference Points 468
   Placing Fixed Points 468
   Measuring Distances 468
   The Right Tools for the Job 469
   3D Scene Points 470
   Tutorial 10.1 CamPoint Setup 470
   Tutorial 10.2 Camera Tracking 474
What to Do if the Camera Won't Track 482

CHAPTER 11 • TEXTURING AND LIGHTING

Texturing 484
    Materials 484
    Maps 485
    The Material Editor 485
    Tutorial 11.1 Texturing the Alien 486
    Tutorial 11.2 Texturing the Rover 494
The Battery Material 514
    Battery Textures 514
    Tutorial 11.3 Texturing the Battery 515
Lighting 523
    Types of Lights 523
    Shadows 525
    3D Light Limitations 526
The Set 526
    Global Illumination 526
    Tutorial 11.4 Lighting the Set 526
Catching Shadows with Materials 535
    Tutorial 11.5 Shadows for the Ground 535

CHAPTER 12 • ANIMATION AND RENDERING

Bringing Files Together 542
    Merging 542
    Cross Referencing 542
    Bringing our Scenes Together 543
    Tutorial 12.1 Putting Together the Final Scene File . 543
Time to Animate 557
    Storyboard Check 557
    Rover Animation 557
    Alien Animation 557
    Safe Frames 558
    Tutorial 12.2 Animating the Rover and Alien 559
Rendering 573
    Rendering for Post production 573
    Video Post 573
    Tutorial 12.3 Rendering the Shots 574

CHAPTER 13 • POST PRODUCTION WITH COMBUSTION

Introduction to Combustion 584
    Combustion Files 584
    Operators 585
Combustion User Interface 586
Rendered 3ds max Files for Combustion 589
    Our RPF Files 589
    Dealing with Shadows 590
    Tutorial 13.1 Footage Import 590
Contrast and Color Adjustment 593
    Contrast Adjustment 593
    Adjusting Colors 593
    Tutorial 13.2 Color Correction 594
Combustion Tracking Techniques 602
    The Tracker 602
    Tutorial 13.3 Tracking the Selections 604
Paint 607
    Tutorial 13.4 Painting the Footage 609
Finishing the Job 625
    Picking Your Battles 625
    Delivering the Job 625
    Backing up the Job 625
    Our Version of the Animation 626

INDEX 627

Preface

3ds max 4 is a PC-based software package from Discreet for modeling, texturing and animation. This book is designed to show you how to take a professional animation project from start to finish using **3ds max 4**.

This book also covers the use of Discreet's **Combustion**, a compositing and painting package fully compatible with **3ds max 4** and ideal for compositing 3D animation with live action footage.

In this book, we start with a premise for a real-life 30-second commercial and take you through preliminary sketches to modeling, texturing and animation in **3ds max 4**, then compositing and rendering the final frames for broadcast in **Combustion**.

WANTED: 3D ARTIST WITH PRODUCTION EXPERIENCE

There is a joke that goes around between animation instructors. A student takes his seat on the first morning of a three-day beginner's class on 3D animation. The instructor starts the lesson by showing the class how to make a sphere. The student wrinkles his nose in distaste and raises his hand. "I don't want to make a sphere," he says. "I want to make a fully articulated character with a skeletal structure, and animate it in a live-action environment with a custom user interface written from scratch. Can you show us how to do that instead?"

Now we can. In the past, when you wanted to know how to go about professional production with **3ds max 4**, the only choices were a long training program or figuring it out yourself through trial and error. Although a smattering of materials can be found on bookshelves and the Internet, we couldn't find a comprehensive reference that takes a new animator through the production process from beginning to end.

Why, if there is such demand, has such a book not been written before? Completing this book required several people with production, teaching and writing experience working together for several months. It was, to put it plainly, a lot of hard work to document each and every step with this level of detail. But we thought it was important to do it for our students and for you, so we could finally answer, "Yes!"

HOW TO USE THIS BOOK

This book contains detailed steps for creating a fully animated production of a 30-second commercial. The chapters are tutorial-based, explaining concepts along the way but mostly focusing on the how-to for production tasks.

The tutorials are intended to be done in sequence. Some experience with **3ds max 4** is helpful when doing the tutorials, but is not required.

WHAT'S INSIDE

The chapters in this book follow a progression. In the early chapters, no knowledge of **3ds max** is assumed. In later chapters the tutorials become more complex, with instructions based on material covered in earlier chapters.

- **Chapter 1** describes the commercial and the project in detail, covering pre-production tasks like scheduling, character design and storyboarding.

- **Chapters 2-4** deal with the modeling of each of the three objects in the scene — the remote control, the alien and the rover car. Each tutorial introduces new modeling techniques and gives detailed instructions on the creation of the object. The tutorials increase in complexity from one chapter to the next.

- **Chapters 5-9** make extensive use of MAXScript, the scripting language built into **3ds max**. **Chapter 5** starts with the basics of MAXScript, assuming no prior knowledge of its use. In **Chapters 6-9** you use MAXScript to build a custom user interface for controlling the alien and the rover car. You'll also set up a skeletal structure, IK controls and facial morph targets for the alien.

- **Chapters 10-12** deal with scene composition. Here you'll track the camera motion from the live footage, apply materials to objects, and set up lighting. You'll also animate the objects in the scene, an easy task with the advanced controls created with MAXScript in previous chapters.

- **Chapter 13** takes you through the compositing process with **Combustion**. Here you'll correct the color on the live footage, paint out tracking markers and render the final animation.

WORKING WITH THE CD

This book comes with a CD containing source files, scenes and the final animation. For tutorials that call for a scene created in a previous chapter, you can use your own scene or load it from the CD. The CD folders are structured as follows:

- **Movies**   Background plates and final animation
- **Scripts**   MAXScripts and supporting images
- **Scenes**   **3ds max 4** files
- **Textures** Bitmaps used in tutorials

To do the tutorials in *Chapters 10-13*, you must copy the file *low-res footage.zip* in the *Movies/Background Plates* folder from the CD to your hard disk. Then you must unzip the file so you can use them in the tutorials. These files, which contain the digitized live footage for compositing, take up approximately 700MB of hard disk space.

It is not necessary to copy the remaining files from the CD to your hard disk. However, it is recommended that you set up a folder structure on your hard disk similar to the one on the CD for saving the files you create.

To see the final animation, view the file *Ionz.avi* in the *Movies/Final* folder.

SUPPORT

If you have questions about tutorials, CD files or any aspect of this book, point your Internet browser to **www.The3DBuzz.com/Mastering**. There you can ask the authors questions and find additional resources and tutorials related to this book.

You can also visit **www.AutodeskPress.com** for information on other books in the Autodesk Press catalog.

ABOUT THE AUTHORS

Jason Busby is the Director of Animation at The Renaissance Center in Dickson, Tennessee, the largest 3D animation training facility in the world. Jason oversees commercial projects in The Renaissance Center's Production Department in addition to teaching as a certified instructor on **3ds max**, Maya and Softimage|XSI. The animation in this book was produced at The Renaissance Center by Jason and his production team. This is Jason's first book.
The Renaissance Center www.rcenter.org

Michele Bousquet is an animator and video producer at Many Worlds Productions in southern New Hampshire. She began her production career as a video editor in 1985, and got into animation in 1990 with 3D Studio DOS R1. Two years at the Australian Broadcasting Corporation and nine years of freelancing later, she currently produces animation and live programming for cable television. Michele, a certified **3ds max** instructor, has previously authored three books on the software.
Many Worlds Productions www.maxhelp.com

ACKNOWLEDGMENTS

Although there are only two names on the cover of this book, it would have been impossible for us to have done all the production and documentation all by ourselves in any reasonable period of time. Fortunately, we had several talented production artists on our team that gave their time and efforts to help us see it through to the end.

Angie Ramsey has been one of the most valuable assets to this entire project. Angie performed numerous technical edits, tirelessly going over each tutorial many times to help us make them as perfect as possible. Her insights on ways to make the tutorials easier to follow were invaluable. If you find a tutorial anticipates all your questions, that's her work talking. Angie, we will never be able to thank you enough for all your input and the time you devoted to this project.

Phillip Prahl, a former **3ds max** student and now one of the most valued employees at The Renaissance Center, was a vital asset to our production team for his technical abilities, attention to detail, and above all his untiring dedication to the job. We want to thank Phillip for all the nights and weekends he gave up to help see this book through. His help in developing the camera tracking, rendering, Combustion, and particularly the scripting tutorials were more than we could ever ask. Without Phillip, the delivery of this book would have been so far off schedule that you might not be reading it now.

David Aguilar, another former student, is now the lead character animator at The Renaissance Center. David's drawing, modeling, and animation skills make him a true gem for any production team. David developed the concept sketches for the alien and rover, the storyboards in *Chapter 1* and the original alien model. He also animated the alien in the final production. We will never be able to thank him enough for all the nights and weekends he gave up to help us make this book happen.

We would like to also thank Joe Clark, one of the most impressive modelers we've ever seen and, we're happy to say, another former student. Joe helped with the cover art, rendering issues and camera tracking, and played an important role in making us see the flip side of what we were doing. His spectacular debates with Phillip were a bonus.

Another Renaissance Center student-turned-employee, Dustin Parsons, deserves a big thanks for his contributions to the Combustion tutorials. Dustin gave up many hours of his personal time to help us out.

We would also like to thank Joaquin Escandon for developing and modeling the original rover, and for working so hard on the cover art.

Another former student and now webmaster at Maxhelp.com, John Paul Casiello, gets our thanks for correcting and adding important details to tutorials and anticipating user questions based on his extensive experience with **3ds max** beginners.

A thanks goes to Jenny Good for layout assistance and proofreading, for working all those odd hours and weekends to accommodate our crazy schedule, and for making us laugh when we needed it most. Additional thanks to David A. Drew Jr. for layout assistance, and to Su Falcon for helping out with proofing.

And what book would ever get published without the tireless efforts of the publishing team? We would like to thank Jim Devoe, John Fisher, Jasmine Hartman, Thomas Stover and David Porush at Thomson Learning for their continued goodwill and support despite having to work with authors like us all day long.

From Michele: A special thanks to my co-author Jason who first talked me into co-authoring this book with him, then went ahead and wrote the bulk of the content himself. Jason, you make me look so good! And last but not least, a huge, enormous can't-make-it-big-enough thanks to my wonderful husband David, who provided a warm and welcoming nest when the computers finally got turned off for the night.

From Jason: A special thanks goes out to my mom and family. Thank you to Angie, Heather, and Meagan for hanging in there and providing a constant stream of support. I'd never had made it without you. You're the greatest and I'll love you each forever. It'll be tough making up for all the lost time, but I'll find a way to do it! And thank you to my mom, Belinda, for your constant encouragement every time something looked like it was going wrong. All our telephone conversations were a blessing in helping me keep my drive going strong. And finally, I'd like to thank my co-author Michele for inspiring me in my career from the first day I met her.

The Job

Congratulations! You've just been awarded your first contract for a complete animated commercial. The client likes your reel, and has given you a general budget range and an overall idea for the commercial. The finished piece will air on prime time television in just a few weeks, giving your company the exposure you've been looking for. It's your big break!

Getting here has been an uphill battle, but it's been worth it. You think of all the hard work to get the equipment and software up and running, plus the endless hours of polishing your reel to showcase what you're capable of doing. You just needed to get your name out there, and here's your opportunity.

The project you just won is a tough one — not that it's something you can't handle, but time is against you. The commercial is slated to air in only three weeks.

Generally speaking, this is the way it is; clients expect animation projects to be delivered in shorter and shorter amounts of time. Regardless of the time given, experienced animators know that time is always against them. Even when you think you have plenty of time to complete a project you still manage to find yourself working around the clock during that last week, trying to outdo your last project with even more spectacular work.

The project in this book assumes you're working alone on the job, with perhaps a few fellow artists pitching in now and again to help out.

This chapter is all about the preparation that must take place before modeling and rendering can be started.

THE JOB

The commercial's purpose is to promote a new battery called Ion Z. This battery uses a completely new technology that makes it last up to ten times longer than an ordinary battery.

Your client's target market is people who use batteries for toys and games. They want an "out of this world" theme to stress the incredible longevity of the batteries, and have suggested that the commercial take place on a distant alien planet.

THE STORY

Through discussions with the client, you decide on a story line for the commercial. The star of the commercial will be an alien "child" playing with his remote control car. When the car stops working, the alien first suspects the remote is the problem, then tries the car and reveals that it's using off-brand batteries. The alien screams for his mother as the screen fades to black. A voiceover asks, "Got the wrong batteries again?" The Ion Z products and logo appear as a voiceover tells you Ion Z is the way to go for longevity.

JOB SPECIFICATIONS

The commercial is intended to be somewhat comical while getting the message across. In addition, the client wants the entire commercial to look completely photorealistic. To television viewers it will be obvious that it couldn't possibly be real, yet it will look so real that viewers will not be able to help watching it and commenting on it to their friends. Both you and the client are excited about this challenge and its possibilities – the client for the soaring battery sales, and you for the attention you'll get from this project.

As for the design of the alien, remote control, and car, the clients have only general descriptions to offer for their appearances. The details are up to you.

PAPERWORK

When starting a new job, it is very tempting to simply jump in and start modeling. Before you spend (and possibly waste) your time on production tasks, your paperwork must be in order. This includes a contract or agreement with the client, a schedule and a budget. You'll also need to get your office in order in preparation for efficient workflow.

CLIENT AGREEMENTS

Before starting any sort of large-scale animation project, you should have every last detail of the client agreement ironed out. Animation is an incredibly involved art with many variables that must be considered. Sometimes the slightest misunderstanding between the artist and the client can result in days or even weeks of lost work.

Overall Agreement

The first document is the overall agreement with the client. So many animators skip the important step of making this agreement as comprehensive as possible, including every aspect of the job. It can take several hours or even a few days to come up with the complete agreement document, but this time is well spent. Disagreements and disputes with the client, perhaps even the cancellation of the job, can all be avoided with good, solid agreements in writing.

The agreement need not be worded in legal terms. The documentation's purpose is not to support potential lawsuits, but to record agreements between you and the client in case someone forgets or misinterprets what was verbally stated. Clients can and will forget what they originally told you, and simply pointing to an agreement resolves disputes in a courteous and professional manner.

The overall agreement should include the following items:

- Project description
- Schedule
- Budget
- Important deadlines and milestones (often tied to payments)
- Payment amount and terms
- Delivery method (videotape or digital format)

It is customary to ask for 1/3 to 1/2 of the total budget before starting a job. You can also ask for payment in full upon delivery of the final product.

When creating this agreement, keep in mind the specifics of the job, including:

- Will the client supply logo art, or will you create it? How strict is the client about colors, fonts and letter sizes in the logo? What trademark and copyright symbols must be used?
- What music will be used? Who will supply it, and who will pay any licensing fees?
- What delivery method will be used? For example, if the job is for a commercial to be aired on television, what video format should the job be delivered on? Does the client want VHS copies as well, and if so, how many?
- Will the client allow you to use the commercial on your own promotional reels when the job is done? How about for competitions and festivals?

Agreements During the Job

There has yet to be an animation agreement that didn't require changes during the production process. The best approach to a clear understanding is to have everything documented and signed. When the client asks for something, document it. Be as descriptive as possible and then have the client sign it after they agree to what they need completed. This way, if there are any questions later, you can always pull out your documentation and show them what they originally agreed to.

Clients will take you more seriously if every detail is documented. In addition, documenting everything will ensure you won't forget any important details over the course of the project.

The Tightrope of Client Happiness

Giving the client what they want while keeping the job within the schedule and budget is something of a tightrope walk. Engaging in lots of open communication and coming up with firm, written agreements that everyone understands are the keys to keeping everyone happy, including yourself.

An important part of the initial agreement is determining when and how the client can look at work in progress and recommend changes. In your documentation, be sure to include specific approval intervals and guidelines for client comments.

The client must evaluate and determine changes in a timely manner in order to keep the schedule. Make sure they know that swift turnaround on approvals is important if the job is to be done on time. In other words, if the agreement calls for the client to look at and approve a segment in two days, but it takes five days to decide that they want a green landscape instead of red, this could cause major problems in your own schedule. Make sure everyone agrees on the approval intervals and timetable.

You'll also need to determine what the client should see. Almost anyone can understand rendered stills with full texturing and lighting. Preliminary previews of animation are excellent for approvals, as long as you educate the client about the limitations of this media, such as the smaller size and limited lighting and materials.

You might also run into a situation where some distant third party with no involvement in the approval process has to okay the job before it can be considered done (and you can get paid). For example, suppose all your direct contact is with the client's Marketing Team. They inform you that the Vice President has to approve the commercial, but they don't want to show it to him until it's all finished. Does this mean the Vice President can dictate changes? If so, you'll need to point out the dangers of this situation and work with the client to come up with a solution.

You could agree that when you complete the job to the satisfaction of the Marketing Team for the current fee and schedule, then you'll make a new, separate agreement with a schedule and billing for the Vice President's changes. You could also insist that they show the storyboard to the Vice President to ensure he likes the general idea.

Make sure the client knows that any major changes they request near the end of the job will be considered out of the original job scope, which means there will be additional charges for extra hired help (or compensation for your lack of sleep). Also make it clear that you, not the client, get to decide what a "major" change is.

Of course you want to keep the client happy, but what might seem like a minor change to the client ("Can you make the character purple instead of blue?") often means rerendering the entire project. Clients sometimes don't understand why rerendering takes so long, especially if they're accustomed to working with print media where only one picture needs to be changed. You should state in the client agreement that you'll give them a cost estimate for each major change requested so they can decide if it's worth it. Clients don't always ask for major changes, but they do it often enough to warrant this clause.

These restrictions can be communicated in a patient and pleasant manner. The client merely needs to be educated as to what's acceptable and what's not within the job scope.

If the client seems bent on making the job impossible for you to complete, you should find this out before making any agreements and simply walk away. But most clients aren't trying to make your life miserable (although it may seem that way sometimes!). Usually it's just a matter of communicating with the client and working out a solution.

In short, it's up to you to make sure you get a fair shake while still delivering what the client wants. Be fair, but don't leave yourself open for catastrophy. The client will rely on you to spot any potential trouble areas so they can be avoided.

THE SCHEDULE

Scheduling is one of the most important and challenging tasks in any job. A production schedule should outline every single task that will need to be completed and its expected completion date. A good schedule allows you to quickly assess the progress of the production.

The production schedule should be given to the client as part of the signed agreement. If there are any additions or changes throughout the project (and there typically are), the schedule should be updated and signed off by both you and the client.

Some elements might not need client approval. For example, the creation of programs used internally for animation is not the client's concern. To avoid confusing the client, you can make two versions of the schedule, one for you and one for them. Just make sure you update both when changes are made to the project.

Here's the production schedule for the Ion Z battery commercial. The dates haven't been filled in yet, but the tasks are all there.

PRODUCTION SCHEDULE   Client: ION Z   Job: ALIEN BATTERIES

	Est. Comp. Date	Date Comp.	Client Signoff
Preproduction			
Character Concepts Object Concepts Scenery lockdown Storyboards Level I Animatics Filming background video			
Modeling			
Modeling alien Modeling remote control Modeling rover			
Technical Setup			
Character setup Rover setup Camera tracking Scene matching			
Lighting and Materials			
Lighting Materials for alien Materials for remote control Materials for rover			
Animation			
Controller scripts Level II Animatics Animating rover Animating alien			
Rendering/Postproduction			
Test Renders and Tweaks Rendering Color correction Masking trees Assembling edit			

Figure 1.1 *The production schedule form*

For the Ion Z project, the client has already provided the voiceover, music and logo elements to be used for the commercial, so the creation or procurement of these elements is not part of our schedule.

BIDDING THE JOB

In your bid for the job, you will naturally include the price of your time plus enough to cover office overhead such as telephone charges and rent.

Be sure to include other items needed for the job. Some you'll have already, while others will have to be purchased or rented.

It's important to take into account everything required to complete the job to ensure you finish on time and make a profit.

For the Ion Z project, you know you're going to need the following items:

ITEMS NEEDED	
Video camera	For shooting the background footage. You can rent a high-end digital camera just for the job, but make sure you have the ability to digitize the footage.
Computer graphics workstation	A little extra memory or hard disk space might be required.
Additional computers for network rendering	Just about any computer on your network can be pressed into service for network rendering. See **CD Tutorial 4** in the **Tutorials** folder on the CD for information on how to set up network rendering.
Digitizing hardware/software	For digitizing the background footage.
Discreet **3ds max 4**	Of course!
Paint program for textures	Adobe Photoshop™ is the authors' program of choice.
Composting software	Discreet **Combustion** integrates with **3ds max 4**. The project in this book includes the use of **Combustion** for compositing.
Output equipment	The client requires output to Beta SP video format. Rental of a Beta SP video deck for one or two days must be included in the budget.
Various media for input, output and backups	Videotapes, backup tapes, Jaz or Zip disks, CD-ROMs. The costs of these items add up quickly!
Organizational supplies	Whiteboard or bulletin board, file folders, tape measure, calculator, string, sticky tape, pens and paper. Start a new binder or folder for the project where you can store the contract, your notes, client feedback, photographs, sketches and other materials.

Figure 1.9 *Items needed for the job*

Keep all your software CDs labeled and stored in a safe place that's easy to access. Software can and will fail, usually at the worst possible moment. There's nothing worse than finding you need to reinstall Photoshop at 3am on a Saturday night just before a deadline, then suddenly remembering you loaned your Photoshop CD to your cousin Martha in Illinois.

Also make sure you have a method for delivering artwork to the client for approval. If you plan to use an FTP site for file transfer, secure a backup site in case the primary one goes down. If you'll be using a courier such as UPS or FedEx, make sure you know where the nearest offices are and their latest pickup times. Write the client's delivery address in block letters on a large piece of paper and tape it to the wall, or better yet, fill out the courier forms ahead of time and keep a stack handy.

Making sure you have everything is very important. If you discover that you left something out, you'll end up paying for it out of funds that would have been profit.

GOOD WORKFLOW AND THE IMPORTANCE THEREOF

Developing a good workflow is another important step in a successful production cycle, especially in this situation where you will get to play all the roles on your team. On this project, you'll be the storyboard artist, modeler, animator, system administrator (in charge of backups and disk/folder management), and technical director (managing and putting together all the elements of the job).

In juggling all these tasks, a good workflow is essential. By a good workflow we mean putting in place and following an administrative system that makes sense for the job and your work habits.

As a one-man (or one-woman) team, it can be very easy to forget to do particular tasks. You might think you'll remember tomorrow, but often you won't, especially if you've been working long hours and your brain is starting to drag a little. A good workflow put into place and actually followed will keep all your tasks in order and make it much easier to make deadlines.

Some examples of good workflow:

- Start a binder or folder for the project. Keep all client communications and other important documents in the binder.
- Use a whiteboard, blackboard or bulletin board to keep track of the steps you've taken and how much farther you have to go to complete tasks. Use the schedule to determine the content for the board. The board is also a good place to post the storyboard.
- Set up an intelligent hierarchy of folders on your main workstation's hard disk. Never just dump files in any old folder. We promise that you'll forget where they are. Not that we've ever done this or anything.
- As you work, keep a written record of filenames, where they're stored and what's in them. Name files intelligently, but keep a written record too. An inexpensive wirebound notebook is excellent for keeping notes of this nature.

- Back up your files, and not just to another network machine. Save the files to independent media such as CD-ROMs or Jaz disks. If all your computers crash, you should be able to restart the job with little delay on another computer with your backups alone.

A good workflow made into a habit makes every aspect of production easier and smoother. Before starting preproduction, put your workflow tools in place.

PREPRODUCTION

The contract has been signed and the schedule approved, and you're ready to start. But before you jump in and start modeling, you have to complete the first step: preproduction.

Since you've worked out your schedule so carefully, you know the steps of preproduction and their sequence. Let's go!

DESIGN APPROVAL

It is very important to make sure you lock down the exact look of the characters, objects and scenery on paper before starting to model. You would think that this was so obvious that no one would ever forget to do it, but the problem isn't one of forgetfulness. You might have a very clear idea of what the characters and objects will look like, but the client needs to know too.

Some clients are offhand about designs, saying that they trust you to do a great job. It's very tempting to just go ahead and start the project based on the client's blind faith, but as often as not, this leads to disaster. We have seen projects like this, where the animators did not insist the client look at the designs and give approval. During later phases of production, the client would see the models for the first time and call for a halt to all work until drastic changes were made. There is no surer way to waste your valuable time than to have to start the design and modeling processes all over again.

It's your responsibility to get the client to look at the designs and sign them off.

SKETCHES

Knowing how to draw is a tremendous asset in the animation profession. An experienced pencil artist can sit with the clients, interview them, and create quick sketches that will help everyone visualize the final work. If you can't draw, you can bring an artistic friend into the project to help with the sketches.

Sometimes you'll have to create additional sketches or even a clay model to get the client to understand the character design. The small amount of extra time spent up front will save you much time later on.

The sketches you show the client are not necessarily the ones you'll use for modeling. Later on, you'll draw specific types of sketches for use as modeling references.

Character Design

Character sketches should show the front and side views of the character, plus a few sketches in poses similar to those in the final project. This approach helps the client get a feel for the character in action.

Include written notes with details such as how the body will move and what types of accessories it wears, particularly if these details aren't obvious from the sketches.

Figure: 2.2 *Concept sketches of alien character*

During an early client meeting you draw some rough sketches based on how the client describes the alien child character. After a few tries, you hit on something that everyone likes. The client's representatives murmur their approval. You take the rough sketch back to your office and begin several high-quality drawings.

After more back and forth with the client, the final design is approved. Good job!

The client is very pleased with your ability to convey the childlike aspect of the alien. After all, who knows what an alien child really looks like? But an experienced artist knows that children's heads, hands and feet are large in comparison to their bodies, and small children have superior flexibility and little pot bellies. You have included all these elements in the alien drawings to make it look like a child. A big floppy nose tube on the character's head contributes a comic element while giving you another tool for expressing the character's emotions.

Inanimate Object Design

Now that we know what the alien will look like, we need to get final approval for the other objects in the scene. In this project, the alien will be interacting with two objects: the remote control and the rover. As with the alien, it is important that these two objects look exactly the way the client has visualized them. So it's back to the drawing board for preliminary sketches, then client feedback and approval.

After showing the client a few designs, they discuss them and finally agree on the ones they like. The approved sketches are shown in figures that follow.

Figure 1.3 *Concept sketch of remote control*

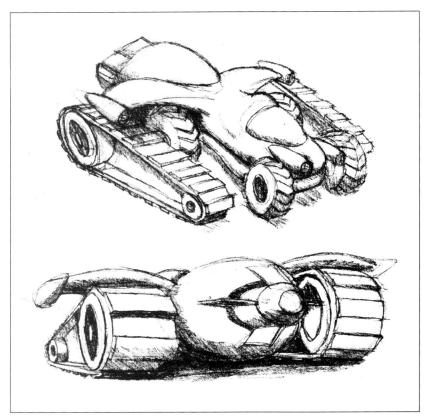

Figure 1.4 *Concept sketches of rover vehicle*

SCENERY

With all of the designs in place for the character and the two objects he will be interacting with, it's time to discuss the environment where everything takes place. You know that the client is looking for photorealistic quality, and time is against you. In this case, you decide that the best course of action is to film the background and give it an alien planet feel in postproduction.

Now, where can you find this soon-to-be-alien landscape? You hop into the car for a pleasant Sunday drive, taking your 35mm camera with you. After a few hours of location scouting (this is work?) you stumble upon a rock quarry.

Figure 1.5 *The near-perfect rock quarry*

The landscape is just about perfect. You snap a few photos for use in later planning steps, write down directions to the quarry, and return home.

STORYBOARD

A *storyboard* is a pictorial representation of an animated sequence on paper, similar to a comic strip. The purpose of the storyboard is to communicate the final animation so everyone knows where it's going before the real work is started. By its very nature, the storyboard forces you to fill in any gaps in the story and work out the basic composition, staging and scale of scene elements.

The storyboard is the most vital part of the preproduction process. A good storyboard saves time and money by giving the client a first hand look at the completed animation, albeit in rough form. At this time, any and all client feedback and requests can be incorporated into the project before you get into the long process of modeling and animating.

You don't have to be a professional artist to create a storyboard, but a good hand at drawing will help you tremendously in getting across your ideas.

Creating a Storyboard

1. Create or find storyboard paper.

   Storyboard paper can be purchased at large art supply stores. Alternatively, you can make your own by drawing 3 to 6 rectangles on a page with space underneath or between each for scene notes.

2. Make a rough outline of your sketches and the order in which the scenes will appear.

   In general, you should plan on 3 to 5 pictures per shot, more if there is a lot of camera or character motion.

3. On the storyboard paper, draw a picture for each point on the outline with light pencil.

   Include brief notes on what is occurring in the scene. Using a variety of camera angles and different viewpoints will also give you a better understanding of the animation. Describe the camera movements on the storyboard with arrows and cues.

4. Go over the rough draft and perfect it.

   It's also a good idea to show the rough draft to an uninvolved person to ensure the storyboard communicates the animation sequence effectively.

5. When the storyboard is in good shape, go over it with heavier pencil or black marker.

6. Deliver the storyboard to the client.

7. Make any changes necessary to get client approval.

   Take care not to include shots that will make the job difficult or impossible. For example, the client might want to add an overhead shot of the scene, but if this is going to require you to rent a helicopter, you'd better talk the client out of it. Unless, of course, the client is willing to pay for it. If so, then go for it! Helicopter rides are great fun.

Here is the final storyboard for the Ion Z battery commercial.

Figure 1.6 Storyboard page 1

Figure 1.7 *Storyboard page 2*

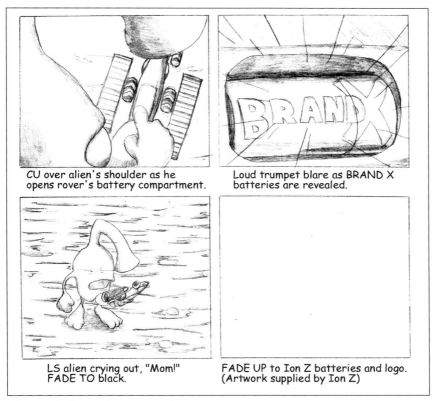

Figure 1.8 *Storyboard page 3*

ANIMATICS

An *animatic* is a series of individual frames edited together to play at the same pace as the final project. An animatic goes a step further in giving both animator and client an idea of how the final sequence will look.

The simplest animatic consists of the storyboard panels, each one held for the length of the shot. Any music or voiceover is also added.

A more comprehensive animatic can be created by panning and zooming on the frames where called for in the storyboard. You can also use scans of the scenery as background images.

By the time the animatic is approved, both you and the client should have a pretty complete idea of what the final project will look like.

STARTING THE PROJECT

The most important things to keep in mind while working on the production are time and money. The goal is to deliver high quality final animation that is both on time and at or below budget. With this in mind, let's begin the production!

Modeling the Remote Control

We'll start our project by making the remote control. The creation of this object will serve as an excellent introduction to complex modeling. The tools used to create the remote control will give you a sound base for additional techniques later in this book.

TECHNIQUES IN THIS CHAPTER

In this chapter, you will learn how to:

- Set up reference images in the scene as a modeling aid.
- Use box modeling tools to sculpt a complex 3D object.
- Work with sub-objects (vertices, polygons and edges).
- Model one half of an object, then mirror it to the other side and attach it without leaving any seams.
- Smooth the appearance of an object.
- Lathe a profile shape to make a 3D object.
- Align an object with a polygon or face.
- Create objects with lofting and change their profiles easily and quickly.

Don't be concerned if some of these concepts don't make sense just yet. Each topic and the steps to use each technique are clearly explained as we go along. Suffice it to say that by the end of this chapter you will know plenty about modeling, and will be able to create all kinds of complex objects on your own.

TUTORIALS IN THIS CHAPTER

The modeling of the remote control is divided into several tutorials, each of which has many subsections. The tutorials are designed to be done in sequence.

- **Tutorial 2.1 Virtual Studio for the Remote Control** will show you how to set up reference images to aid in the modeling process.

- **Tutorial 2.2 Remote Control Body** makes up the bulk of this chapter, showing you how to use box modeling techniques to create the remote control body from scratch, step by step. A variety of tools are used, many of which will come up in later chapters when modeling the car and alien.

- **Tutorial 2.3 Joystick Assembly** uses the Lathe modifier to create the joysticks, and now-familiar box modeling techniques to form the spherical bases.

- **Tutorial 2.4 Antenna and Hand Grip** uses lofting and deformation to create these two elements.

Each tutorial will introduce new modeling techniques and practices. The focus is on not only on creating a great scene, but also on learning tools that you can use to create your own scenes.

REFERENCE IMAGES

The first thing you'll need when modeling is some sort of reference picture or real-life example of what you plan to model. This is absolutely essential. Going into a modeling assignment without an example to look at while working can be difficult if not disastrous. Please don't be like so many students we've had who have disregarded this advice only to learn the hard way. Refusing to use references does not make you a "real" artist, it makes you a struggling artist who creates models that somehow don't look right. Even seasoned professionals use references all the time.

Pictures and drawings are more readily available than real-life objects or sculptures, so these are the types of references you'll use most often. If you're able to draw reference pictures yourself or if you have someone around who can draw them for you, then you're in great shape. A photograph of the object or a similar object is just as useful. If you can't get drawings or photographs, you can use encyclopedias, picture books and the Internet as a source for references.

There will be times when you have to settle for a picture that isn't entirely what you had in mind, such as a very rough sketch. This is still far better than nothing. Always start with one or two photographs or drawings.

THE VIRTUAL STUDIO

Once you have your reference pictures, the best approach is to create a *virtual studio*. This setup uses reference materials displayed as viewport backgrounds or applied to flat objects in the scene.

The model is then drawn on top of the reference pictures, allowing you to use the pictures both for modeling and for checking the accuracy of your work.

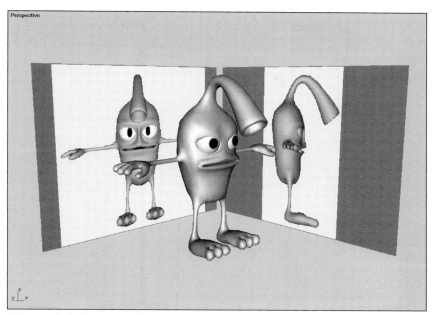

Figure 2.1 *Virtual studio in practice*

A common setup for a virtual studio uses two drawings or photographs of the object to be modeled, usually a side view plus a top or front view. The two images are mapped onto two intersecting planes. You then swap back and forth between views to build the model.

Two reference drawings for the top and side views of the remote control are supplied on the CD. These drawings will be used to set up a virtual studio for modeling.

PRACTICE MAKES PERFECT

A virtual studio eases the modeling phase, but don't think this technique will automatically make you into the next modeling genius. That will come only with a lot of practice and hard work.

Modeling also takes a lot of patience. If you don't have this attribute, find it! You can't expect to sit down and finish modeling something with any amount of detail in just a few minutes. We've all known artists who have worked on a complex model hour after hour, week after week.

If your first or second effort doesn't quite look right, keep working at it until you get the result you want. Hours of practice will reward you with skill and speed.

TUTORIAL 2.1 VIRTUAL STUDIO FOR THE REMOTE CONTROL

The first step in making our virtual studio is to create and place the planes to hold the reference images. For this task, you will use two images from the CD, shown below. These images show the top and side of the remote control, respectively.

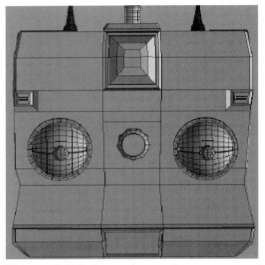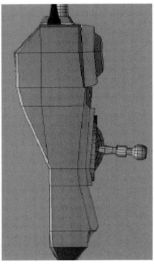

Figure 2.2 *Reference images for remote control*

Creating a Plane

A plane is a flat object similar to a box. Although a plane has no actual thickness, it is shaded and rendered like a 3D object. Planes make excellent objects for holding reference images for a virtual studio.

1. Reset **3ds max 4**.
2. Activate the **Top** viewport and press the **<G>** key on the keyboard.

   This turns off the grid in the Top viewport to make our scene easier to work with.

3. On the **Create** panel, click **Plane**.
4. In the Top viewport, click and drag to create a plane of any size.
5. With the newly created plane still selected, go to the **Modify** panel. Set the following values on the Parameters rollout:

Length	300
Width	300
Length Segs	1
Width Segs	1

Figure 2.3 *Plane*

6. At the top of the **Modify** panel, enter the name **Top_Reference** for the object.

    We want to place the plane so its center is at the 0,0,0 position where the three viewport grids intersect. We'll use the Transform Type-In dialog to accomplish this. This dialog is accessed by right-clicking on the selected transform button, such as **Select and Move**.

7. Click **Select and Move** on the Main Toolbar, then right-click **Select and Move**. The Move Transform Type-In dialog appears. Under the Absolute column, enter **0** for each of the **X**, **Y**, and **Z** values.

8. Close the Move Transform Type-In dialog.

Assigning the Top Reference Material

To put the reference image on the plane, you must make a material that uses the image, then apply the material to the plane. The first step is to make the material.

1. Click the **Material Editor** button on the Main Toolbar to open the Material Editor.

2. Verify that the top left sample slot is selected. The material is currently named 1-Default. Replace this name with the name **mTopRef.**

    Next you'll assign the top picture of the remote control top to the Diffuse Color channel, which defines the color or pattern that you see on the object.

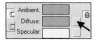
3. On the Blinn Basic Parameters rollout, locate the Diffuse color swatch. Click on the small shortcut button next to the Diffuse color swatch.

    The Material/Map Browser appears.

☼ TIP ☼

Always name everything you create as you go along, whether they are objects or materials. In this book, we use a lower case first letter for materials and an upper case first letter for objects. This will help you quickly identify materials as opposed to objects.

Figure 2.4 *Material/Map Browser*

4. Select **Bitmap** and click **OK**.

   The Select Bitmap Image File dialog appears.

5. Select the file **Remote_Top.jpg** from the **Maps/Remote Refs** folder on the CD, and click **Open**.

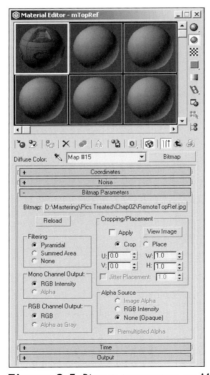

Figure 2.5 *Bitmap parameters on Material Editor*

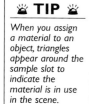

TIP

When you assign a material to an object, triangles appear around the sample slot to indicate the material is in use in the scene.

The Material Editor displays to show the bitmap parameters, and the bitmap appears on the sphere in the first sample slot.

6. Click and drag from the sample slot to the plane.

   The material is now assigned to the object.

   In your viewports, you can't see a picture on the plane just yet. Displaying the picture in viewports is what makes it possible to use the image as a reference while modeling. We'll get the picture to appear in the Top viewport with the next step.

7. To enable the map to show in the viewport, click the **Show Map in Viewport** button on the Material Editor toolbar.

   Our goal is to be able to see the picture in the Top viewport so that we will have a reference picture to draw over when we create the remote control. Currently, the Top viewport is in wireframe mode. To see material bitmaps assigned to objects in a viewport, you need to make the viewport a shaded view.

8. To change the Top viewport to a shaded view, right-click on the viewport label to access the pop-up menu. The Wireframe option is currently selected. Choose **Smooth + Highlights** from the menu.

   The remote control reference becomes visible in the Top viewport.

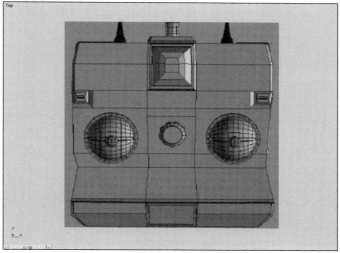

Figure 2.6 *Top view of reference image on plane*

9. In the same way, change the Front and Left viewports to **Smooth + Highlights** mode as well.

Converting the Plane to an Editable Mesh

The plane could be left as a standard primitive, or it can be converted to an Editable Mesh. An Editable Mesh is a very useful type of object. To work with an Editable Mesh's vertices, there is no need to apply an Edit Mesh modifier as the commands are already right there on the Modify panel. We'll convert the plane to an Editable Mesh so we can take advantage of this feature later on.

1. In the Top viewport, right click on the plane to open the Quad menu.

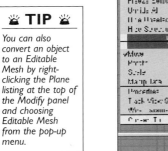

> ☸ **TIP** ☸
>
> You can also convert an object to an Editable Mesh by right-clicking the Plane listing at the top of the Modify panel and choosing Editable Mesh from the pop-up menu.

Figure 2.7 *Quad menu, convert to Editable Mesh*

2. In the lower right quadrant, select **Convert To**, then **Convert to Editable Mesh**.

There is no visible change to the object as a result of the conversion. We will see the results later on when we edit the object.

Assigning Mapping Coordinates

Mapping coordinates tell a bitmap how to lie on an object. It's obvious to you and me that you want the bitmap to lie on the plane and retain its original aspect ratio (length/width ratio), but not so obvious to **3ds max**. You'll have to tell it so with mapping coordinates.

> ☸ **TIP** ☸
>
> To quickly find a modifier on the list, press its initial letter on the keyboard repeatedly until the modifier appears at the top of the list. Then press <Enter> to apply the modifier..

Mapping coordinates are represented by a *gizmo*, or outline, that defines the area and orientation for the map. The tools for defining this area and orientation are accessed by applying a UVW Map modifier to the object.

1. Make sure the plane is still selected.
2. On the **Modify** panel, locate the Modifier List drop-down list arrow. Click on the arrow and scroll through the list to locate the UVW Map modifier. Choose the **UVW Map** modifier to apply it to the plane.

The UVW Map modifier appears on the modifier stack just above the base object Editable Mesh. The UVW Map gizmo appears around the plane as an orange border. By default, the Mapping type is set to Planar on the Parameters rollout. This setting places the bitmap flat on the object, which is exactly what we want.

3. Click on the **Bitmap Fit** button in the Alignment section of the command panel. The Select Image dialog appears. Select the file **Remote_Top.jpg** file again.

>
> **TIP**
>
> Many primitives automatically generate mapping coordinates for themselves when they are created, including the Plane. In practice, you will rarely use a primitive's default mapping coordinates, and will usually need to apply a UVW Map modifier to adjust the mapping coordinates to exactly the way you want them.

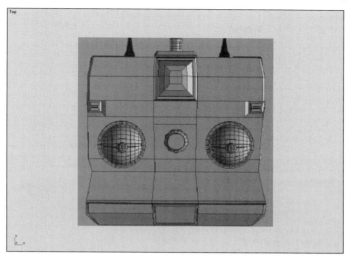

Figure 2.8 *Reference image mapping coordinates adjusted*

This adjusts the gizmo to the aspect ratio of the bitmap so when the picture lies on the object, it retains its original ratio. The picture now appears correctly. The change is subtle, but it will make a difference in helping us create the remote control with exactly the same proportions as the reference image.

Creating the Side Reference Plane

Now that we have our top reference in place, we can use it to quickly generate a side reference plane by copying and rotating the first plane.

1. Click **Select and Rotate** on the Main Toolbar.

2. Click the **Angle Snap Toggle** at the bottom of the screen.

3. Locate three text fields at the bottom of the screen labeled **X**, **Y** and **Z**.

These are the transform type-in fields. They display the current values for the transform being used. In this case, they'll display the rotation angle as you rotate the object.

> **TIP**
>
> Because the Angle Snap Toggle is turned on, the plane rotates by five degree increments, making it easier to rotate the plane exactly 90 degrees.

4. In the Front viewport, hold down the **<Shift>** key while you click and drag on the Top_Reference plane's Z axis. Watch the **Z** transform type-in field and rotate until the number reaches 90, then release the mouse.

By holding down the <Shift> key while you perform a transformation, you generate a clone (copy).

The Clone Options dialog appears.

Figure 2.9 *Clone Options dialog*

5. Leave the Object selection as **Copy**, the default selection. Enter **Side_Reference** in the Name field, and click **OK**.

   A new plane has been created at a 90-degree angle to the original plane.

6. Turn off the **Angle Snap Toggle**.

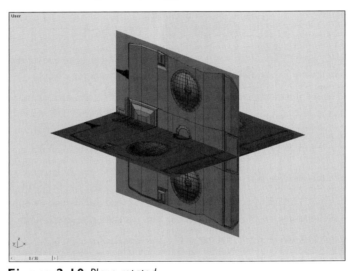

Figure 2.10 *Plane rotated*

Creating the Side Reference Material

Now we need a new material for the new plane. Because the two materials will be very similar, we can use the top reference material as a base for the side reference material.

In the Material Editor, note that the first sample slot has triangles in all four corners. The second slot does not. This indicates that the first slot is "hot", meaning it has been assigned to one or more elements in the scene.

TIP

The practice of copying a material from one sample slot to another is a great time saver when you have a material that already closely resembles one you need to create.

1. In the Material Editor, click and drag the sample slot that contains the top image over to the next empty slot.

   This creates a copy of the material which we can use as a base for the side reference material.

2. Change the material name to **mSideRef**.

3. On the Material Editor, scroll down to the Maps rollout.

   Note that the Diffuse Color button in the Map column currently shows a map assigned to it, the map for the top reference. This is the result of clicking the small button next to Diffuse on the Blinn Basic Parameters rollout for the original material. We will need to change this bitmap to the side reference image.

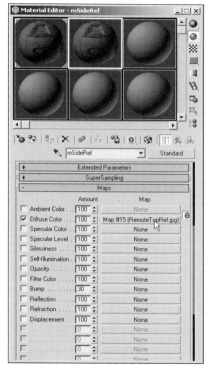

Figure 2.11 *Top reference image in Diffuse Color Map slot*

4. On the Maps rollout, click on the button across from **Diffuse Color** under the Map column.

   This will change the Material Editor display to show the bitmap parameters.

   On the Bitmap Parameters rollout, you will see the label **Bitmap** with a long button next to it. The text on the button is the path to the old image that is currently assigned to this bitmap map.

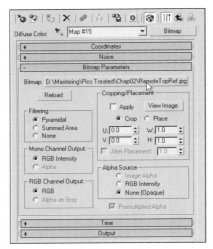

Figure 2.12 *Bitmap button*

5. Click the long button next to **Bitmap**.
6. Select the file **Remote_Side.jpg** from the **Maps/Remote Refs** folder on the CD, and click **Open**.

    The image on the sample sphere changes to show the side reference image.

7. Select the **Side_Reference** plane in any viewport.
8. On the Material Editor toolbar, click **Assign Material to Selection**.

    The side reference material has been assigned to the new plane.

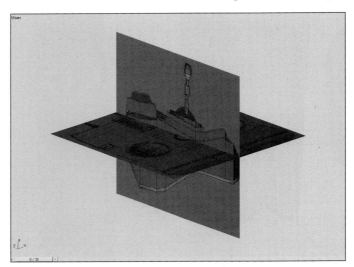

> **☼ TIP ☼**
>
> *Using the Assign Material to Selection button is another way to assign a material to an object.*

Figure 2.13 *Side reference image on plane*

By default, only one side of the geometry will show in the Perspective view, meaning you might have to rotate the view with **Arc Rotate** to see the new map.

Adjusting Mapping Coordinates

The side image displays as squashed. We'll fix this by applying mapping coordinates and fitting them to the bitmap's aspect ratio (length/width relationship).

1. With the **Side_Reference** plane selected, go to the **Modify** panel.

   Note that the UVW Map modifier has already been assigned, as this plane was copied from the top reference, which already had the modifier applied. The modifier is currently set up to fit the bitmap to the same aspect ratio as the top reference, which is not correct for the side bitmap.

2. Click the **Bitmap Fit** button on the command panel. From the Select Image dialog, select the *Remote_Side.jpg* image and click **Open**.

   The bitmap on the plane is resized to the same aspect ratio as the side reference image.

   The image also tiles (repeats) on the Side_Reference plane. This can become distracting during the modeling process.

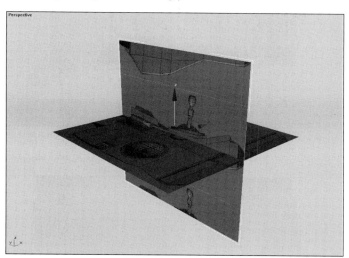
Figure 2.14 *Side reference with fitted mapping coordinates*

3. To correct the tiling problem, go to the Coordinates rollout in the Material Editor for the **mSideRef** material. Under **Tile**, uncheck the checkbox across from **U**.

   The side reference image no longer tiles.

Collapsing the Modifier Stack

To keep the memory usage as low as possible, we can collapse the modifier stack. Collapsing the stack compresses the base object and all modifiers into one object. You won't be able to go back and change any of the modifiers later, but you will save memory and keep your computer's speed as fast as possible.

1. Select one of the planes.

2. On the **Modify** panel, right-click in the modifier stack window and select **Collapse All** from the pop-up menu.

   A warning appears explaining the dangers of collapsing the stack.

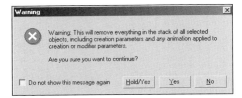

Figure 2.15 *Warning message*

3. Click **Yes** to collapse the stack.

   The stack view shows that the object has been collapsed to an Editable Mesh.

4. Repeat this process for the other plane.

Finalizing the References

For the best working environment, we want the planes moved apart to give us room to model, and we want the bitmap to appear on both sides of the plane.

1. In the Front viewport, move the **Side_Reference** plane to the right until it is just past the edge of the top reference panel.

2. In the Front viewport, move the **Top_Reference** plane downward until it is just past the bottom of the side reference.

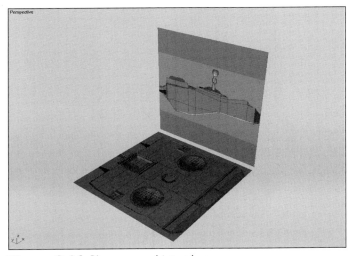

Figure 2.16 *Planes moved into place*

We can make the bitmap appear on both sides of each plane by turning on each material's 2-Sided option. On both materials, you are currently at the child (map) level of the material in the Material Editor. The 2-Sided option is available only at the parent level.

3. In the Material Editor, select the sample slot for one of the reference materials.

4. On the Material Editor toolbar, click **Go To Parent**.

5. At the top of the Shader Basic Parameters rollout, check the **2-Sided** checkbox.

6. Repeat this process for the other material.

    We don't need the default viewport grids for this project. Let's turn them off in each viewport.

7. For each viewport other than the Top viewport, click in the **Viewport** to activate it, then press the **<G>** key on your keyboard to turn off the grid.

    The virtual studio is now set up and ready to use.

8. Save your scene with the filename **Remote Virtual Studio.max**.

BOX MODELING

In the next tutorial, the main body of the remote control will be created with a technique called *box modeling*. Box modeling starts with a primitive such as a box, cylinder or cone. You then push and pull the primitive into the shape of the final object. Afterwards, a smoothing modifier can be applied to give the object a smooth, rounded appearance.

Box modeling is extremely useful for modeling odd-shaped objects and organic objects such as characters. In fact, you will use this technique in later chapters to model the car and the alien.

The next tutorial makes use of many basic box modeling tools, each described in detail as it is introduced. Learning how to use these tools will help you greatly in modeling the more complex objects later in this book.

SUB-OBJECTS

Box modeling works with *sub-objects*. A sub-object is a component that makes up an object. Sub-objects include:

- **Vertex** - A point on an object
- **Edge** - A segment or line that connects vertices
- **Face** - A surface surrounded by three edges
- **Polygon** - A surface surrounded by three or more edges, made up of one or more faces

The plural of vertex is *vertices*.

Sub-objects are dependent on one another. For example, if you delete all the edges around a face, the face is also deleted.

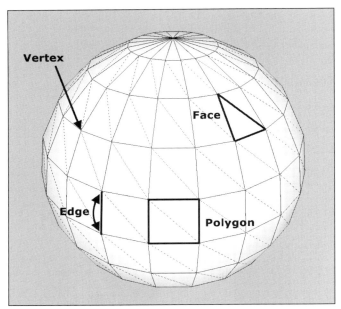

Figure 2.17 *Vertex, edge, face, polygon*

With an object that has been converted to an Editable Mesh, you have direct access to sub-objects.

Accessing Sub-Object Levels

There are several ways to access a sub-object level.

- You can right-click on the object to display the Quad menu, then choose **Sub-Object** and select the appropriate sub-object level.
- On the modifier stack, you can click the [+] next to Editable Mesh and highlight the sub-object level.
- You can click the appropriate picture button on the Selection rollout:

    Vertex
    Edge
    Face
    Polygon
    Element

Selecting Sub-Objects

In this chapter, at almost all times when you work with sub-objects, you will be required to select and change several sub-objects at a time. When selecting sub-objects, you can use a dotted bounding area, or *marquee*.

To use a marquee, click any selection tool from the Main Toolbar such as Select object, then click and drag on the screen to draw a rectangular marquee. The sub-objects inside the marquee will be selected.

To select with a marquee, you can also use a selection tool on the Main Toolbar that also transforms, such as **Select and Move**. With this type of tool, you must take care to begin dragging in an area where there are no sub-objects. If you click and drag on a sub-object, you will transform the sub-object rather than simply selecting it.

Always use marquee selection when selecting vertices. When selecting other sub-objects such as polygons, you can use a marquee, or you can click on each one while holding down the <Ctrl> key.

About Faces and Polygons

Polygons are always made up of one or more faces. A face always has three sides, while a polygon can have any number of sides. In this book, you will be working primarily with polygons.

Face edges are not always visible, but edges around polygons are always visible. The edges you see in wireframe mode in viewports are edges around polygons. It is very common to find a four-sided polygon made up of two faces, as shown in Figure 2.17.

Faces and polygons are always flat. An object appears rounded when it has sufficient faces to fool the eye into thinking the object is rounded. The shading performed by **3ds max** also plays a part in making an object look smooth.

Many reference materials use the terms *face* and *polygon* interchangeably.

THE ANATOMY OF THE REMOTE CONTROL

Box modeling requires you to transform vertices, edges, faces and polygons. In **3ds max**, to *transform* something means to move, rotate or scale it. In making the remote control, you will also cut, slice, collapse, attach, and weld various sub-objects.

Many of the transformations and other operations will be performed on selections of vertices. For this tutorial to go as smoothly as possible, we need to set a few guidelines so you can locate and select rows and columns of vertices when called for.

Naming the Sides of the Remote Control

The remote control will start out its life as a simple box. A box has six sides. In the steps that follow, we will refer to the box's sides in the following way:

- **Front** is the side holding the joysticks and knob.
- **Back** is opposite the Front side.
- **Left Side** and **Right Side**, **Top** and **Bottom** are as shown in the following illustration.

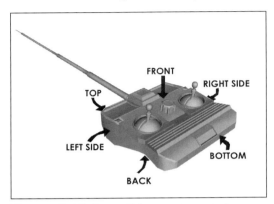

Figure 2.18 *Parts of the remote control*

TUTORIAL 2.2   REMOTE CONTROL BODY

The first step in creating the remote control is creating a basic box from which the object will be modeled.

Setting up the Box

1. Load the file **Remote Virtual Studio.max** from the last tutorial. If you didn't create this file, load this file from the **Scenes/Virtual Studios** folder on the CD.

2. Go to the **Create** panel and click **Box**. In the Top viewport, create a somewhat flat box with roughly the same dimensions as the Top_Reference plane.

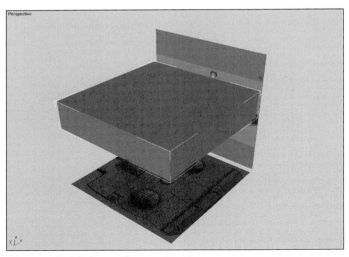

Figure 2.19 *Box in Perspective view*

Now we'll adjust a few of the box's settings.

2. On the **Modify** panel, set the following properties for the box:

Name	Remote_Body
Height	45
Length Segs	6
Width Segs	3
Height Segs	1

   In the Top and Left viewports, the new box blocks out the reference pictures. We need to be able to see the box and the pictures at the same time. There are several ways to solve this problem, but the easiest is to make the box semi-transparent (see-through). This will allow you to see through the box to the reference pictures.

You can also make an object transparent by checking the See-Through option on the Display panel while the object is selected.

3. With the box still selected, and press **<Alt-X>** on your keyboard to make the box semi-transparent.

   With the box in see-through mode, it is very difficult to tell where all the box's edges are, and seeing these edges is very important while modeling. Changing the display to Edged Faces will cause the wireframe outline to appear over the shaded object.

You can also activate Edged Faces by right-clicking the viewport label and choosing Edged Faces from the pop-up menu.

4. Activate the Top viewport and press the **<F4>** key to enable Edged Faces.

   We are now able to see a wireframe outline on the semi-transparent object.

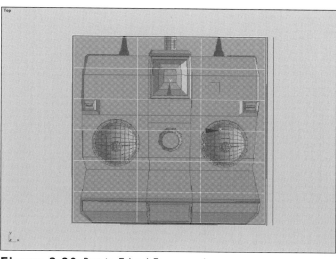

If using Edged Faces makes the object look too transparent, choose Views/ Shade Selected from the menu.

Figure 2.20 *Box in Edged Faces mode*

5. Repeat this step for the Front, Left and Perspective viewports.
6. In the Top and Left viewports, move the box around until it is centered on the view of the reference image in both viewports.

Accessing the Vertex Sub-Object Level

1. Convert the box to an Editable Mesh. To do this, right-click on the **Box** and from the Quad menu, select **Convert To** then select **Convert To Editable Mesh**.

   The parameters that were available on the **Modify** panel have changed. We can now access the Editable Mesh's sub-objects so we can begin box modeling.

2. Go to the **Modify** panel.
3. Right click on the box. On the Quad menu, select **Sub-objects** then **Vertex**.

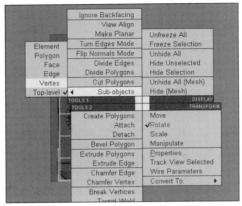

Figure 2.21 *Switching to Vertex sub-object mode from the Quad menu*

Transforming Vertices

In this tutorial, all vertices will be referred to by row or column numbers to make it clear which vertices you're supposed to select.

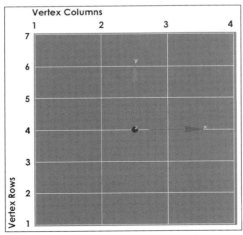

Figure 2.22 *Counting rows and columns in the Top viewport*

When viewing the remote body in the Top viewport (top reference picture), you can count seven lines running horizontally from the bottom of the viewport to the top. These will be referred to as rows 1 through 7.

When referring to the vertical columns of vertices in the Top viewport, they are counted as columns 1 through 4 running vertically from left to right. When working with the box in the Left viewport (side reference picture), the same rows are counted from left to right as 1 through 7.

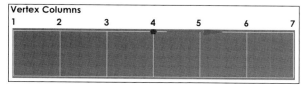

Figure 2.23 *Column numbers for side vertices in Left viewport*

You will often be required to move an entire row or column of vertices. Make sure you select with a marquee unless instructed otherwise. If you just click to select vertices, you could accidentally move only one or a few vertices rather than the entire row or column.

Refer to these illustrations as necessary to help you select the correct row or column of vertices.

1. On the Selection rollout, make sure the **Ignore Backfacing** option is unchecked. This ensures that you can select vertices on the back and front at the same time.

2. Click **Select and Move** on the Main Toolbar.

3 In the Top viewport, use a marquee to select the top row of vertices, horizontal row 7.

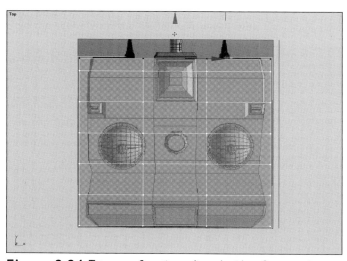

Figure 2.24 *Top row of vertices aligned with reference picture*

4. Move the selected vertices downward along the Y axis to align with the top of the main body of the remote control in the reference picture.

5. Use a marquee to select the next row of vertices, horizontal row 6.

6. Move the selected vertices downward along the Y axis so they align with the area of the reference picture where the remote control begins to slant inward.

7. Position the remaining rows to align with the reference lines as shown in the following figure.

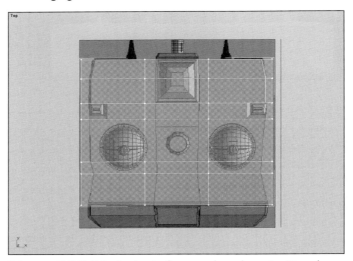

TIP
When moving vertices to match the reference pictures, you don't have to match the pictures exactly. Just get the vertices as close as you can.

Figure 2.25 *Top viewport after rows have been positioned*

Note that the lowest row of vertices falls short of the bottom of the remote control body. This area of the object will be created with a different technique, which we will get to later.

Scaling Vertices in the Top View

1. Choose **Select and Non-uniform Scale** from the Main Toolbar. To do this, click and hold the **Select and Uniform Scale** button on the Main Toolbar until the flyout appears. Choose **Select and Non-uniform-Scale** from the middle of the flyout.

2. In the Top viewport, select horizontal row 7 and scale the vertices on the X axis until they match the width of the reference picture.

3. Repeat the above scaling operation for all remaining rows.

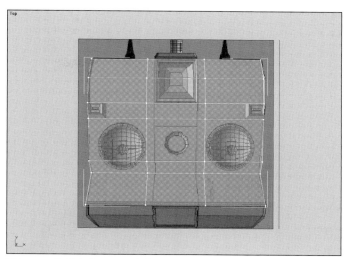

Figure 2.26 *Rows scaled*

Shaping the Side View

Next you are going to shape the side view of the remote control in the Left viewport.

1. Click **Select and Move** on the Main Toolbar.
2. In the Left viewport, use a marquee to select the first four bottom vertices in columns 1 to 4.
3. Move these vertices downward to line up with the nearest reference line on the reference picture.
4. Use a marquee to select each set of vertices along the top of the box, and move them up or down in the Y axis as required to line up with the reference picture as shown in Figure 2.27.

The remote control body should look similar to the following figures in the Left and Perspective viewports. Move any other vertices as necessary to make the remote control look like the figures.

> **TIP**
>
> Keep in mind that the reference pictures are not displayed at their full resolutions, and might not match each other exactly for this reason. Just match up the vertices as best you can, but don't worry too much about exact placement of vertices.

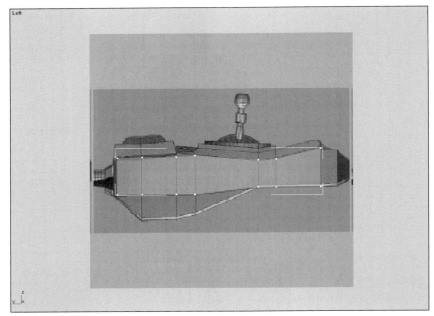

Figure 2.27 *Box and reference picture in Left viewport*

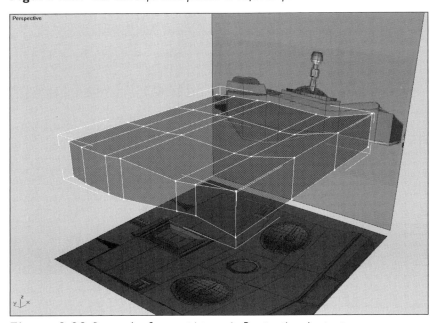

Figure 2.28 *Box and reference pictures in Perspective viewport*

Selecting Polygons for Extrusion

Next, we'll use a box modeling technique called *extruding*. Extruding takes a face or polygon and pushes it away from or into the object, creating new polygons on the sides of the extrusion. Extruding is a powerful way to add additional detail to an object.

1. Use **Arc Rotate** to rotate the Perspective view so that you are looking at the back of the remote body (the part facing the top reference picture).

2. Switch to the Polygon sub-object level. You can do this from the Quad menu, or by clicking the **Polygon** button on the Selection rollout.

3. Activate the Perspective viewport, and press the **<F2>** key on the keyboard to cause selected polygons to appear as shaded in red.

   This will make it much easier to see which polygons you're selecting.

4. In the Perspective view, select the two rows of four polygons at the back that are closest to the top of the remote and its outer edge, as shown in the following figure. You should select eight polygons in all.

> **TIP**
> To select multiple polygons, hold down the <Ctrl> key as you click on each one.

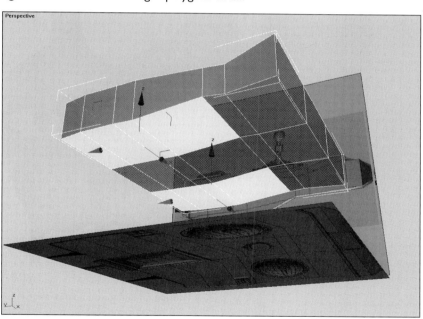

Figure 2.29 *Selected faces on bottom of remote control*

Extruding the Back

To shape the back of the remote, we can't simply move our current selection of polygons downward as this would destroy the shape that we've developed. Instead, we will shape the back of the remote by extruding new polygons and working with the new vertices generated by the extrusion.

1. On the **Modify** panel, expand the Edit Geometry rollout if necessary.

   On this rollout you will find the Extrude button with a text field and spinner to the right of it. The button allows you to interactively control the extrusion by clicking and dragging on the screen, while the text field and spinner set a specific amount of extrusion with a numeric entry.

   At the bottom of this section is the Normal setting with two choices, Group and Local. These settings determine the direction the extruded polygons relative to face normals (direction perpendicular to each face plane). Don't concern yourself with understanding exactly what these selections do. For now just follow the instructions, and the use of these selections will become clear with time and practice.

2. Select the **Local** option.
3. Click the **Extrude** button.

   The Extrude button is depressed and its color has changed.

4. Place your cursor over any of the selected polygons, then click and drag.

   The cursor changes to the extrusion cursor, which looks like a stack of polygons. When dragging on the selection, you can see that moving the cursor in one direction extrudes the selected polygons outward, while the other direction pushes them inward.

5. Press **<Ctrl-Z>** on your keyboard to undo the practice extrusion you just performed.

6. Click and drag on the selected polygons in the Perspective view while watching the Left viewport. Extrude them until they are aligned with the lowest part of the remote control reference picture.

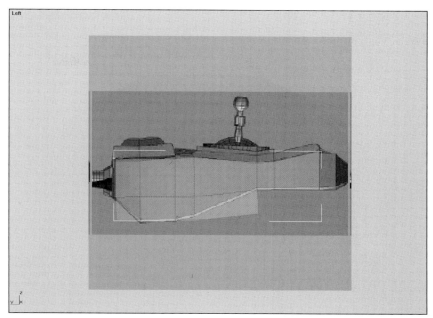

> **TIP**
>
> Click and drag just once to extrude the polygons. Otherwise, you will end up with more than one set of extruded polygons, which will lead to difficulties in shaping the remote control in later steps.

Figure 2.30 *Extrusion aligned with lowest part of side reference picture*

Don't worry about all the extruded polygons not lining up yet. We will fix this next.

7. Click the **Extrude** button again to turn it off.

Setting up Snap Tools

With the new polygons extruded, we have several new vertices that we can move around to achieve the shape of the remote control. In some cases, we will want to move a set of vertices directly on top of another set of vertices. This can be accomplished with *snap tools*. These tools can be used to force a vertex to snap (move exactly to) another vertex.

1. Switch to the **Vertex** sub-object level.

   In order to use a snap tool to snap vertices, we first have to set up the snapping options to work with vertices rather than other screen components.

2. Choose **Customize/Grid and Snap Settings** from the menu.

   The Grid and Snap Settings dialog appears.

TIP

You can also access this dialog by right-clicking any of the snap tools at the bottom of the screen.

Figure 2.31 *Grid and Snap Settings dialog*

3. Check the **Vertex** option, and uncheck all other options.
4. Close the dialog by clicking the **X** at its upper right corner. The settings will be retained even after the dialog is close..

5. Click **3D Snap** at the bottom of the screen to turn on this snap tool.

6. Click **Select and Move** on the Main Toolbar.

If you move your cursor near a vertex, a blue icon appears over the vertex. This indicates that snapping is in effect. The blue icon is called the *snap marker*.

Snapping Vertices

In the Left viewport, the left and right ends of the extrusion need to be tapered upward to the upper part of the remote body. We will use our snap tool to move the vertices

1. In the Left viewport, use a marquee to select the vertex at the bottom left corner.

   This will also select all vertices behind the visible vertex. Note that these vertices are not over any part of the remote control reference picture.

2. In the Left viewport, click and drag on the selected vertices, and move them to the vertices directly above them.

   When the selected vertices are close to the new location, they will snap to the new location and the snap marker will appear.

3. Select the bottom vertex in the 5th column.

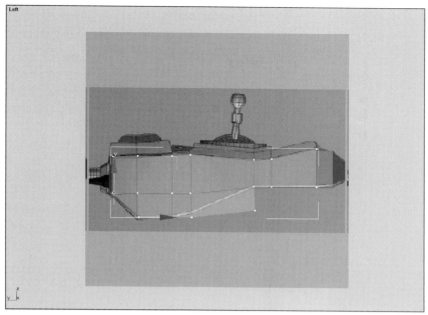

Figure 2.32 *Vertices in column 1 snapped to vertices above*

4. Move and snap these vertices to the ones just above them.

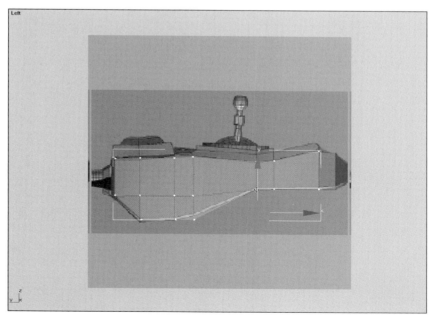

Figure 2.33 *Vertices in column 5 snapped to vertices above*

5. Turn off **3D Snap**.

Moving the Remaining Vertices

1. Use a marquee to select the vertices at the bottom of column 4 and move them to align with the reference picture.
2. The vertices at the bottom of the remaining columns should already be aligned with the reference picture from the extrusion you performed earlier. If they are not lined up, move them into place now.

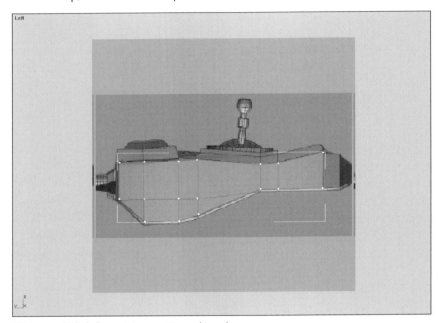

Figure 2.34 *Remaining vertices aligned*

Welding Vertices

The remote control is starting to take shape. However, now we have vertices on top of other vertices.

Where two or more vertices are sitting on top of one another, we can *weld* the vertices together. This will replace multiple vertices in the same place with one vertex. This will not change the appearance of the object in its current state, but will help preserve the surface's integrity if you modify it later on.

1. In the Left viewport, use a marquee to select all the vertices at the lower left corner of the remote control object.
2. On the **Modify** panel, expand the Edit Geometry rollout.

The Edit Geometry rollout has a section labeled Weld. Clicking the **Selected** button in this section will look for all selected vertices within the unit threshold distance from one another, and weld them into a single vertex.

The unit threshold is provided in the entry field next to the Selected button The default value of 0.1 unit should work fine for welding the snapped vertices together.

3. Click the **Selected** button to weld the vertices.
4. Use a marquee to select the vertices at the bottom of column 5, and click **Selected** again to weld the vertices together.

Extruding and Beveling the Bottom

The bottom of the remote control body is shorter than the reference picture. We will use extrusions to pull out the bottom. We will also use another tool called Bevel to shape the extrusions.

1. In the Left viewport, use **Region Zoom** to zoom into the right end of the reference picture.

2. Switch to the **Polygon** sub-object level.

3. Select the three polygons at the bottom-end of the remote control object.

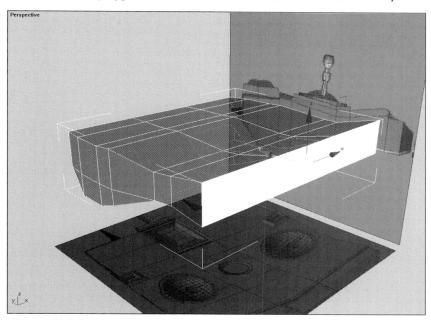

Figure 2.35 *Three selected polygons at bottom of remote control body*

This time we will be using a combination of the Extrude and Bevel tools. The Bevel tool scales selected polygons. To achieve the rounded end of the remote, we will alternate between extruding and beveling.

4. In the Left viewport, use **Region Zoom** to zoom into the right end of the remote control.

5. While watching the Left viewport, click and drag on the spinner to the right of the **Extrude** button until the extruded polygons have reached the next major division line in the reference picture.

6. With the new polygons still selected, click and drag downward on the **Bevel** spinner until the new faces have scaled down to fit the reference.

7. Use **Extrude** and **Bevel** again to create another section for the very bottom of the remote control.

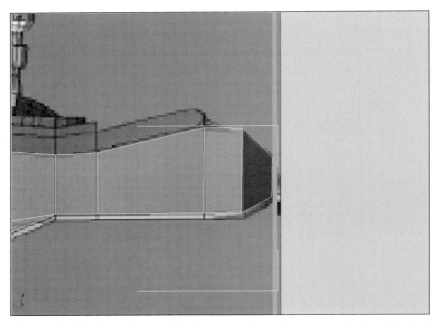

Figure 2.36 *Zoomed view of Left viewport shows bottom extruded and beveled*

8. Switch back to the **Vertex** sub-object level. In the Top and Left viewports, move vertices as necessary to make them conform to the reference picture.

   The rough shape of the remote control is complete. We've done a substantial amount of work here, so it's a good idea to save the scene.

9. Save your scene with the filename *RemoteControl02.max*. You can do this automatically by choosing **Save As** from the **File** menu and clicking the **[+]** button on the Save File As dialog.

Extruding the Bottom Front Detail

We could leave the remote control with its rough shape and use it in the scene. However, the remote will look much more realistic with small details added. These details will be added using the same techniques we have been using so far, extrusion and beveling.

The extrusions and bevels you will use to create the details will be much smaller than the ones used to sculpt the overall shape.

The first bit of detail we'll be creating is a rectangular groove at the bottom front of the remote control. Our goal is to create a small lip that extrudes outward a little before extruding inward to add some extra detail. If you are not seeing your selected polygons as a shaded red color, don't forget to press the **<F2>** button.

 1. Switch back to the **Polygon** sub-object level. In the Perspective view, select the two polygons on the front closest to the remote bottom.

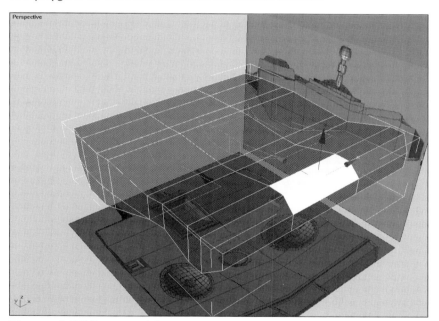

Figure 2.37 *Bottom-middle two polygons selected on front of remote body*

 2. On the Edit Geometry rollout, set the Normal option to **Local**.

3. Use the **Extrude** spinner to extrude out a small lip of approximately 2 units.

We want to make the lip a fairly hard edge with a 90 degree turn headed inward. To do this we will need to extrude again, but by a very small amount.

4. Click the **Extrude** spinner up arrow one time to make an additional extrusion of a very small size.

5. Use the **Bevel** spinner to scale the polygon by -1.0.
6. Click and drag downward on the **Extrude** spinner to extrude into the remote control a little past the original starting point.
7. Click the **Extrude** spinner down arrow one time to create another small extrusion.
8. Using the **Bevel** spinner, scale the face in just a little.
9. Using the **Extrude** spinner up arrow, extrude the face forward just enough to create a small groove running around the inside of the newly detailed area.

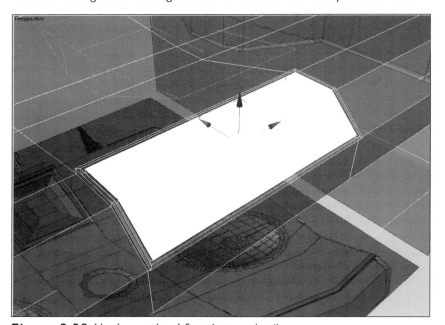

Figure 2.38 *Newly completed front bottom detail*

Extruding the Middle Front Detail

If you were to count the rows of polygons running horizontally in the Top viewport, you would see that there were 8. Keep in mind we are not talking about rows of vertices this time, instead we are talking about rows of polygons. Armed with this information, you can complete the next series of steps:

1. In the Perspective view, select all of the horizontal polygons on the front of the remote body in rows 3, 4, 5, and 6, as shown in the figure that follows.

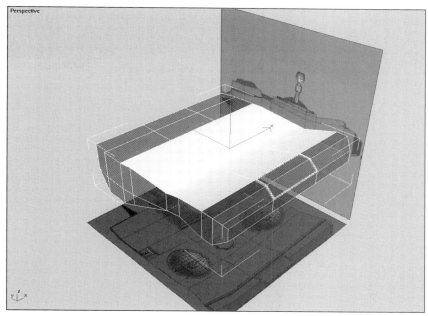

Figure 2.39 *Polygon rows 3-6 selected on front of remote control body*

2. While watching the Left viewport, click and drag upward on the **Extrude** spinner until the extruded polygons are about halfway to the next major division line on the reference image.

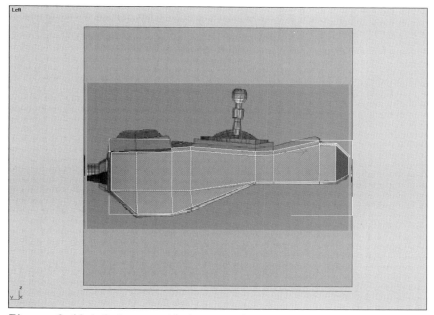

Figure 2.40 *Left viewport with the new polygons extruded*

3. Switch to the **Vertex** sub-object level.

 4. Use **Region Zoom** to zoom in on the top vertex in the third column.

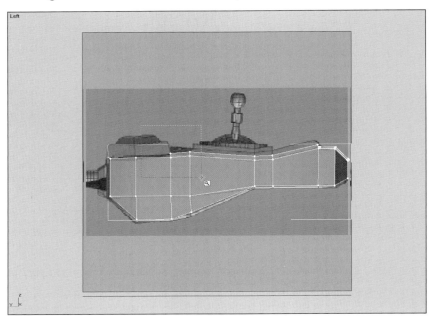

Figure 2.41 *Zooming into top of third column*

 5. Turn on **3D Snap**.

6. Select the vertices at the top of the third column, and move them so they snap to the vertices just below them.

 7. Turn off **3D Snap**.

8. Without changing the selection, use the Y axis arrow to move the vertices back up so they make a straight line with the vertices to their immediate right.

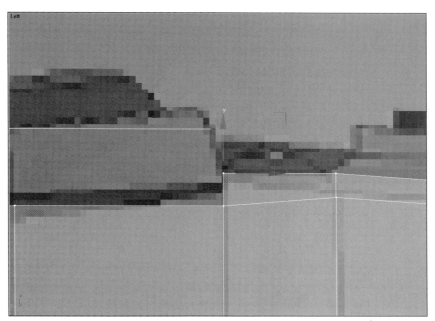

Figure 2.42 *Vertices moved straight up*

 9. Click **Zoom Extents** to zoom out the view.

Cutting the Screw Indentation

Now we will create the small indentation for the square-shaped screw.

From this point forward we will work only on one side of the remote control, the left side. When we're done putting in all the detail, we'll mirror and copy the left side to the right. Having to work only with one side will speed up the modeling process.

To create the indented area for the screw, we will have to make some cuts to our edges so that we'll have the required faces for manipulation and extrusions. The Cut tool on the Edit Geometry rollout can be used to place new edges, and thus new polygons, on the object.

 1. Switch to the **Edge** sub-object level.

 2. On the Selection rollout, check the **Ignore Backfacing** option.

This will enable us to work with the front of the remote control without accidentally affecting edges on the back.

3. On the Edit Geometry rollout, click **Cut** to activate it.

To use the Cut tool, you will click on two edges to place a new edge between them. Remember that you will be working only with the screw hole on the left as viewed in the Top viewport.

4. In the Top viewport, click at the lower right corner of the screw hole on the left, then click at the screw hole's upper right corner. Right-click to end the cut operation, then right-click again to turn off the **Cut** button.

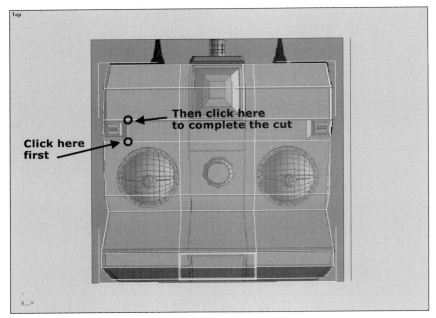

TIP

If you cut and don't see a new edge created, undo the cut and perform the clicks in the opposite direction.

Figure 2.43 *Click points for cut*

5. If the cut line is not straight, switch to the **Vertex** sub-object level and move the new vertices as necessary to make the cut line straight.

Locating and Removing Isolated Vertices

Cutting and slicing edges and polygons will sometimes leave isolated vertices. These are vertices that are just sitting by themselves on an edge and are not required for the geometry to maintain the its shape. The last few cutting and extruding operations may have left some isolated vertices.

1. Switch to the **Vertex** sub-object level.

2. In the Perspective and Top viewports, inspect the front lip of the newly added screw hole detail. Look for any vertices sitting on an edge that don't really need to be there.

3. If you do have isolated vertices, turn on **3D Snap**. Move the isolated vertex and snap it to the closest vertex on the same edge. Select both vertices and press the **Collapse** button on the Edit Geometry rollout. Be sure to turn off **3D Snap** when you're done.

Creating the Screw Indentation

1. Switch to the **Polygon** sub-object level.

   In the Perspective view you will find all of the polygons that you had selected earlier still selected. You will also see that there are some new faces in the upper left corner, at the location of the two cuts you performed.

2. Hold down **<Ctrl>** and click the polygon over the screw hole to deselect it.

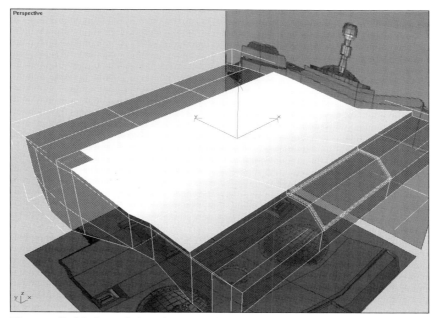

Figure 2.44 *Screw hole polygon deselected*

3. Use the **Extrude** spinner to extrude the selected polygons until the they extend all the way to the highest major division line in the side reference

   This will leave an indented area for the screw hole as we didn't extrude this polygon with all the other ones.

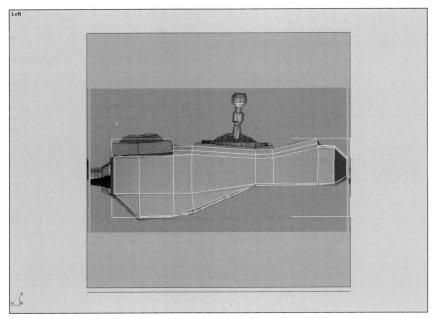

Figure 2.45 *Left viewport showing newly extruded polygons*

4. Switch to the **Edge** sub-object level.

   We will now slant the polygons surrounding the indented area by moving the edges around the polygons.

5. Use **Arc Rotate** to spin the Perspective view so you are looking toward the bottom right of the remote control. Zoom in so you can see the screw area clearly.

6. Select the edge at the inside of the wall of the indentation.

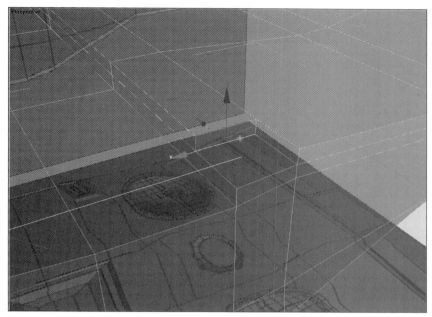

Figure 2.46 *Inside indentation edge selected*

7. Move this edge along the X axis toward the center of the screw hole by about -5 units. Watch the X field in the Transform Type-In area at the bottom of the screen to see how far the edge is moving.

Figure 2.47 *Edge moved on X axis*

8. Select the edge at the bottom side of the screw, and move it toward the center of the screw hole by about 5 units.

Figure 2.48 *Edge moved on Y axis*

Creating the Screw Hole

1. Press **<Alt-X>** to turn off semi-transparent mode. This will make it easier to make the screw hole.
2. Switch to the **Polygon** sub-object level.
3. Select the polygon at the bottom of the indented area.
4. Extrude the polygon upward by about 4.5 units so it is just below the highest edge of the remote body when looking in the Left viewport.
5. Use the **Bevel** spinner to uniformly scale the new polygon inward by -1.0.
6. Click the **Select and Non-Uniform Scale** on the Main Toolbar. In the Perspective or Top viewport, scale the polygon inward on its Y Axis by 50%.
7. Click the **Extrude** spinner up arrow once.
8. Use the **Bevel** spinner to bevel the new polygon by -1.5 and create a top lip for the screw hole.
9. Use the **Extrude** spinner again to extrude the face down into the screw body by -4.0 units.

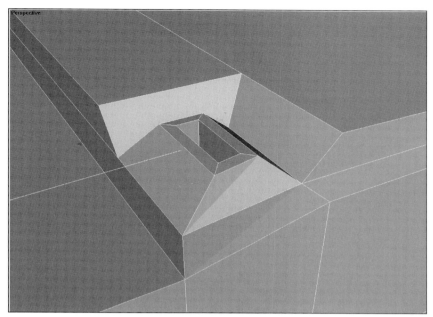

Figure 2.49 *Final shape of square screw detail*

Check the screw hole in the Top viewport to see how it looks. If necessary, go into the **Vertex** sub-object level and move vertices around to straighten out the sides of the screw indentation or hole, and make it match more closely to the reference picture.

Making Room for Joysticks

1. Press **<Alt-X>** to make the remote control semi-transparent again.

2. Switch to the **Vertex** sub-object level.

3. Uncheck **Ignore Backfacing** on the Selection rollout.

4. In the Top viewport, select the two center vertical columns of vertices. Scale these vertices on the X axis until they line up with the two-vertical lines that divide the two joystick areas on the reference picture.

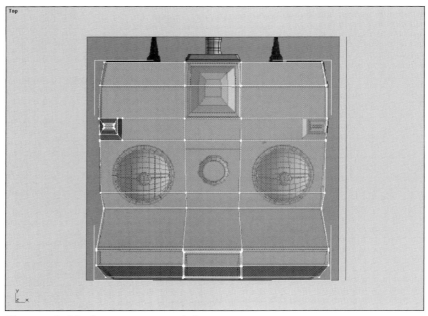

Figure 2.50 *Center vertices scaled*

Extruding the Joystick Area

1. Switch back to the **Polygon** sub-object level.

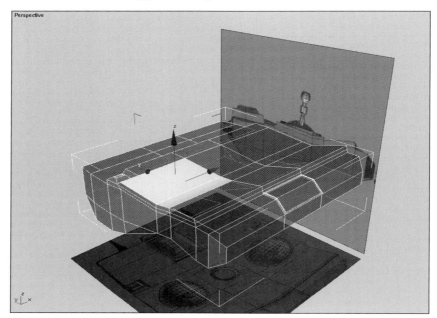

Figure 2.51 *Polygons selected for joystick area*

2. Select the top two polygons on the left where the joystick hole will eventually be, as shown in the previous figure.

3. Watching the Left viewport, use the **Extrude** spinner to extrude the polygons upward until they reach the base of the joystick in the side reference picture.

4. If necessary, switch to the **Vertex** sub-object level and move vertices in the Left viewport to more closely match the reference picture.

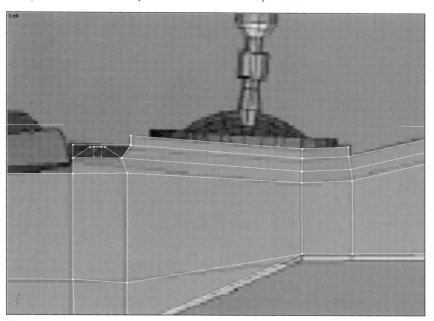
Figure 2.52 *Joystick area extruded and adjusted*

Extruding Faces for Monitors

The remote control will have two areas for displaying monitor information. You will create one monitor on the left side.

1. If necessary, switch back to the **Polygon** sub-object level.

2. Select the top two polygons at the upper left corner of the remote body's top, as shown in the following figure.

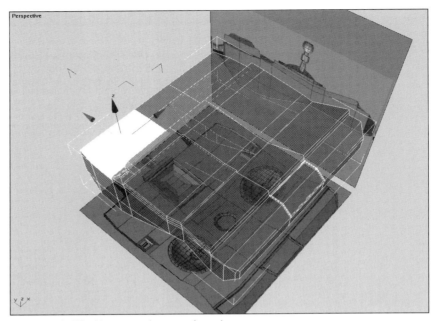

Figure 2.53 *Monitor polygons selected*

As with our other details, the monitor will be created with the Extrude and Bevel tools. By now you should be familiar enough with these tools that you don't need as precise instructions as you did earlier. After all, you won't always have us here to tell you how many units to extrude by! So use your judgment in these next steps, using the next figure as a guide.

3. Use the **Extrude** spinner to extrude the polygons upward very slightly.
4. Use the **Bevel** spinner to scale the polygons inward. The goal to to create a slight slant on the new outer edges.
5. Click the **Extrude** spinner one time. This new polygon will be the top lip of the monitor.
6. Using the **Bevel** spinners, scale the new face inward creating the rim of the monitors.
7. Use the **Extrude** spinner again to extrude the polygon inward creating a hollowed out area for the monitors.
8. Extrude inward a small amount more.
9. Bevel the last extruded face in just a little.

The last Extrude and Bevel operations will give the inside of the monitor some nice edges for a slightly rounded, more realistic appearance.

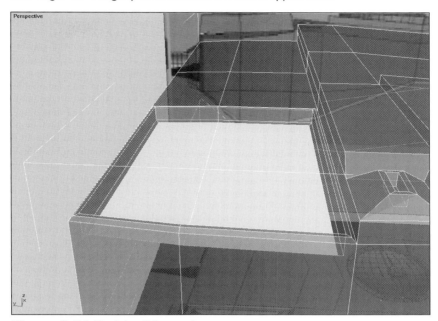

Figure 2.54 *Final detail for monitor*

Creating the LED

The remote will also have an LED (light-emitting diode, a type of light display) for giving information about the remote control operation. The LED is at the top center of the remote control.

1. In the Perspective view, select the two center polygons located on the front of the remote body up at the top, as shown in the next figure.
2. On the Edit Geometry rollout, click **Make Planar** button to make the polygons coplanar.

    This flattens the polygons so they lie in the same plane.
3. Watching the Left viewport, use the **Extrude** spinner to extrude the polygons upward until they line up with the corresponding height on the side reference picture.

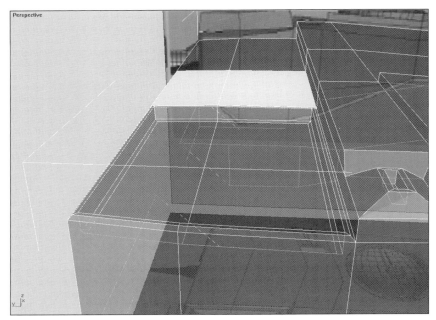

Figure 2.55 *LED polygons selected*

 4. Switch to the **Vertex** sub-object level. In the Left viewport, position the new vertices so the extrusion matches the reference picture.

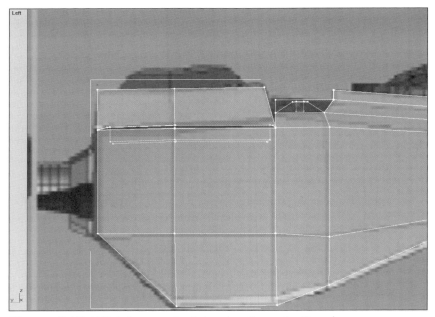

Figure 2.56 *LED polygons extruded and vertices repositioned*

5. Switch to the **Polygon** sub-object level. The two previously selected polygons should still be selected.

6. In the Top viewport, use **Select and Non-Uniform Scale** to scale the polygon down to about 75% on the X axis.

7. Use what you've learned about **Extrude** and **Bevel** to make the LED detail as shown in the following figure. Note that there is a hollow gap between the outer and inner structures.

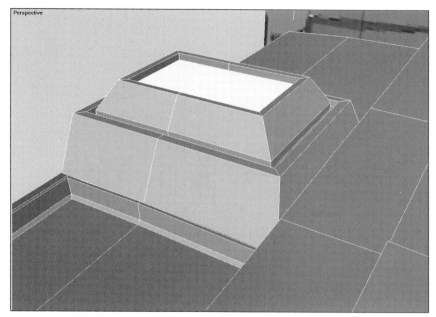

Figure 2.57 *Final LED detail*

Cutting Edges for the Ramp and Grooves

There is a raised ramp section between the two joystick chambers. We will create it by using the Cut tool again. There are also four grooves near the bottom of the remote control that will be made with the Cut tool. We may as well do all our cuts at once.

1. Press **<Alt-X>** to make the remote control semi-transparent.
 2. Switch to the **Edge** sub-object level.
3. Check **Ignore Backfacing** on the Selection rollout.
4. Click **Cut**.
5. In the Top viewport, place a cut on the two edges between the joystick chambers on the reference line. To place the cut, click on one edge, then the other edge. Right-click to end the creation of the cut.

6. Create four additional cuts near the bottom of the remote, as shown in the following figure. Be sure to right-click between each cut.
7. Right-click again to turn off **Cut**.

> ☼ **TIP** ☼
>
> *If you cut across in one direction and don't see a new edge created, undo the cut, then cut in the opposite direction. If you do not undo and try to cut again, there will be at least one isolated vertex.*
>
> *If you end up with isolated vertices, you can get rid of them using the method described under Locating and Removing Isolated Vertices on page 56.*

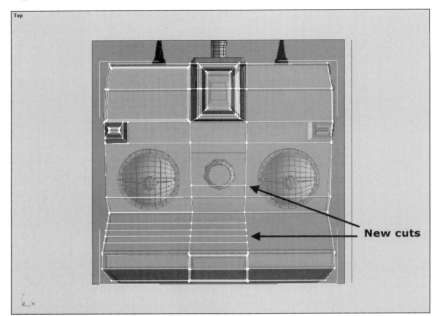

Figure 2.58 *New cut between joystick chambers, and four new groove cuts*

Extruding the Ramp

1. Press **<Alt-X>** to turn off semi-transparent mode.
2. Switch to the **Polygon** sub-object level.
3. Select both polygons on either side of the cut, as shown in Figure 2.59.
4. Extrude the selected polygons upward to a height just below that of the joystick chambers.
5. Switch to the **Vertex** sub-object level.
6. Turn on **3D Snap**.

Modeling the Remote Control 69

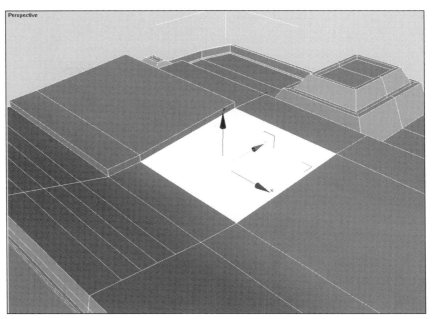

Figure 2.59 *Two center polygons selected*

7. Click and drag on one of the selected vertices and move it downward to snap to the vertices below them.

   This creates a small ramp.

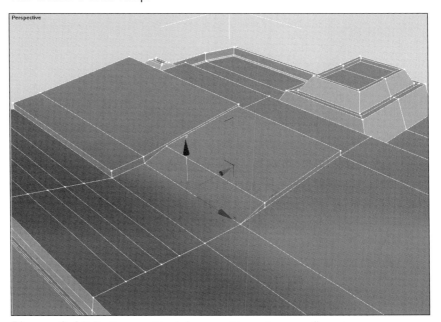

Figure 2.60 *Vertices snapped downward to make a small ramp*

8. Turn off **3D Snap**.
9. Select the snapped vertices, and click **Selected** under Weld to weld the vertices together.

Creating the Grooves

To create the grooves, we will use another tool called Chamfer. The Chamfer tool divides an edge and makes new edges.

 1. Switch to the **Edge** sub-object level and select the new groove edges that you just created.
2. On the Edit Geometry rollout, use the **Chamfer** spinner to subdivide each edge into two closely spaced edges.

Figure 2.61 *Chamfered edges*

 3. Switch to the **Polygon** sub-object level.
4. Select the four rows of polygons in between the chamfered edges.
5. Extrude the polygons into the remote body by a small amount.
6. Use the **Bevel** spinner to make a slight inward bevel.

Creating the Antenna Notch

The remote control body is almost complete. All we need is a notch for the antenna. The actual antenna will be created later.

1. Select the center polygon at the top of the remote control body.
2. Use a combination of extrusions and bevels to create an interesting area for the antenna. Be creative!

   Our version of the antenna notch is shown in the following figure.

   We've done a fair amount of work here, so it's a good time to save.
3. Save your scene with the filename **RemoteControl03.max**.

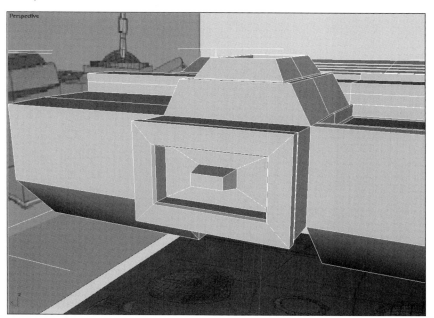

Figure 2.62 *Antenna notch detail*

Cutting the Shape for the Joystick Chambers

Here we'll use a new tool, ShapeMerge, to cut a circular polygon that can then be extruded for the joystick chamber.

One of the reasons we saved our work is that ShapeMerge is a Compound Object feature. These tools are sometimes hard to undo. You should always save your work before using a Compound Object so you can return to an earlier saved version if the object doesn't turn out right.

The first thing we need when using ShapeMerge is a shape to work with. In this case, we're going to use a circle.

1. Press **<Alt-X>** on the keyboard to make the remote control semi-transparent.
2. On the **Create** panel, click **Shapes**, then click **Circle**.
3. In the Top viewport, click and drag to create a circle that matches the left side cylinder chamber in the reference picture.
4. If necessary, use **Select and Move** to position the circle exactly over the circle in the reference picture. If the size isn't right, go to the **Modify** panel and change the **Radius** parameter for the circle.

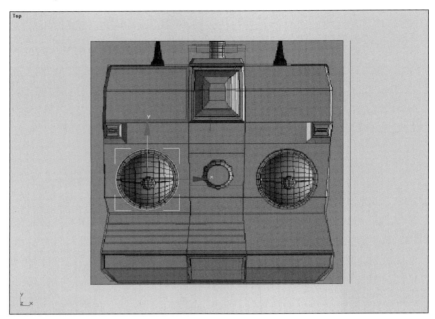

Figure 2.63 *Circle to match joystick chamber*

5. On the **Create** panel, click **Geometry**, and choose **Compound Objects** from the pulldown menu.

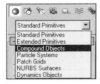

Figure 2.64 *The Geometry pulldown list*

6. Select the remote control object, and click **ShapeMerge** on the command panel.

7. On the Pick Operand rollout, click the **Pick Shape** button, and click the circle. Click **Pick Shape** again to turn it off.

   There is now a circular shape cut into the remote body.

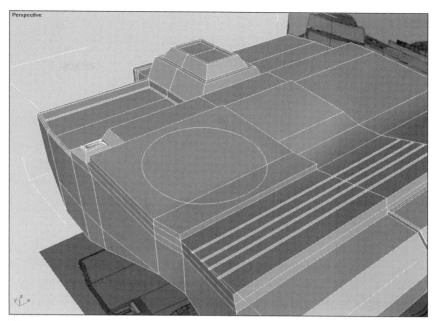

Figure 2.65 *Circle shape merged into remote body*

The remote control in the figure above is shown in non-transparent mode for clarity. Your remote control should still be semi-transparent.

8. If necessary, right-click to turn off the **ShapeMerge** tool.
9. Select the original circle shape, and delete it by pressing **<Delete>** on the keyboard.

Extruding the Joystick Chamber

Using the ShapeMerge tool has turned our object into a ShapeMerge object rather than an Editable Mesh. The first order of business is to turn the object back into an Editable Mesh.

1. Select the remote body and go to the **Modify** panel. Right click on the modifier stack and select *Convert to Editable Mesh*.

2. Switch to the **Polygon** sub-object level.
3. You should find that the two new polygons that make up the circle are already selected. If not, select them now.
4. Click **Make Planar** to make the circle polygons coplanar.
5. In the Extrude section, change the Normal setting to **Group**.

   This will prevent the polygons from coming apart when you extrude them.
6. Extrude the face to the height of the chamber wall as seen in the Left viewport.

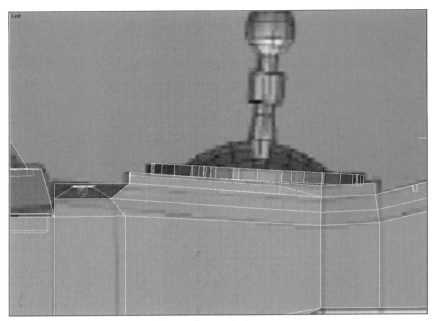

Figure 2.66 *Joystick chamber wall extruded*

7. Press **<Alt-X>** to turn off semi-transparent mode.
8. Use the **Extrude** and **Bevel** tools to create the joystick chamber as shown in the following figure.

If the circle starts to come apart when you extrude it, select the Group option on the Edit Geometry rollout, then extrude again.

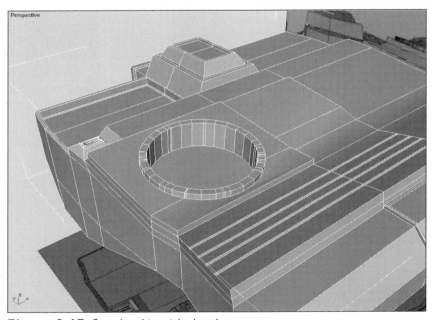

Figure 2.67 *Completed joystick chamber*

Creating Bevels for Highlights

In the real world, it is very rare that an object has all hard edges. Usually, objects have slightly beveled edges. Adding beveled edges to your models will greatly enhance the final renders as the edges provide places off which specular highlights can reflect.

To make the remote control body a little more interesting to look at when it is rendered, we are going to bevel most of the edges running around its side.

1. Switch to the **Edge** sub-object level.
2. Select the hard edges around the outside of the remote body, and the edges around the joystick chamber. See the figure below.

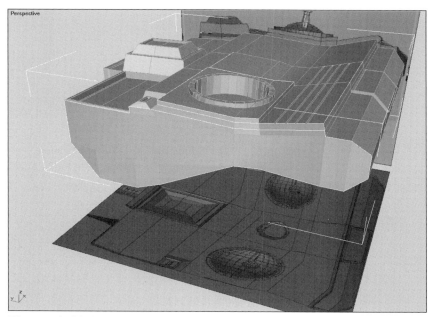

Figure 2.68 *Selected outer edges*

3. Using the **Chamfer** spinner, apply a slight chamfer (about 0.3) so the selected edges become slightly beveled.

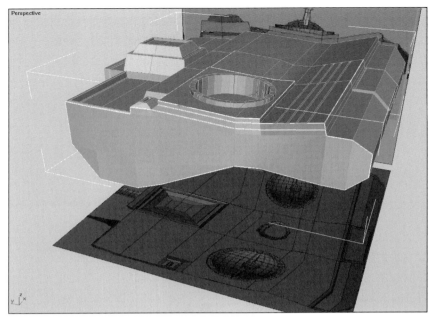

Figure 2.69 *Hard ddges after being chamfered*

The detail on the remote control body is now complete! We just have to mirror the detailed half over to the other side, and then we're done.

4. Save the scene with the filename **RemoteControl04.max**.

Slicing the Remote Control

It is now time to slice the remote in half, delete the side without all of the detail and then mirror the detailed side. Finally, we will attach and weld the two sides together so the remote body becomes one seamless piece.

1. Turn off the **Edge** sub-object level.
2. Click the Modifier List down arrow and choose the **Slice** modifier.

   The slice plane gizmo appears in the scene. This plane determines where the slice will be made.

3. On the modifier list, click the **[+]** next to the Slice modifier, and select the **Gizmo** sub-object level.

4. Turn on the **Angle Snap Toggle** at the bottom of the screen.

5. Click **Select and Rotate** on the Main Toolbar.
6. In the Front viewport, rotate the plane 90 degrees around the Z axis.
7. If necessary, move the slice plane gizmo so it is as close to the center of the object as possible. You don't have to have it exactly at the center, but the closer the better.

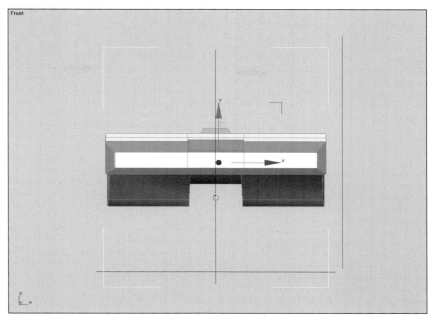

Figure 2.70 *Slice plane through center of remote control*

Having the object as an Editable Poly will help later on when mirroring and attaching the two halves of the remote control.

8. Choose the **Remove Top** or **Remove Bottom** option on the command panel, whichever deletes the undetailed side while leaving the detailed side.

9. Right-click the modifier stack and choose **Collapse All** from the pop-up menu. Answer **Yes** to the warning dialog to collapse the stack.

   This should have converted the object to an Editable Poly. If it did not, right-click the modifier stack and choose **Convert to Editable Poly** from the pop-up menu.

Mirroring the Detailed Side

1. With the remote control body selected, click the **Mirror Selected Objects** button on the Main Toolbar.

   The Mirror dialog appears.

Figure 2.71 *Mirror dialog*

2. On the Mirror dialog, set the Mirror Axis to **X**, and choose the **Copy** option. Click **OK**.

3. If necessary, move one or both of the halves so the two centers line up as exactly as possible.

Attaching the Two Halves

Before we can weld the center vertices together, we need to attach the pieces so that they will become one object again.

1. With one of the remote halves selected, click **Attach** on the Edit Geometry rollout, then click on the other half of the remote control to attach it.

   The two halves are now part of the same object.

2. Click **Attach** to turn it off.

3. Switch to the **Vertex** sub-object level.

4. In the Top viewport, use a marquee to select the vertices down the center of the object.

5. On the Edit Geometry rollout, click **Selected** in the Weld section to weld the vertices together.

   If a message appears stating that no vertices are within the weld threshold, increase the threshold value next to Selected and try again.

   Mirroring the two halves has given us a set of edges down the center of the remote control that we don't really need. Because we're working with an Editable Poly, we can delete these edges while keeping the overall polygon structure intact. This is why we are working with an Editable Poly now and not an Editable Mesh. If you delete edges on an Editable Mesh, you are left with gaping holes where edges used to be.

6. Switch to the **Edge** sub-object level.

7. Select the edges down the center of the remote control.

8. Press **<Delete>** on the keyboard to delete the selected edges.

   The unnecessary edges have been deleted while the overall structure is retained.

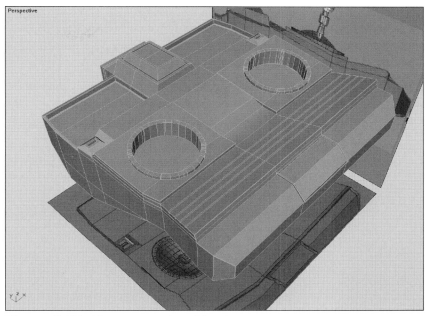

Figure 2.72 *Unnecessary edges deleted*

Although this technique gives you a nice object with no visible seam, it does leave isolated vertices down the middle of the object. Leaving the vertices there won't cause any problems in this particular case, but if you really want to get rid of them, follow the instructions under **Locating and Removing Isolated Vertices** earlier in this chapter.

9. Save the scene with the filename **RemoteControl05.max**.

Smoothing Groups

Smoothing groups provide a way to generate smooth edges or hard edges.

Some areas on the left and right sides of the remote body appear smooth and other areas have very distinct edges, particularly when the scene is rendered. Although a smoothed or faceted appearance is determined to some degree by the angles between polygons, it has a lot more to do with smoothing groups.

Every polygon can belong to a number of smoothing groups, to which it can be assigned via an Editable Mesh or the Smooth modifier. When **3ds max** displays or renders an object, it looks at each polygon's smoothing groups and the smoothing groups of the polygons around it. If two contiguous polygons share a smoothing group, **3ds max** makes a smooth transition between the two, giving the object a rounded appearance at that place. If two contiguous polygons do not share a smoothing group, the polygons are displayed or rendered with a hard edge between them.

To assign smoothing groups to polygons, simply select the polygons and click a smoothing group number on the Surface Properties rollout.

1. Switch to the **Polygon** sub-object level, and uncheck **Ignore Backfacing** if necessary.
2. Select all the polygons on the object, and click **Clear All** on the Surface Properties rollout.

   This removes all smoothing groups from the selected polygons.

3. Select all polygons around the sides of the remote control body. Do not select the polygons created from the Chamfer operation. Selecting the polygons might be easier if you hide the reference planes by selecting them and clicking **Hide Selected** on the **Display** panel.

4. On the Surface Properties rollout, click the **1** button under Smoothing Groups.

   This assigns all the selected polygons to smoothing group 1.
5. Select the sides of the outer walls of the joystick chambers, and assign them smoothing group **2**.
6. Select the all the polygons on the plate at the bottom front of the remote control and assign them smoothing group **3**.
7. Select the inner areas of the monitors and assign them smoothing group **4**.
8. Select the polygons in the inner area of the top of the LED and assign them smoothing group **5**.

   Feel free to experiment and try out different combinations of smoothing groups. The rendered result of the smoothing groups described above is shown in the following figure.

Figure 2.73 *Rendered remote control with smoothing groups assigned*

Modeling the Remote Control **81**

 If you have hidden the reference planes, be sure to unhide them when you are finished by choosing **Unhide All** on the **Display** panel.

If you render the model, you will find that the sides of the remote no longer appear faceted, and that the outer joystick chambers are smooth and round.

Cleaning up Turned Edges

Turning edges can make a big mess, so be sure to save the scene before continuing.

After an object has been box modeled and smoothing groups have been assigned, there are sometimes issues with edges. An edge can be turned the wrong direction, causing the face to look strange.

To determine if you have problems with edges, rotate the Perspective view and carefully inspect the remote control body. Look for faces or polygons that appear indented or pushed out for no apparent reason. If you find a face or polygon like this, follow this procedure to try and solve the problem.

1. On the **Display** panel, uncheck the **Edges Only** option.
2. On the **Modify** panel, access the **Edge** sub-object level.
3. Click the **Turn** button on the Edit Geometry rollout.
4. Click on an edge that looks puckered, pinched or pushed out.
5. Look carefully at the model to see if turning the edge solved the problem. If it didn't solve the problem or made it worse, press **<Ctrl-Z>** to undo the change.
6. When you have finished turning edges, click the **Turn** button again to turn it off, and check the **Edges Only** option on the Display panel.

The remote control body is now complete.

7. Save the scene with the filename *RemoteControl06.max*.

THE JOYSTICK AND ITS BASE

In the next tutorial you will be putting together the joysticks and their spherical bases. The bases will be made from ordinary spheres converted to Editable Meshes, while the joysticks will be created with a profile shape and the Lathe modifier.

TUTORIAL 2.3 JOYSTICK ASSEMBLY

Creating the Spherical Base

1. You can continue working with the remote body that you modeled in the last tutorial or open the file *RemoteControl06.max* from the **Scenes** folder on the CD.

2. On the **Create** panel, click **Geometry**. Choose **Standard Primitives** from the pulldown menu.
3. Click **Sphere**.
4. In the Left viewport, create a sphere.

5. Go to the **Modify** panel and set the following parameters for the sphere.

Name	Left_Joystick_Base
Radius	31
Segments	24
Slice On	checked
Slice From	90
Slice To	-90

> ☼ **TIP** ☼
>
> *The slice settings cut the sphere into a hemisphere.*

6. In the Top viewport, move the hemisphere over the left joystick chamber.

7. In the Left viewport, move the hemisphere down so it sits inside the joystick chamber.

Extruding the Joystick Slot

1. Convert the sphere to an Editable Mesh.
2. Switch to the **Polygon** sub-object level.
3. Select the two rows of center faces running horizontally.
4. Extrude polygons to make a slot for the joystick, and bevel edges to catch highlights.

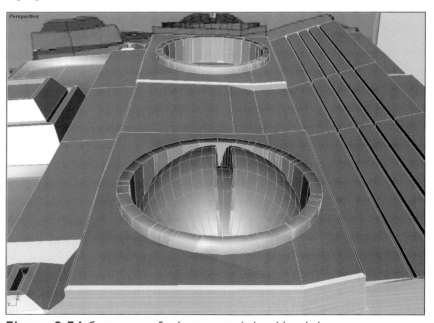

Figure 2.74 *Center rows of polygons extruded and beveled*

5. Turn off the **Polygon** sub-object level.
6. In the Left viewport, rotate the hemisphere slightly so the opening aligns with the reference picture.

Creating the Joystick

We will use an outline shape and the Lathe modifier to create the joystick.

1. On the **Create** panel, click **Shapes**.
2. Click **Line**.
3. In the Left viewport, draw a rough profile of one side of the joystick handle. Draw the profile as if the joystick were standing upright. Click to set each vertex, then right-click when you have finished. Use no more than 9 or 10 vertices.

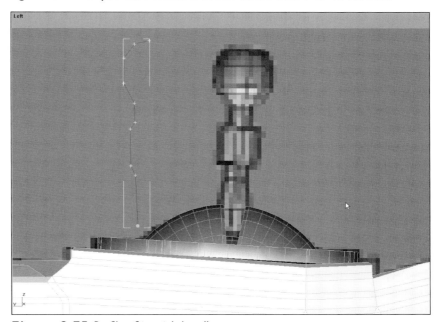
Figure 2.75 *Profile of joystick handle*

4. Go to the **Modify** panel.

5. Choose the **Vertex** sub-object level.
6. Select all the vertices on the line. Right-click any selected vertex and choose the **Smooth** option from the upper left quadrant of the Quad menu.
7. Move and edit the vertices that you've placed so that the profile matches the joystick shape as accurately as possible.

> **TIP**
>
> A vertex can be one of four types: Bezier-Corner, Bezier, Corner, or Smooth. Convert the vertices to any of the four vertex types as needed to get the joystick shape you want. Bezier and Bezier Corner vertices have two handles that you can move around to control the segment on each side of the vertex.
>
> Vertex types are discussed in detail in Chapter 4.

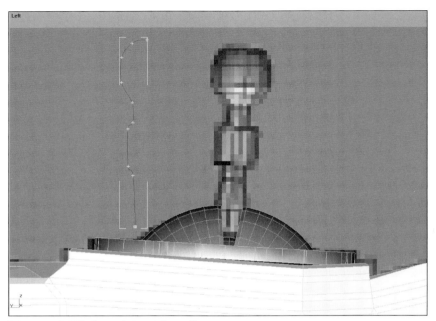

Figure 2.76 Smooth joystick profile

Lathing the Joystick

1. With the line selected, click the down arrow for the Modifier List and choose the **Lathe** modifier.

    The line is lathed, but it looks funny. The lathe center axis is not properly placed.

2. If you created a left-profile as shown in Figure 2.76, click the **Max** button on the Parameters rollout to push the axis to the right of the shape. If you created a right-side profile, click **Min** to push the axis to the left.

    The axis jumps to the rightmost or leftmost side of the shape, and the joystick looks more normal.

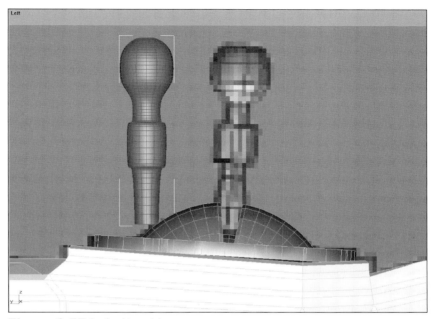

Figure 2.77 *Lathed joystick*

TIP

A normal is an imaginary arrow pointing outward perpendicular to a face. The direction of the normal determines the "outside" direction of the face, which is the side that is shaded and rendered.

Look carefully at the joystick in your Perspective viewport. Does it appear to be inside-out? If so, then your surface normals are pointing the wrong way. The Lathe Modifier has an option that allows you to flip the directions of the normals.

3. Check and uncheck the **Flip Normals** checkbox to see which way the joystick looks better.

4. If your joystick is too thick or too thin, you can return to the **Line** level of the stack and adjust the vertices until the joystick looks right.

5. When you have finished adjusting the joystick, convert the object to an Editable Mesh.

6. Name the object **Left_Joystick**.

7. Move and rotate the joystick so it aligns with the side reference picture and it fits into the slot on the sphere.

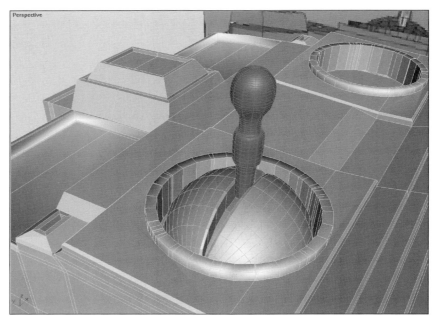

Figure 2.78 *Joystick complete and placed*

8. Make a copy of the joystick and sphere, and place them in the other joystick chamber. After you copy them, name them **Right_Joystick** and **Right_Joystick_Base**.

 TIP

If you like, you can rotate the sphere on the right so the slot runs vertically instead of horizontally.

Don't spend too much time on these details as the face of the remote control will be visible only for a few moments in the final scene.

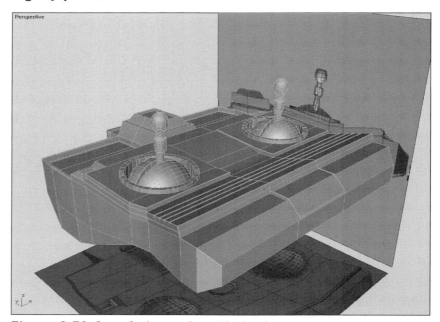

Figure 2.79 *Copy of sphere and joystick placed*

Creating the Center Knob

The center knob will be created with a cylinder.

1. Select the remote control and press **<Alt-X>** to make it semi-transparent.

2. On the **Create** panel, click **Cylinder**. In the Top viewport, create a cylinder over the center knob in the reference picture.

3. On the **Modify** panel, set the following parameters.

Name	Knob
Radius	15
Height	20
Height Segments	1
Cap Segments	1
Sides	18

4. Convert the cylinder to an Editable Mesh.
5. In the Top viewport, position the cylinder over the center knob in the reference picture.
6. If the cylinder is inside the remote, move it up and out so you can work with it.

Creating the Knob Fins

1. Access the **Polygon** sub-object level.
2. Select every other polygon on the sides of the cylinder.

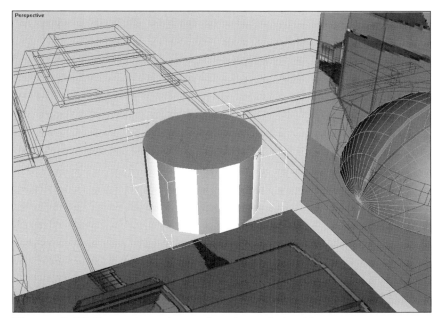

Figure 2.80 *Every other polygon selected on cylinder sides*

3. While watching the Top viewport, use the **Extrude** spinner to extrude the polygons outward to match the size of the gears in the reference picture. Bevel the faces inward by a small amount.

4. In the Front viewport, move the selected polygons downward so they meet the bottom of the knob.

5. Select the polygon on top of the cylinder.

6. Extrude the face up a small amount, then bevel it inward.

7. Click the **Extrude** spinner up arrow one time to create a rim, and bevel the polygon inward to define the rim.

8. Extrude the face downward, and bevel inward again so that the inner portion of the knob has a slight flare.

Aligning the Knob

The knob must be rotated slightly to follow the incline of the polygons on which it rests. You could rotate it manually with Select and Rotate, but it can be difficult at this point to figure out exactly how much to rotate it.

Instead, we are going to use a secret weapon called Align Camera. This tool forces the selected object's Z axis to point at a specified face or polygon.

This tool was originally designed to cause a camera to point at a particular face or polygon, but we can also use it to make an object's rotation align with a specified polygon.

1. Turn off the **Polygon** sub-object level.

2. Use **Region Zoom** to zoom into the knob area of the Top viewport.

3. With the knob selected, choose **Align Camera** from the **Align** flyout on the Main Toolbar.

4. In the Top viewport, click and drag the cursor over the polygon on which the knob sits and look for the blue line or dot. Position the blue line or dot over the center of the knob and release the mouse.

This aligns the knob with the polygon underneath it. The change will not be obvious in the Top viewport, but you will be able to tell in the Left and Perspective viewports that the knob has been rotated.

Figure 2.81 *Knob rotated to align with polygon underneath*

5. If necessary, use **Select and Move** to move the knob vertically so it sits just above its polygon.

6. Save the scene with the filename *RemoteControl07.max*.

FINISHING TOUCHES

The remote control only needs two more objects to make it complete: an antenna for controlling the car, and the hand grip for easy toting into the field.

So far we have used box modeling almost exclusively to create the remote control and joystick assembly. To make the antenna and hand grip, we will use a different technique called *lofting*.

Lofting sweeps or pushes one shape along a path, leaving behind a surface. For example, you could make a pipe by creating a path in the shape of the pipe, then passing a circle shape along the path.

Once an object is lofted, you can easily scale different parts of it and perform other modeling tasks on the object.

Since you have not been supplied with a reference picture for the antenna, you will be using your imagination and judgment to determine the way the antenna will look.

Over the course of creating the antenna and hand grip you will be given some specific instructions, but you will also have the opportunity to experiment and make your own personalized antenna for the remote control.

TUTORIAL 2.4 ANTENNA AND HAND GRIP

Lofting requires two shapes, one for the path and another to pass along the path.

Creating the Loft Path

The first order of business is to create the path for the lofted object. A path is an ordinary shape, in this case a straight line. In order to ensure the line comes out perfectly straight, we will use snap tools to snap to the viewport grid.

1. Right-click the **3D Snap** button.

   This displays the Grid and Snap Settings dialog. Earlier, you accessed this dialog by choosing *Customize/Grid and Snap Settings* from the menu.

   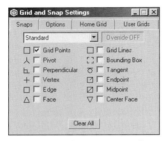

   Figure 2.82 *Grid and Snap Settings dialog*

2. Uncheck all checkboxes, then check **Grid Points**. Close the dialog.

   This will enable us to snap to grid points when a snap tool is turned on.

3. Use **Zoom** on the Top viewport to zoom out so the remote control takes up only about 1/4 of the viewport.

4. Use **Pan** to pan the view so the remote control is sitting at the bottom center of the viewport.

5. Press the **<G>** key on the keyboard to activate the grid for the viewport.

6. Turn on **3D Snap**.

7. On the **Create** panel, click **Shapes**, then click **Line**.

8. Create a line with two vertices to represent the total length of the antenna. Click once near the base of the antenna and once at the far end, then right-click to end creation of the line. Be sure to snap to grid points to ensure the line is straight.

9. Turn off **3D Snap**.

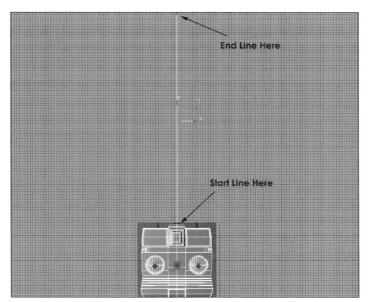

> **TIP**
> It might be difficult to see the line right after you create it. If in doubt, look for the transform gizmo (XY axes). If they appear at the center of the area where you created the line, this means the line is there. You can also make the line easier to see by pressing <G> to turn off the grid.

Figure 2.83 *Click points for path in Top viewport*

Creating the Loft Path

> **TIP**
> If you draw a very small circle, the circle might not be drawn. if this happens, make a larger circle then change the Radius afterward.

Next, you must create a shape to be passed along the path. A circle is perfect for this shape.

1. On the **Create** panel, click **Circle.**
2. In the Top viewport, draw a small circle that represents the general thickness of the antenna.
3. On the panel, change the circle's **Radius** to 8.

Checking for Corner Vertices on the Path

When creating the path line, it's easy to accidentally click and drag a little rather than just click to set the vertex. Simply clicking creates a Corner type vertex, but clicking and dragging creates a Bezier vertex.

When you create a line to be used as a straight-line path for a loft, it is important that the vertices at each end be Corner vertices. Otherwise, the lofted object might pucker or pinch at the ends.

It is difficult to tell if the vertices are Corner or Bezier type without going through these brief steps, and it will save you many headaches if you just test the vertices right after making the path.

1. Select the line that represents the antenna path.

2. On the **Modify** panel, access the **Vertex** sub-object level.

3. Right-click each vertex and see which vertex type is checked on the Quad menu. If the vertex is a Corner type, leave it as is. If it is a Bezier type, change it to a Corner type by choosing **Corner** on the Quad menu.

   If your path is too long or short, you can move one of the vertices to change its length at this time. Just be sure to move the vertex only along the Y axis in the Top viewport to ensure the path stays straight.

4. Turn off the **Vertex** sub-object level.

Lofting the Antenna

To use the loft tool, you must first select one of the shapes to be used, then pick the other shape after you access the loft tool.

Although you can loft with either the path or profile shape selected first, you should keep in mind that the first selected shape determines the location of the lofted 3D object. We'd like the 3D antenna to appear where the path is, so we'll start out with the path selected.

1. Select the line that represents the antenna path if it is not already selected.

2. Go to the **Create** panel and click **Geometry**. From the pulldown menu, choose **Compound Objects**.

3. Click **Loft**.

4. Click **Get Shape**, then click on the circle profile shape.

   The loft operation creates a cylindrical object around the path.

☼ TIP ☼

Make sure you turn off Get Shape before proceeding. If you forget, it is very easy to click on another shape (such as the loft path) and end up with a mighty strange looking object, with no idea how you got it.

If this happens, press <Ctrl-Z> to undo, then turn off Get Shape before continuing. And don't say we never told you!

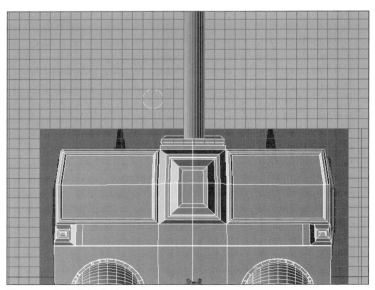

Figure 2.84 *Lofted antenna*

5. Right-click to turn off **Get Shape**.
6. Name the object **Antenna**.

Accessing Scale Deformation

You may well wonder, what's the big deal with lofting when all we did was make a cylinder?

One important difference between loft objects and standard primitives is that lofts have a host of tools for deforming the mesh. For instance, it is a simple matter to change the scale of a lofted object at different points along the path to create a complex shape. The deformation can be performed in real time without having to select vertices or polygons. You will learn how to use this feature in the steps that follow.

1. Go to the **Modify** panel.
2. The lofted object should still be selected. If it is not, select it now.

3. Expand the Deformations rollout at the bottom of the command panel.

   The options on this rollout are available only on the **Modify** panel, and only for loft objects.

4. Click **Scale**.

   The Scale Deformation window appears.

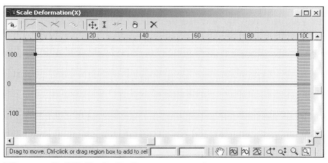

Figure 2.85 *Scale Deformation window*

The line on the graph represents the path, while its value represents the thickness of the lofted surface. The numbers at the left represent the percentage of the original profile shape. The line is currently at the 100 mark all the way across, which means the thickness of the object is at 100% of the original profile shape all along the path.

The values across the top of the window represent the percentage of the total path length. There are currently two control points on the line, one at each end. The left end of the line (under the 0 column) represents the first point you created on the path, while the other end represents the last point. You will add more control points so you can change the shape of the line.

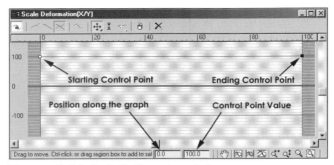

Figure 2.86 *Scale Deformation window areas*

Deformations work with the concept of XYZ axes, but not the ones you're accustomed to using in viewports. With a loft deformation, the direction of the path is the local Z axis, and the XY plane is considered to lie on the profile shape as it passes up the path.

The line on the Scale Deformation window represents the Z axis along the path. As we change the shape of this line, the lofted object's width (X and Y axes) will change at those points to a percentage of the original width. In a sense, you will be changing the lofted object's profile.

In the Scale Deformation window, you will work with the loft object's local X and Y axes to change the width of the object at various points along the path.

5. In the Scale Deformation window, click **Display XY Axes**.

This will ensure that we are working with both axes at the same time.

Shaping the Antenna

In order to make changes to the line and thus shape the antenna, we need to add more control points to the line.

1. On the Scale Deformation window, click **Insert Corner Point**, then click on the line. This adds a control point to the line.

2. Click **Move Control Point** and click and drag on the point to move it.

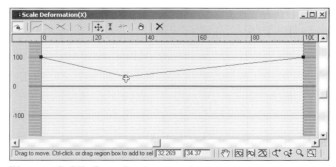

Figure 2.87 *Control point moved*

As you move the control point, the shape of the antenna changes accordingly. You have the choice of placing Corner or Bezier control points on the line. Bezier point allow you to change the shape of the line around them with handles. To place a Bezier control point on the line, click and hold the Insert Corner Point button and choose Bezier Control Point from the flyout, then click on the line to create the point. Use Move Control Point to move the handles.

After creating a control point, you can change its type by right-clicking the point and choosing a point type from the pop-up menu.

3. Create and move control points to shape the line so it looks similar to the figure below.

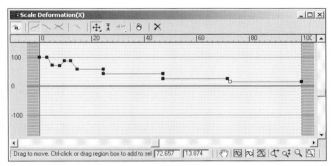

Figure 2.88 *New antenna profile in Scale Deformation window*

4. Reduce the number of faces on the loft object by reducing the **Shape Steps** and **Path Steps** values on the Skin Parameters rollout. It is usually possible to reduce these values to 1, 2 or 3 while still retaining the shape of the object.

The resulting antenna should look similar to the following figure.

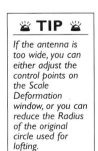

☀ **TIP** ☀

If the antenna is too wide, you can either adjust the control points on the Scale Deformation window, or you can reduce the Radius of the original circle used for lofting.

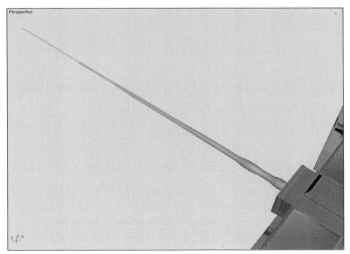

Figure 2.89 *Final antenna*

5. If necessary, move the antenna so it comes out of the top center of the remote control.

6. When you are completely satisfied with the antenna, delete the path used to make it. Leave the circle in the scene for later use in lofting the hand grip.

Creating the Hand Grip

You can create a hand grip for the remote control using a technique similar to the one just used to loft and deform the antenna. You're already familiar with the technique so we'll just give you the basics.

1. On the **Create** panel, click **Shapes**. In the Top viewport, create a **Line** shape in the rough shape of the hand grip.

2. On the **Modify** panel, access the **Vertex** sub-object level and adjust vertices as needed.

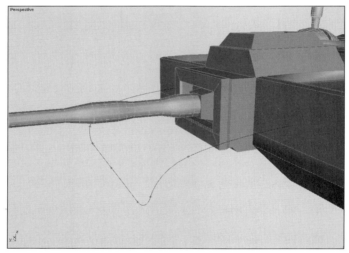

TIP
You don't want the hand grip to pass through the antenna, so you'll need to move some of the path vertices toward the remote control's back.

Figure 2.90 *Loft path for hand grip, with reference planes hidden for clarity*

3. With the path selected, go to the **Create** panel and click **Geometry**. Click **Loft**.

4. Click **Get Shape** and click the circle used for the antenna loft. Right-click to turn off **Get Shape**.

5. On the **Modify** panel, click **Scale** on the Deformations rollout.

6. Create and move control points to set the hand grip profile.

Our profile shape and the resulting appearance of the hand grip ends is shown in Figure 2.91. The window is zoomed into the left end so you can see the shape of the control points in detail. The right end of the window is a mirror image of the left end.

You can make your hand grip look like ours, or tap into your native creativity and make it look any way you like.

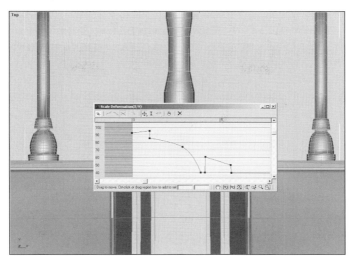

Figure 2.91 *Suggested scale deformation and resulting hand grip detail*

The remote control is now complete.

Figure 2.92 *Rendered image of remote control in all its glory*

When you later merge the remote control into the main scene, it will be easier to merge if it has been grouped.

7. Select all objects in the remote control assembly (body, joysticks and bases, antenna and hand grip).
8. From the **Group** menu, choose **Group**. Name the group **RemoteControl**.
9. Save the scene as **Remote Control Model.max**.

SUMMARY

In this chapter, you have learned how to:

- Set up a virtual studio with reference images
- Convert an object to an Editable Mesh or Editable Poly
- Extrude and bevel polygons to sculpt an object
- Cut new edges in an object for additional detail
- Slice and mirror an object, and attach one side to the other seamlessly
- Apply smoothing groups to faces or polygons
- Use the Lathe modifier to create a 3D object from a shape
- Align an object's Z axis with a polygon
- Loft an object from two shapes
- Change a lofted object's profile with Scale Deformation

Congratulations on making it through these modeling techniques! Now you're ready to go on to the next task, modeling the alien.

In this chapter we will model the alien character for our commercial from start to finish. We'll use box modeling techniques to make its body, head and limbs all in once piece. Special smoothing tools will be used to smooth out the rough shape of the body.

TECHNIQUES IN THIS CHAPTER
In this chapter, you will learn how to:

- Display a smoothed version of the model while working with the boxy structure underneath.
- Create additional detail where needed by subdividing vertices, edges and polygons in a variety of ways.
- Sculpt hands, feet and facial features with box modeling.
- Use an instanced object as an aid in modeling.

TUTORIALS IN THIS CHAPTER

The tutorials in this chapter focus exclusively on box modeling.

- *Tutorial 3.1 Alien Torso* shows you how to create the main body of the alien model with an Editable Poly using NURMS subdivisions.

- *Tutorial 3.2 Arms and Hands* and *Tutorial 3.3 Legs and Feet* use numerous box modeling tools to sculpt the alien's limbs.

- *Tutorial 3.4 Finishing the Model* completes the alien model by adding the eyes and attaching the left side to the right.

EDITABLE POLY FEATURES

In *Chapter 1 Modeling the Remote Control*, you learned a number of box modeling techniques. Throughout most of the chapter you edited the remote control as an Editable Mesh, then you converted it to an Editable Poly only at the end when you wanted access to a few specific Editable Poly features.

In this chapter you will use a lot of the same box modeling tools, but will be working on an Editable Poly right from the start. We will do this because an Editable Poly has a superior smoothing feature that we can use to make our alien look great with a minimum of work. This feature is called called *NURMS subdivision*.

NURMS SUBDIVISION

NURMS stands for *Non-Uniform Rational MeshSmooth*. NURMS subdivision automatically adds faces to an object and adjusts the faces to make the object appear very smooth and rounded.

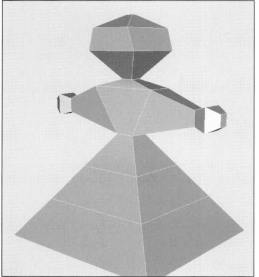
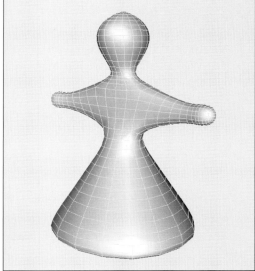

Figure 3.1 *Simple figurine before and after NURMS subdivision is applied with 2 Iterations*

When you use NURMS subdivision, the number of faces on the object might double or triple, but the amount of smoothness you get for this level of detail is outstanding.

NURMS subdivision is enabled on the Surface Properties rollout at the Editable Poly root level by checking Use NURMS Subdivision. The Iterations parameters in the Display and Render sections set the number of times NURMS subdivision is applied in viewports and the rendering respectively. In order for you to see smoothing in viewports, Iterations under the Display section must be set to at least 1.

You could achieve the same smoothing effect with the MeshSmooth modifier. In fact, a common approach is to use NURMS subdivision while modeling, then turn it off and apply MeshSmooth after the fact as a separate modifier. But during the modeling process, NURMS subdivision gives you an additional feature: The ability to edit the original vertices while seeing the results on the smoothed surface.

MESH CONTROL AT SUB-OBJECT LEVELS

Once you have enabled NURMS subdivision and set Iterations for the display, accessing a sub-object level (such as Vertex) displays a "cage" around the smoothed object representing the original boxy object. You can then move, rotate or scale the vertices on the cage to automatically adjust the smoothed version of the object, enabling you to see your changes in real time.

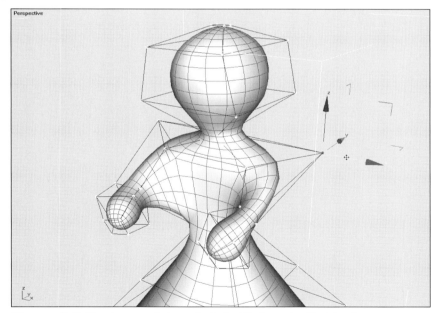

Figure 3.2 *Figurine cage being adjusted at Vertex sub-object level*

At the Vertex sub-object level, you can set different weights for vertices to push the smoothed mesh inward or outward at different points.

NURMS subdivision is an excellent tool for modeling characters such as the alien.

TUTORIAL 3.1 ALIEN TORSO

We'll start our modeling session by creating the alien torso. Before we get started, we'll need to set up a virtual studio to help us create the alien correctly.

Setting up the Virtual Studio

If you simply want to load the virtual studio and get started with modeling, load the file **Alien Virtual Studio.max** from the **Scenes/Virtual Studios** folder on the CD that comes with this book.

If you want to create your own virtual studio, basic instructions are given here. More detailed information on setting up a virtual studio can be found in **Chapter 2 Modeling the Remote Control**.

1. Create a plane in the Front viewport.

2. Create a material with the **Diffuse Color** map set to the file **Alien_front.jpg** from the **Maps/Alien Refs** folder on the CD. Assign the material to the plane and click **Show Map in Viewport** on the Material Editor toolbar.

Check the 2-Sided option for the material to ensure it shows up on both sides of the plane.

3. Apply a **UVW Map** modifier to the plane. Adjust mapping coordinates as necessary.

4. Copy and rotate the front plane to make a plane facing the Left viewport. For the side plane, create and apply another material with **Alien_side.jpg** as the **Diffuse Color** map. Display the map and adjust mapping coordinates.

For the side material, you will need to uncheck the Tile option on the Coordinates rollout at the Bitmap level to prevent the side image from tiling.

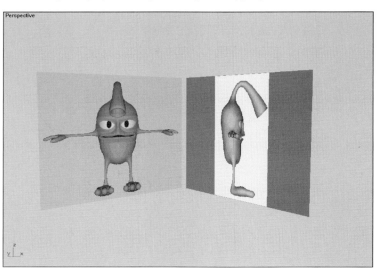

Figure 3.3 *The alien virtual studio*

It's always a good idea to freeze the planes so you won't select them by accident during the modeling process.

5. Select the two reference planes.

6. On the **Display** panel, under the Display Properties rollout, uncheck **Frozen in Gray**.

7. On the Freeze rollout, click **Freeze Selected**.

   Now we're ready to start creating the alien body.

Preparing the Box

The alien torso will be created with box modeling. Box modeling always begins with a simple object such as a box. Through experience you will know how large to make the box and how many segments to give it before you start modeling. We'll help you out here by giving you the exact dimensions and detail for the box.

1. Create a **Box** in the Top viewport with the following settings:

Name	Alien
Length	110
Width	110
Height	160
Length Segs	2
Width Segs	2
Height Segs	1

2. Press **<Alt-X>** on the keyboard to make the box semi-transparent.

3. Click in each viewport and press **<F4>** to turn on Edged Faces in each one.

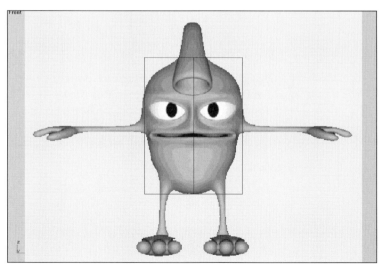

Figure 3.4 *Box lined up with main part of alien torso in Front viewport*

4. In the Front viewport, move the box until it is positioned directly in line with the alien torso in the front reference picture.

5. Right click on the box. On the Quad menu, choose **Convert to**, then **Convert to Editable Poly**.

   As you saw in **Chapter 2**, an Editable Poly is similar to an Editable Mesh, but it has a few properties that an Editable Mesh doesn't. We will make use of these properties in modeling the alien.

6. On the **Modify** panel check the **Use NURMS Subdivision** checkbox on the Surface Properties rollout.

7. On the Surface Properties rollout, in the Display section, set **Iterations** to 1.

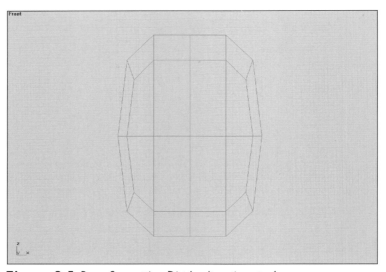

Figure 3.5 *Box after setting Display Iterations to 1*

Using NURMS subdivision creates additional polygons on the box and rounds it out, giving us an egg shape that approximates the shape of the alien's torso. It will also give us some special modeling tools for further shaping of the box.

Shaping the Box to Match the References

1. Access the **Vertex** sub-object level.

   Note the "cage" around the object. We will use this cage to shape the box to match the reference images.

Modeling the Alien **105**

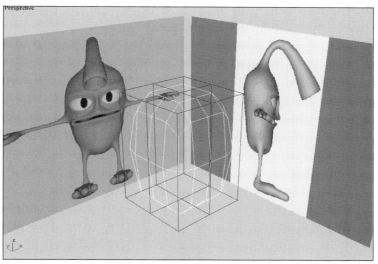

Figure 3.6 *Cage around object at Vertex sub-object level*

2. In the Front viewport, use a marquee to select the top row of vertices.

3. Use **Select and Move** from the Main Toolbar to move the selected vertices on the X axis so that they align with the top of the main part of the head (not the alien's snorkel).

4. Use **Select and Non-Uniform Scale** to scale the size to match as well.

5. Repeat this operation for the bottom row of vertices as well.

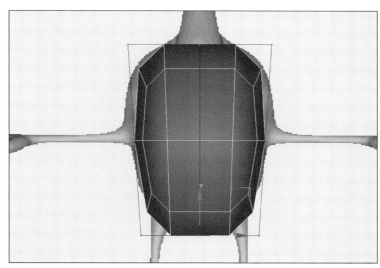

> ☀ **TIP** ☀
>
> Note that you are not moving the vertices up or down on the Y Axis in the Left viewport as this would adversely affect the changes you made in the Front viewport.

Figure 3.7 *Vertices moved and scaled to match front reference*

6. In the Left viewport, move and scale both the top and bottom rows of vertices until the box matches the side reference.

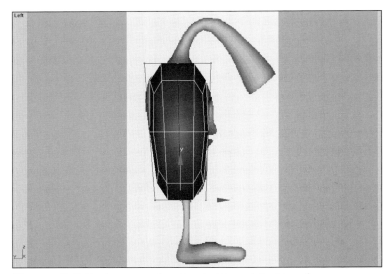

Figure 3.8 *Vertices moved and scaled to match side reference*

Creating New Edges for the Mouth

1. Switch to the **Polygon** sub-object level.
2. Activate the Perspective viewport and press the **<F2>** key to cause selected polygon to be shaded in red.
3. In the Left viewport, use a marquee to select all the center polygons. Verify that they are selected by looking in the Perspective viewport for the red polygons.

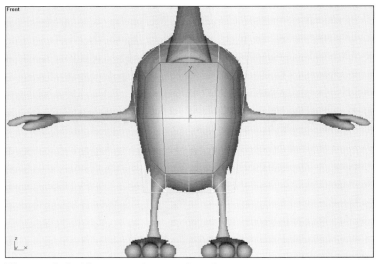

Figure 3.9 *All center polygons selected*

4. On the Edit Geometry rollout, click **Slice Plane**.

   This will give you a slice gizmo that you can now position in preparation for a slice.

5. In the Front viewport, move the slice plane gizmo so it cuts horizontally through the center of the mouth.

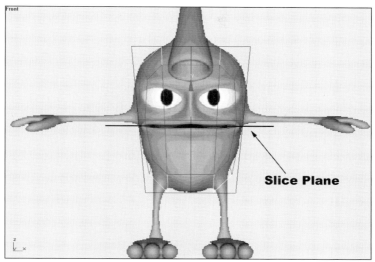

Figure 3.10 *Slice plane positioned at the center of the mouth*

6. Click **Slice** to slice new edges in the object.
7. Click **Slice Plane** to deactivate it.

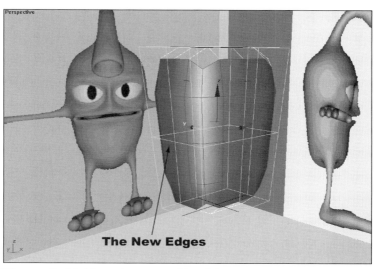

Figure 3.11 *New edges created by slicing*

8. Switch to the **Edge** sub-object level.

   The new edges are already selected by default.

9. While watching the Front viewport, use the **Chamfer** spinners to divide the single row of edges into two rows of edges.

   Your goal is to chamfer these edges until the new top edge is halfway between the bottom of the eyes and the top of the mouth. Try a Chamfer setting of 11 for this purpose.

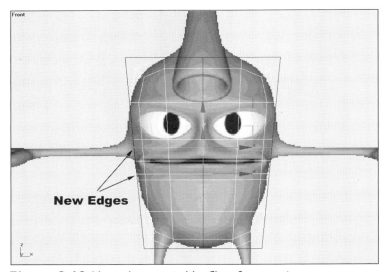

Figure 3.12 *New edges created by Chamfer operation*

Adjusting the Mouth Vertices

Let's take a moment to consider the newly created detail.

Through the Slice and Chamfer operations, we have created two new rows of edges. This of course gave us two new rows of vertices. In the steps that follow, we will be changing only vertices in the new top row.

In the Front viewport, you can see that each of the new row has three vertices: one on the left, one on the right, and a vertex in the center.

1. Switch to the **Vertex** sub-object level.
2. In the Front viewport, use a marquee to select the top center vertex.
3. Move the vertex upward until it is aligned with the bottom of the whites of the eye.
4. Use a marquee to select the vertices on both the right and left ends of the row. Move these vertices to the middle of the arm openings on the reference picture.

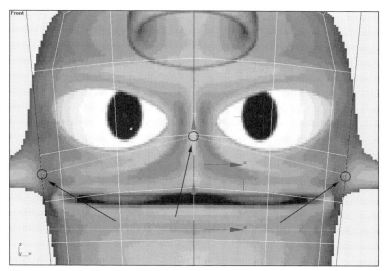

Figure 3.13 *New positions for vertices*

5. Save the scene with the filename *Alien01.max*.

Dividing Edges for the Eyes

1. Switch to the **Edge** sub-object level.

2. On the Edit Geometry rollout, click the **Divide** button.

3. Activate the Perspective viewport. Use **Arc Rotate** to spin the view so you can see the front of the alien. If the alien torso is currently semi-transparent, press **<Alt-X>** to make it solid.

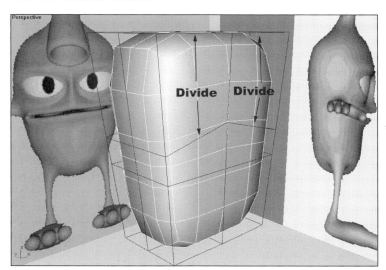

Figure 3.14 *A guide to the divide locations*

We are going to divide two of the edges on the very last row of edges at the top of the alien torso. The front edges are already divided in two. Currently, the alien has a left side and a right side when viewed from the front.

4. With the **Divide** button active, click on the middle of the top left-front edge, then click directly below it on the middle of the second row down.

5. Repeat for the right side, clicking on the top front right edge and the next row down.

6. Click **Divide** again to turn it off.

Adjusting the Eye Vertices

Dividing the edges has given us new vertices in the eye area. These vertices are now going to be used to aid in cutting the edges.

1. Switch to the **Vertex** sub-object level.

2. On the Edit Geometry rollout, click **Cut**.

3. In the Perspective view, move the cursor over the new top left vertex until it changes to crosshairs. Click to start the cut. Next, move your mouse over the new vertex directly below it (second row, left side) and when the pointer changes, click again. Right click to end this cut.

   You have just created a new edge between the two vertices.

4. Repeat this operation for the new vertices on the right side as well.

5. Click **Cut** again to deactivate it.

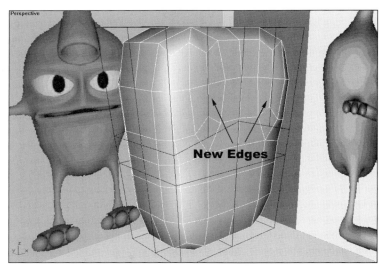

Figure 3.15 *The new edges that were created as a result of the cuts*

6. Save the scene with the filename **Alien02.max**.

Creating the Mouth

 1. Switch to the **Polygon** sub-object level.

   We are now going to begin working with the mouth region that we set up earlier.

2. Select the two polygons in the front middle area where we created the outline for the mouth.

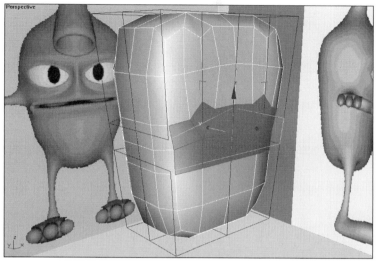

Figure 3.16 *Two selected polygons in the mouth region*

In the Left viewport. note that the alien's lips protrude out from the torso.

3. Extrude the polygons outward to half the distance of the lip in the reference.

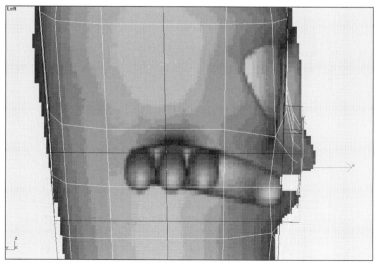

Figure 3.17 *Polygons extruded by half the distance of upper lip in reference*

Next you will use the Outline tool. This tool works the same way as the Bevel tool you used to model the remote control in **Chapter 2**.

4. Use the **Outline** spinner to uniformly scale the polygons inward. Use a value of approximately -2.5 units.

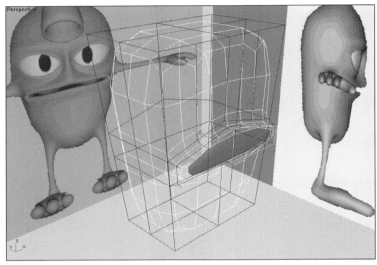

Figure 3.18 *Polygons outlined inward*

5. Extrude the polygons again, this time to the end of the upper lip in the reference.

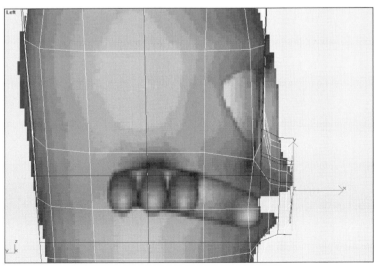

Figure 3.19 *Extrusion matches tip of upper lip in reference picture*

6. Use the **Outline** tool to scale the polygons inward by about -3.2 units.
7. Click the **Extrude** down spinner one time, and use the **Outline** tool to scale the new polygon inward by approximately -3.0 units.

8. Extrude to approximately -18.0 units, then change it again to about -8.0 units.
9. Use the **Outline** tool to scale the polygon outward by approximately 10.0 units.

> ☼ **TIP** ☼
>
> Now we have some lips to work with, but the mouth is too large and doesn't really resemble the reference picture. To achieve the look we want, vertices will have to be moved.

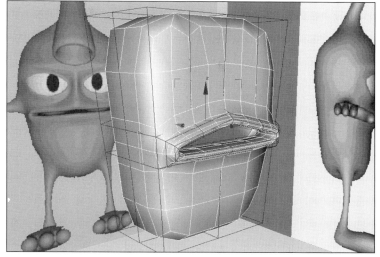

Figure 3.20 *Extruded lips*

Shaping the Mouth

1. Switch to the **Vertex** sub-object level.
2. On the upper lip, use a marquee to select the left and right column of vertices next to the center row. Refer to the next figure.

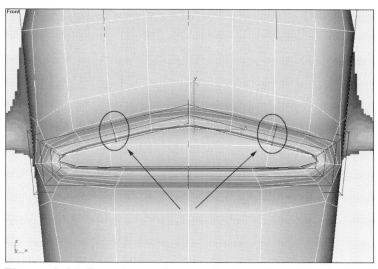

Figure 3.21 *Two columns of vertices closest to the center column*

3. Click **Select and Non-uniform Scale**. In the Front viewport, scale the selected vertices inward on the X axis to approximately half the distance from the starting point to the center vertices.

4. Use a marquee to select the three middle columns of vertices on the top lip.

   This selection includes the two columns that you scaled earlier plus the center column.

5. Press **<Alt-X>** to make the object semi-transparent.

6. Move the selected vertices downward to line up with the lip in the reference.

   You may need to scale the selection upward on the Y axis to better fit the reference.

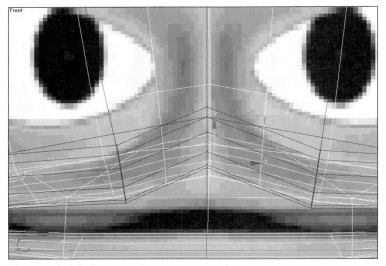

Figure 3.22 *Focus on the center upper lip*

At this point you are ready to go out on your own in manipulating vertex positions. The goal is to make the mouth geometry closely match the picture reference.

Here are a few suggestions to help you achieve this:

- Save the scene as **Alien03.max** before you start so you can return to an earlier version of the model if necessary.
- Toggle in and out of semi-transparency as necessary to help in the positioning. Transparency is turned off and on with **<Alt-X>**.
- Pay close attention to the Left viewport to ensure you have the correct profile.
- To remove the straight-lip look, select all the vertices that make up the outer edges of the mouth (eight on each side) and move them backward on the X axis in the Left viewport. After moving them back, scale them inward on the X axis in the Front viewport.

- Be very careful not to allow vertices from the inner extruded polygons to protrude through the torso as you shape the mouth. If they do, use a marquee to select them and scale them in toward one another.

- Move some of the vertices on the outer edges upward a slight bit to create a hint of a smile.

- After every major movement or scaling of vertices, use **Arc Rotate** to look around your perspective view and inspect the effect the change had on your geometry. Sometimes you might move the wrong vertices and not realize it until much later, when it is very hard to fix. A quick inspection can help identify any problems before it's too late to undo them. You can undo changes by pressing **<Ctrl-Z>** on the keyboard.

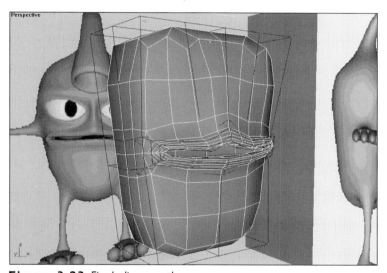

Figure 3.23 *Final alien mouth*

Rounding the Top of the Head

In the next few steps we are going to round the head and prepare to extrude the nose.

1. Rotate the Perspective viewport so that you can see all of the vertices on top of the head.
2. Select all four of the corner vertices.
3. In the Perspective viewport, scale them down on the Y axis.

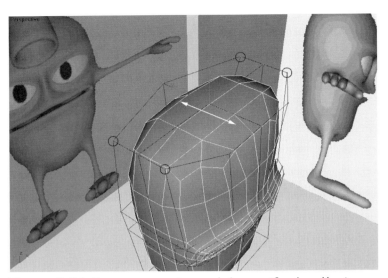

Figure 3.24 *Outer corner vertices scaled non-uniformly on Y axis*

4. Select the top middle vertex and move it up. Use the Front viewport to determine how far up to move it. You might find being in semi-transparent mode helpful.

5. In the Front viewport, use a marquee to select the top row of vertices. Do not select the one that you moved up in the previous step.

6. In the Front viewport, position the selection of vertices so that the head matches the reference picture.

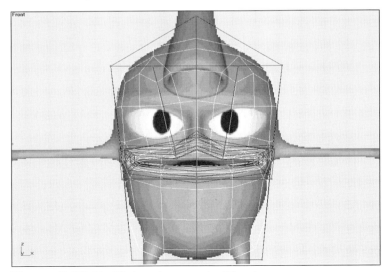

TIP

Uniform and non-uniform scaling are great ways to position entire rows of selected vertices.

Figure 3.25 *Modified vertices on top of head*

7. Switch back and forth between the Front and Left viewports and manipulate the vertex positions to make the geometry better represent the reference pictures.

Creating Edges for Nose Extrusion

The alien has a long snout or snorkel that we call the nose. This nose will be extruded from the top of the head. Currently, there are not enough edges to properly perform the extrusion.

If you look at the vertex that we pulled upward to the very tip of the head earlier, you will see that it has only four edges coming out of it. If you look at the last row of vertices, you will see that there are four vertices on the other ends of those edges, and that there are six vertices not generating edges to the top vertex. We will create edges between these six vertices and the top middle vertex. This will give us sufficient geometry to extrude the nose.

1. Turn on the **Cut** tool.
2. In the Perspective view, click on one of the vertices that's on the top row of vertices and then click on the center vertex. This will create an edge between the two vertices. Right click to complete the cut.
3. Repeat this operation until all six vertices have edges that run up to the top middle vertex. When complete, every outer vertex on the last row (at the top of the head) will have and edge leading to the center top vertex.

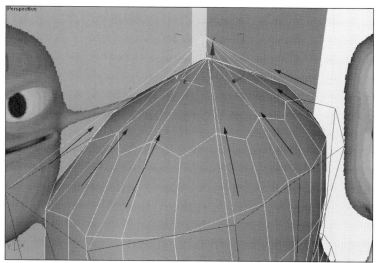

Figure 3.26 *Top of head connects to every vertex on next row*

4. Turn off the **Cut** tool.
5. Use a marquee to select around the vertex at the top middle of the head.

On the Selection rollout, the number of vertices currently selected appears next to the word Vertex. This information is very handy if you're not sure whether multiple vertices are on top of one another.

6. If there is more than one vertex selected, click **Collapse** on the Edit Geometry rollout.
7. Recheck the number of selected vertices. There should be only one.

Creating the Nose Extrusion Polygon

1. With the single vertex still selected at the top of the head, use **Chamfer** to split it into an area large enough to accommodate the first nose extrusion.

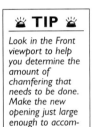

TIP

Look in the Front viewport to help you determine the amount of chamfering that needs to be done. Make the new opening just large enough to accommodate the diameter of the nose's first extrusion.

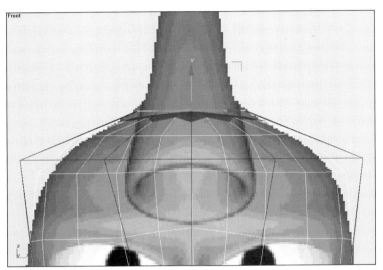

Figure 3.27 *Front view of the new chamfered vertices, wide enough for nose opening*

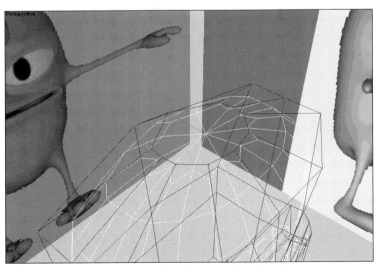

Figure 3.28 *Top of the head with vertex chamfered to create circular shape*

2. Select the center vertex and all the new vertices around the newly chamfered polygon. In the Front viewport, move the vertices downward to flatten out the head a little.

3. In the Top viewport, position the vertices around the top circular area until they more closely resemble the shape of a slightly wide ellipse. The closer you get to a perfect ellipse, the better the final nose will look.

TIP

This is a very important step, so take a few moments to arrange the vertices as closely as possible into a perfect ellipse.

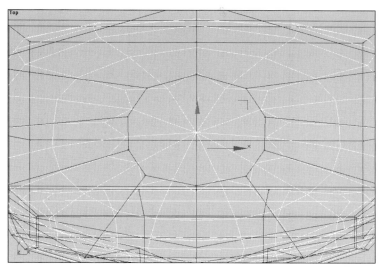

Figure 3.29 *Vertex pattern arranged into an ellipse*

4. Switch to the **Polygon** sub-object level.

5. Select the elliptical polygon. On the Edit Geometry rollout, click **Make Planar** to flatten it out.

6. Save the scene with the filename *Alien04.max.*

Extruding the Nose

You will now use a combination of extrusions, outlines, and rotations to create the nose. Use the Left viewport reference as you do each of these operations.

The top elliptical polygon should still be selected.

1. Extrude the polygon approximately 20 units upward.

2. Use the **Outline** spinner to scale the face inward by about -3 units.

3. In the Left viewport, use **Select and Move** and **Select and Rotate** to reposition and rotate the polygon so it is aligned with the reference picture.

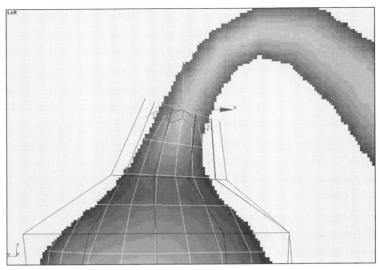

Figure 3.30 *Side view of first nose extrusion*

4. Perform another extrusion, this time roughly 18 units upward.
5. Use the **Outline** spinner to scale the face inward another -3 units.
6. Rotate and reposition the face to align it with the reference picture.

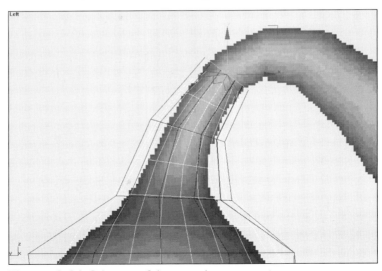

Figure 3.31 *Side view of the second nose extrusion*

Continue this process (extrusion, outline, rotate, position) until you have reached the end of the nose.

7. Save the scene with the filename **Alien05.max** before continuing.

8. At the end of the nose, use a combination of inward extrusions and inward outlines to extrude the polygons back into the nose. You may need to do this a couple of times to get comfortable with the process. If this process goes wrong, return to the saved version of the file and try again.

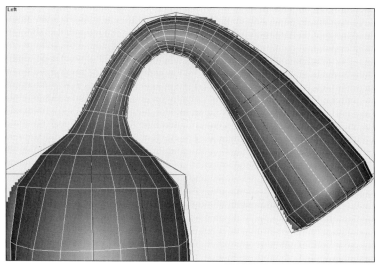

Figure 3.32 *Side view of the completed nose extrusion*

Cutting and Mirroring the Alien

As we've seen with previous modeling tasks, when modeling a symmetrical object it saves time to model one half of the object then mirror and attach it to the other side. For the alien, we're going to use a similar technique.

In this case, we will create an instance (dependent copy) of one side and mirror it to the other side. An instanced object changes when its source object is changed.

Here you will create an instance of the left side and place it on the right. As you make changes to the left side, the right side will change accordingly. At the end of the modeling process, we will attach the two halves together.

1. Save the scene with the filename *Alien06.max*.

2. Switch to the **Polygon** sub-object level.
3. From the Front viewport, use a marquee to select all of the polygons on the right side of the alien (your right, not the alien's).

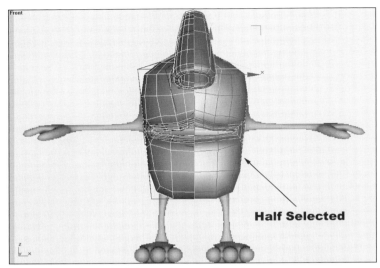

Figure 3.33 *Half of alien polygons selected*

4. Press the **<Delete>** key. When asked if you want to delete isolated vertices, answer **Yes**.

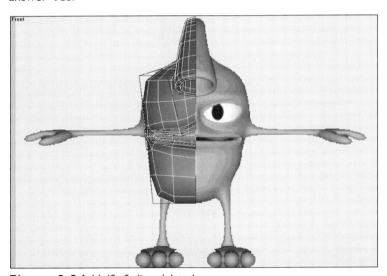

Figure 3.34 *Half of alien deleted*

 5. Turn off the **Polygon** sub-object level.

 6. Click the **Mirror Selected Objects** button on the Main Toolbar. On the **Mirror** dialog, choose the following settings, then click **OK** to mirror the object.

Mirror Axis	X
Offset	0
Clone Selection	Instance

Cutting Polygons for the Eye Socket

As of right now there is not enough detail to create the eye, so the first thing we need to do is to create more edges. We will accomplish this by cutting another edge.

When looking at the front of the alien, you may remember how we divided the head edges and the top of the lip edges with the Divide tool. We then cut between the new vertices that were created. This gave us the starting detail for the eye. This time we are going to cut horizontally across from the middle to the outer edge (right over the edge we cut in earlier). We are going to do this without using the Divide tool.

1. Switch to the **Edge** sub-object level.
2. Click the **Cut** tool to turn it on.
3. Cut from the center edge (between the top of the lip and the top of the head) to the outside edge of the alien. Do not click on the edge in the middle. The Cut tool will automatically divide this area and add a vertex.

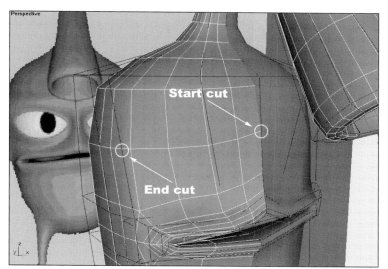

Figure 3.35 *Start cut and End cut points*

4. Click the **Cut** tool again to to finish.

The eye area now has four polygons that make up a large square shape.

Extruding the Eye Sockets

1. Switch to the **Polygon** sub-object level.
2. Select all four of the new square polygons.

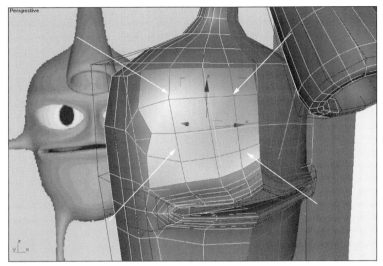

Figure 3.36 *The four selected polygons*

3. Extrude the polygons outward approximately 4.5 units.
4. Use the **Outline** spinner to scale them inward by approximately -8.0 units.

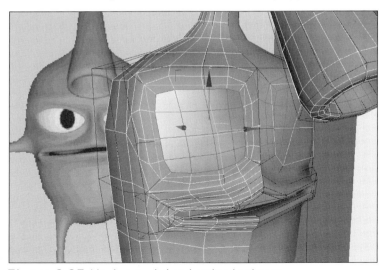

Figure 3.37 *Newly extruded and outlined polygons*

5. With semi-transparency on, use **Select and Move** in the Front viewport to move the selected polygons in toward the center and downward.

Shaping the Eye Socket

There are a total of eight vertices around the eye and one in the middle. We will move these to better shape the eye.

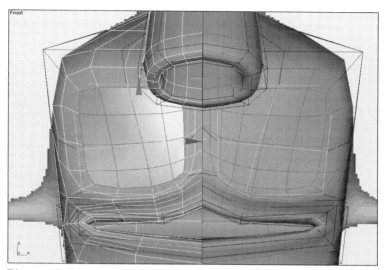

Figure 3.38 *Position of new polygons*

1. Switch to the **Vertex** sub-object level.
2. Using the Front viewport, move the vertices into the shape of the eye, but do not move them all the way to the edges. Leave a little bit of space. Shape the vertices so that the edges resemble the shape of the reference picture.

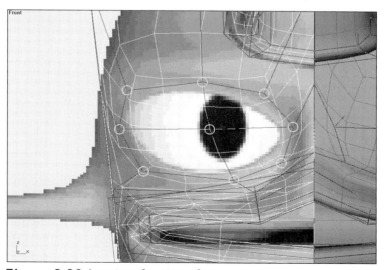

Figure 3.39 *Location of vertices after positioning*

3. Switch to the **Polygon** sub-object level. The same polygons should still be selected. If they aren't, select them now.
4. Extrude by approximately 3.0 units.
5. Use the **Outline** spinner to scale the vertices inward by approximately **-3.6** units until the new edges line up with the edge of the actual eyeball.
6. Extrude the selected polygons back into the head by approximately -25.0 units.

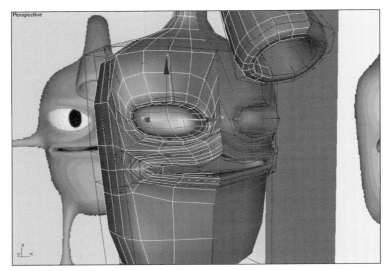

Figure 3.40 *Eye starting to take shape*

Adjusting the Eye Detail

Now that you have created eye sockets, you can adjust vertex positions so the eyes look like those in the reference picture.

1. Switch to the **Vertex** sub-object level.
2. Carefully adjust the position of each vertex around the eye until you are satisfied with final look. A little time spent now will reward you with a great-looking, expressive model.

Some tips to help you along:

- Pull the vertices out that are at the top of the eye in the eyebrow area. This will help create the raised ridge that is seen in the reference.
- Try to avoid any hard creases that may form in the bottom inner corners of the eye. There are a lot of vertices that are close together in this area.
- Continuously check the outcome of each transform in the Perspective viewport.
- In the Left viewport, don't forget to pull the outer edge of the eye back. Otherwise, the eyes will run almost straight across the head.

Final Adjustments to Torso

You are almost finished with the alien torso. At this point you may want to spend some time adjusting the positions of some of the torso vertices. You may also discover the need for a few more cuts.

Inspect the model from the eye all the way around to the back. While they are not required, additional cuts here will give you more geometry for achieving a proper shape. As you add additional detail, the surface will get smoother.

As you continue to add more detail, it will become more difficult to figure out what you're selecting at any given moment. You might find it helpful to uncheck the **Use NURMS Subdivision** checkbox on the Surface Properties rollout to work with the model in its boxy shape. After you have made the additional cuts, be sure to turn **Use NURMS Subdivision** back on.

When you have finished making adjustments, save the scene with the filename *Alien07.max*.

TUTORIAL 3.2 ARMS AND HANDS

Now that the alien torso has been created, we can create the arms and legs as extrusions from its torso.

Creating an Opening for the Arm

1. Select the two vertices on the side of the alien as seen in the following figure.

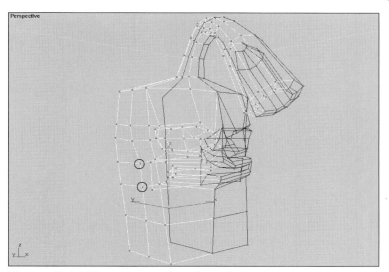

Figure 3.41 *Two vertices to be collapsed*

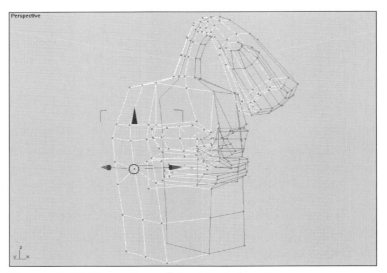

Figure 3.42 *Two vertices collapsed to single vertex*

2. Press the **Collapse** button to combine the two vertices into a single vertex.
3. In the Front viewport, move the newly created vertex up or down until it is positioned in the center of the base of the arm.

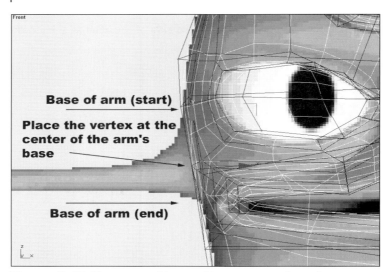

Figure 3.43 *Vertex moved into position at center of arm base*

4. While viewing the Perspective view, use the **Chamfer** spinner to create an opening that is about the size of the arm base. A value of approximately **10** units will work well.

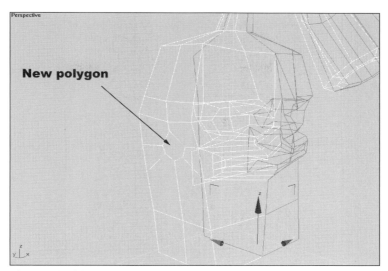

Figure 3.44 *New face resulting from vertex chamfer*

5. From the Left viewport, manipulate the vertices to create an evenly spaced hexagon. Use the following figures as a reference.

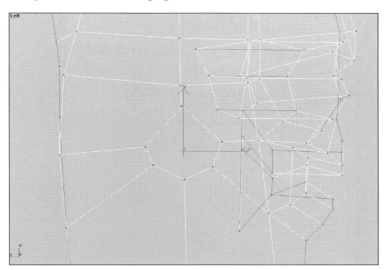

Figure 3.45 *Before the vertices have been properly positioned*

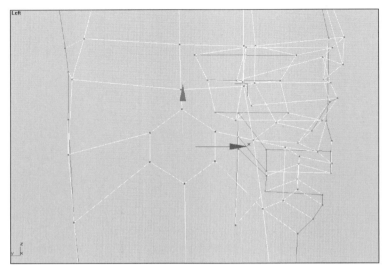

Figure 3.46 *After the vertices have been properly positioned*

Extruding the First Arm

1. Switch to the **Polygon** sub-object level and make sure this new polygon is selected.
2. Extrude the selected face out approximately 4 units.
3. Use the **Outline** spinner to scale the face inward approximately -2.8 units.
4. Extrude another 4 units, and outline inward -1.5 units.

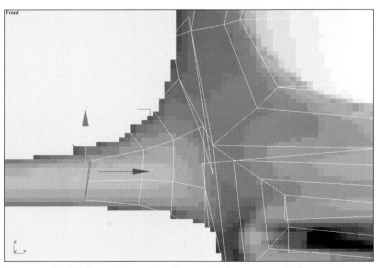

Figure 3.47 *Front view of newly extruded geometry before vertices are adjusted*

5. Extrude again, this time use approximately 7 units, and outline inward -1.3 units.

6. Switch to the **Vertex** sub-object level and position the vertices so that the new geometry matches the reference picture in Front viewport. Use the before and after figures shown here as a guide.

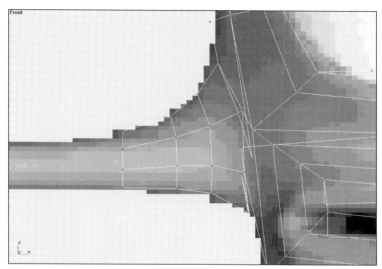

Figure 3.48 *Front view of new arm detail after vertices are adjusted*

7. In the Left viewport, adjust the position of the vertices so that the arm is extending straight out.

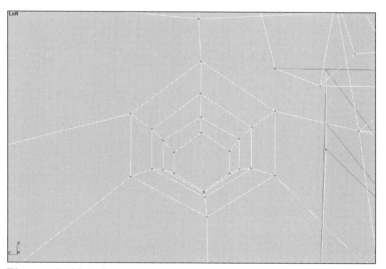

Figure 3.49 *Left viewport close-up of the arm*

8. Switch to the **Polygon** sub-object level and verify that the last extruded polygon is selected.

9. Click in the **Left Viewport** to make it the active view and click the **View Align** button on the Edit Geometry rollout. This will align the selected face with the Left viewport.

   We performed this alignment for two reasons. First, the face is now planar; secondly, any extrusions will stay perpendicular to the Left viewport. The means they will extrude straight and align with the arm.

Placing Detail at the Elbow

1. While using the Front viewport as a reference, extrude the selected polygon to the approximate location just prior to the elbow.

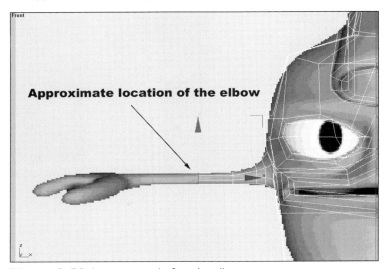

Figure 3.50 *Location just before the elbow*

2. For the arm to be able to Deform properly, we need some additional detail (segments) at the elbow. Perform two more extrusions to create this detail. Below is a close up of the elbow region. The vertical center edges are at the location of the bend.

Figure 3.51 *Front view of elbow region*

3. Extrude again (one time) until you reach the base of the alien's hand.
4. Save the scene with the filename *Alien08.max*.

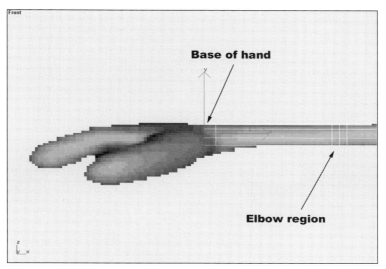

Figure 3.52 *Arm extruded to base of hand*

Creating the Hand

1. Verify that the polygons at the end of the arm are still selected.
2. Extrude the face by approximately 3.5 units
3. In the Top viewport, use **Select and Non-uniform Scale** to scale the new face in the Y axis direction.

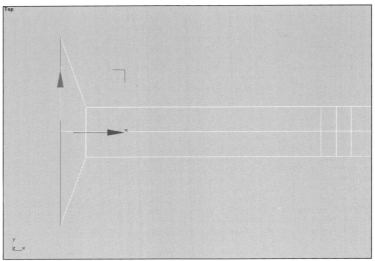

Figure 3.53 *Top view of newly extruded and scaled face that will begin the hand*

4. Extrude again by approximately 15.5 units.

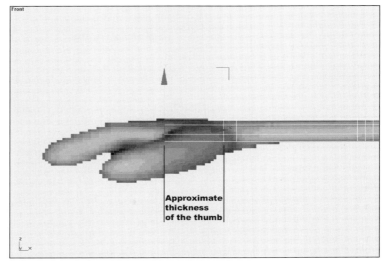

TIP

Use the Front viewport as a reference to make sure the extrusion is long enough to accommodate the next extrusion for the thumb.

Figure 3.54 *Front view of extrusion to be used for the thumb*

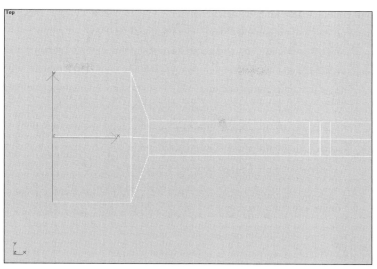

Figure 3.55 *Top view of the hand in its current state*

5. Extrude again by approximately 5 units. This will be the end of the palm and the starting point of the fingers.

Creating the Thumb

1. On the side of the hand, select the polygon that would represent the starting location of the thumb.

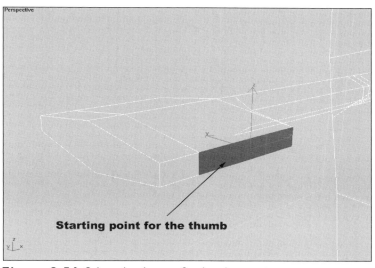

Figure 3.56 *Selected polygons for thumb extrusion*

2. Extrude the selected face approximately 6 units.

3. Click the **Select and Rotate** button on the Main Toolbar and change the **Reference Coordinate System** to **Local**.

   This will allow us to rotate and scale the selected polygon on its local coordinate system.

4. Rotate the thumb so that the next polygon extrusion will have the thumb going in the anatomically correct direction (for an alien).

5. Extrude again, approximately 3.5 units, and outline inward: -0.5 units.

6. Extrude again, approximately 6.0 units, and outline outward 4.5 units.

7. Extrude again, approximately 6.5 units.

8. Collapse the selected polygon. You should have something that looks similar to the following figure.

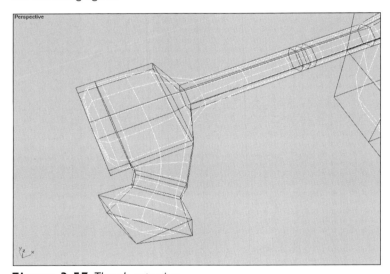

Figure 3.57 *Thumb extrusion*

Adjusting the Thumb

1. Switch to the **Vertex** sub-object level.
2. In the Front viewport, move the thumb vertices around to match the thumb with the reference picture.

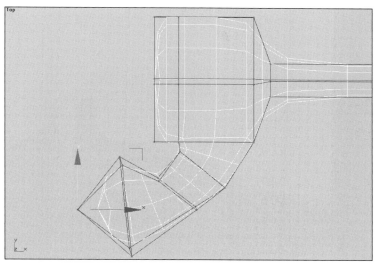

Figure 3.58 *Top view of the new thumb shape after vertices are moved*

Now that you have the thumb matched to the reference picture, you can use it as a base reference for filling out the shape of the palm.

3. Adjust the vertices of the palm so the thumb appears to be the correct size.

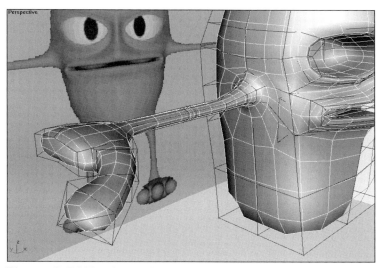

Figure 3.59 *The shape of the palm after matching to the thumb shape*

Making Openings for Fingers

1. In the Left viewport, position the vertices at the end of the palm into the shape of a rectangle.

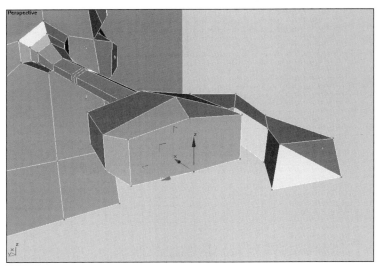

Figure 3.60 *Vertex positions for end of palm*

2. In the Top viewport, move the row of vertices on the end of the palm so that they create the shape shown in the following figure.

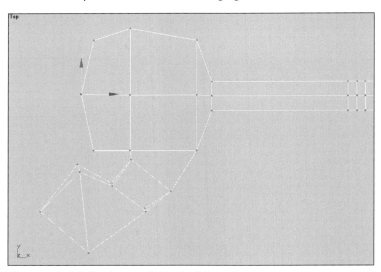

Figure 3.61 *Vertices at the end of the palm*

3. Move the center vertices toward the side of the palm that does not have a thumb.

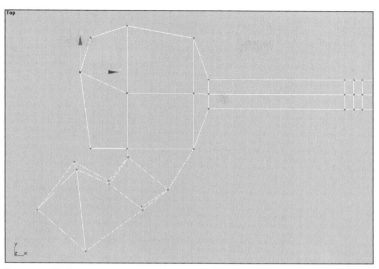

Figure 3.62 *Center-end vertices moved toward the palm side without the thumb*

4. Activate the **Cut** tool. In the Perspective view, Cut between the two center vertices to create an edge.

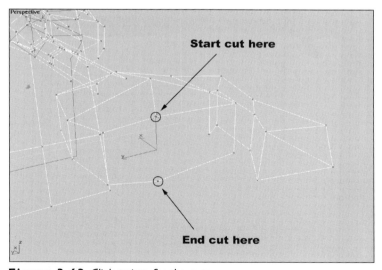

Figure 3.63 *Click points for the cuts*

5. Switch to the **Edge** sub-object level and activate the **Divide** tool. Divide the edges at the location shown in the following figure.

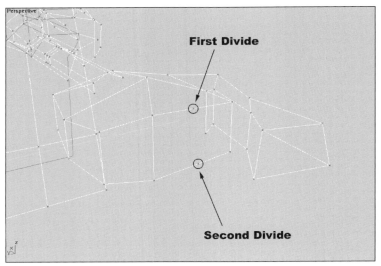

Figure 3.64 *Click points for the Divide operations*

 6. Switch to the **Vertex** sub-object level and activate the **Cut** tool. Make four cuts around the hand in the order indicated by the next two figures, beginning at the top-center vertex and ending at the bottom center vertex.

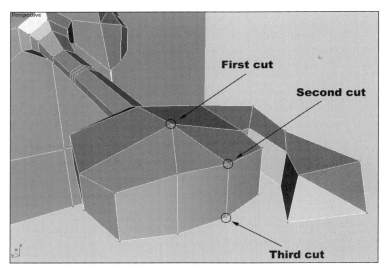

Figure 3.65 *Location of the first three cuts on palm top*

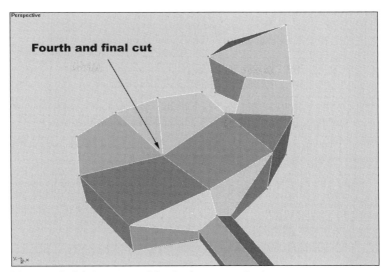

Figure 3.66 *Location of the final cut on palm bottom*

Be sure to make the cuts in the order show in the figures.

7. Deactivate the **Cut** tool.
8. Adjust the positions of the vertices so that the polygons on the end of the palm will have an even spacing. Make sure to look at all views when you do this, pay special attention to the Top viewport.

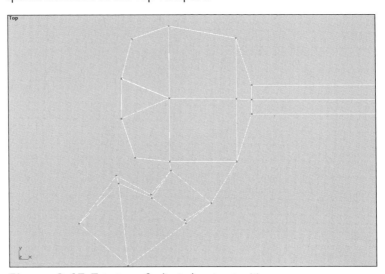

Figure 3.67 *Top view of adjusted vertex positions*

You now have three individual polygons that you can select and extrude into fingers.

Extruding the Fingers

1. Switch to the **Polygon** sub-object level.
2. Select the three polygons.

3. Change the Extrusion Type to **By Polygon**.

   This will extrude the polygons separately in the directions in which they are angled. This method allows us to extrude all three fingers at the same time.

   The next few steps will have you do a combination of extrusions and outlines to create the fingers. The values provided are approximate.

4. Extrude the polygons 5 units, and outline the polygons inward -0.5 units.
5. Extrude again, 5 units. and outline outward 2.5 units.
6. Extrude again, 6.5 units.
7. Press the **Collapse** button to collapse each selected polygon into a single vertex.

Fine-Tuning the Fingers

1. Switch to the **Vertex** sub-object level.
2. In the Top viewport, manipulate the vertices of the new fingers until you have something similar to the following figure.

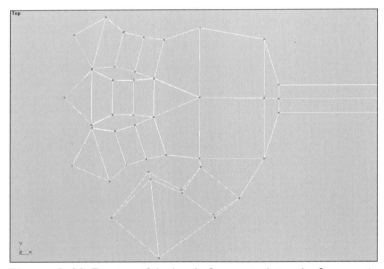

Figure 3.68 *Top view of the hand after manipulating the finger vertices*

3. In the Front viewport, use a combination of moves and rotations to position the vertices so that the fingers match the finders in the reference picture.

   It is best to work with entire selection of vertices at the same time, but sometimes it can be difficult (if not impossible) to use a marquee to select vertices with rectangular selection. This is where fence selection can come in handy.

To use fence selection, click and hold **Rectangular Selection Region** on the Main Toolbar and choose **Fence Selection Region** from the flyout. Click and drag in a viewport to start the selection. Continue clicking to create the fence. The cursor will change to crosshairs when you move your pointer back over the point where you started the fence. Click to finish the selection.

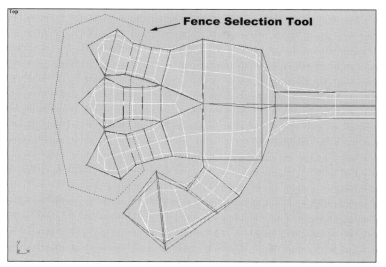

Figure 3.69 *Fence selection providing an easy means for complex selections*

Using fence selection and any other means of moving vertices, complete the hand so it looks similar to the figure that follows.

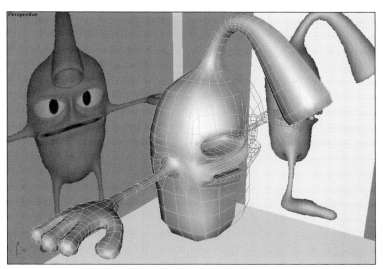

Figure 3.70 *The completed hand*

4. Save the scene with the filename **Alien09.max**.

TUTORIAL 3.3 LEGS AND FEET

We will use the same techniques to model the legs as we did for the arms. We assume you've got the idea by now, so we'll keep it short and to the point.

Preparing the Torso for the Legs

 1. Switch to the **Vertex** sub-object level. In the Front viewport, move vertices to make the model match the reference picture as closely as possible.

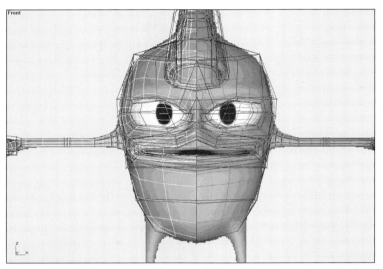

Figure 3.71 *Vertices moved to match the reference image in the Front viewport*

2. In the Perspective view, select the bottom middle vertex on the alien's side.

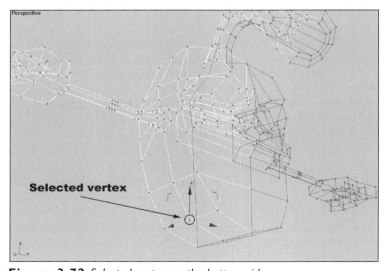

Figure 3.72 *Selected vertex on the bottom side*

3. Chamfer the selected vertex approximately 12 units.

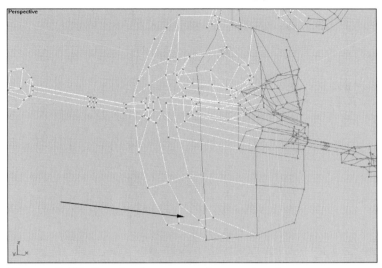

Figure 3.73 *Chamfered vertex creating a series of new vertices*

4. Switch to the **Polygon** sub-object level and select the new polygon.
5. Activate the Top viewport and press the **<B>** key on your keyboard.

    This will change the Top viewport to a Bottom viewport, looking up at the model from the opposite direction.

6. Click **View Align** to align the selected polygon with the Bottom viewport.
7. In the Front viewport, move the face down on the Y axis.

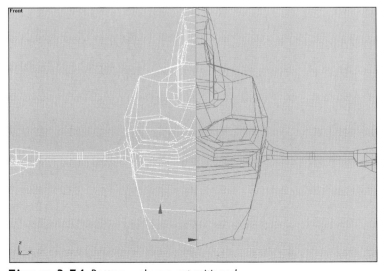

Figure 3.74 *Bottom polygon repositioned*

8. Switch to the **Vertex** sub-object level.
9. In the Bottom viewport, shape the vertices so that they make an even diamond shape. See the before and after figures that follow.

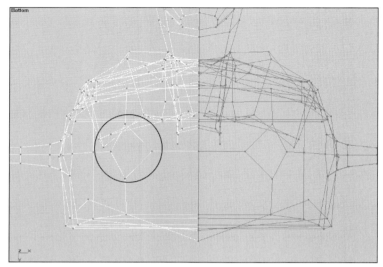

Figure 3.75 *Bottom viewport, chamfered vertices before positioning*

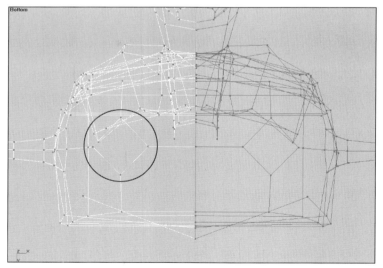

Figure 3.76 *Bottom viewport, chamfered vertices after positioning*

Four vertices are not going to be enough to give the leg a good shape. To increase the number of vertices, we are going to subdivide the edges so we will have eight vertices to work with.

Subdividing the Leg Extrusion Area

1. Switch to the **Polygon** sub-object level, and verify that the newly created polygon is selected.

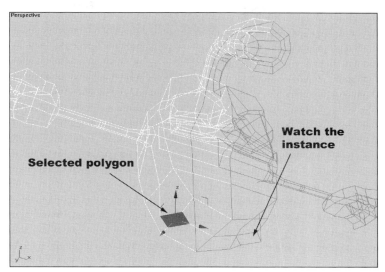

Figure 3.77 Before tessellation is applied

2. On the Subdivide rollout, verify that the Tessellation method is set to **Edge**. Click **Tessellate**.

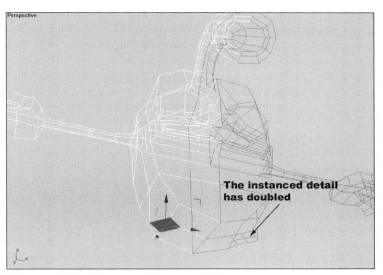

Figure 3.78 After tessellation is applied

3. Switch to the **Vertex** sub-object level.

4. In the Bottom viewport, select all four of the new vertices created by the tessellation.
5. Use **Select and Uniform Scale** to scale the selected vertices outward and create a circular shape as shown in the following figures.

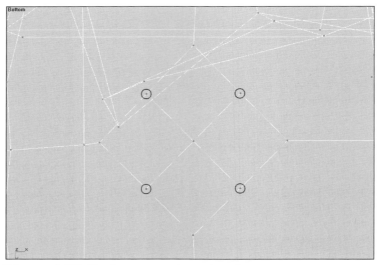

Figure 3.79 *The four new vertices selected*

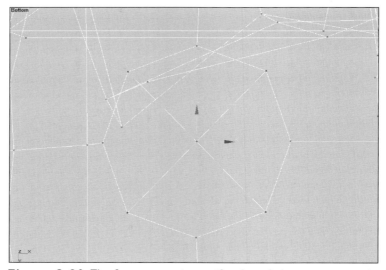

Figure 3.80 *The four new vertices uniformly scaled up to create a circular pattern*

6. Switch to the **Polygon** sub-object level.
7. In the Left viewport, align the four new polygons to match the corresponding area in the reference picture.
8. Use **Select and Uniform Scale** to scale the polygons to match the reference.

Extruding the Legs

1. On the Edit Geometry rollout, set the Extrude Type to **Local Normal**.
2. Extrude approximately 5 units, and outline the polygons down -3.0 units.
3. Extrude another 5 units, and outline inward -2.0 units.
4. Extrude 17 units to just above where the knee will be.
5. Extrude 4 units to just below where the knee will be.

    This will create a little bit of detail so the knees can bend.

6. Extrude 21 units to the ankle.

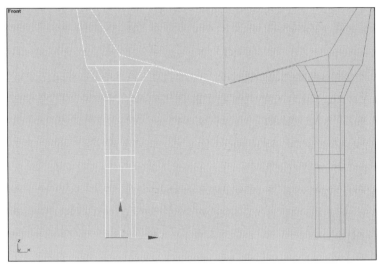

Figure 3.81 *Leg after applying a series of extrusions and outlines*

7. Switch to the **Vertex** sub-object level.
8. Move and scale the vertices in both the Front and Left viewports to align with the references pictures.

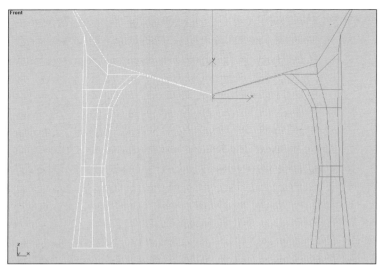

Figure 3.82 Leg reshaped to match reference picture

Shaping the Foot

The foot will be made through a series of extrusions, rotations, and non-uniform scaling.

1. Switch to the **Polygon** sub-object level.
2. In the Left viewport only, use a series of extrusions, non-uniform scaling and rotations to create the shape of the foot as shown in the following figure.

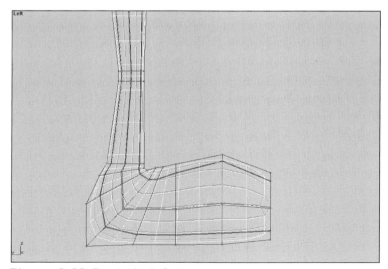

When extruding, set the Extrude Type to Local Normal to make your work easier.

Figure 3.83 Foot in the Left viewport

Extrude to the base of the toes, but don't create the toes themselves. We will create them in a few minutes.

Look at the foot in the Perspective view. It looks like a golf club! This won't do.

3. Switch to the **Vertex** sub-object level.
4. In the Left viewport, select the front half vertices that make up the foot.

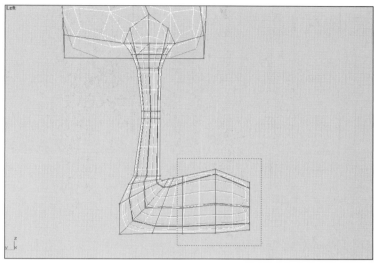

Figure 3.84 *Selecting three columns of vertices*

To increase the volume of the foot, we are going to use a tool called *soft selection*. With soft selection, transformations fall off in effect over a predetermined distance.

In this case we have selected the first three rows of vertices. These vertices will receive 100% of the scaling that we are going to apply. By using soft selection, the fourth row could receive 80% of the scale, fifth row 65%, and so on.

5. Expand the Soft Selection rollout and check the **Use Soft Selection** checkbox.
6. While watching the Left viewport, increase or decrease the **Falloff** spinner on the Soft Selection rollout.

As you change the Falloff value, the vertices change color depending on how much they are affected by the selection.

7. Set **Falloff** to 16.
8. Use the Front viewport to align the vertex selection to the reference image. You will most likely have to move the selection to the left on the X axis.
9. Adjust the vertices of the foot until you are happy with the look.

Our foot still looks a bit like a golf club, but we'll fix that in a moment by adding toes.

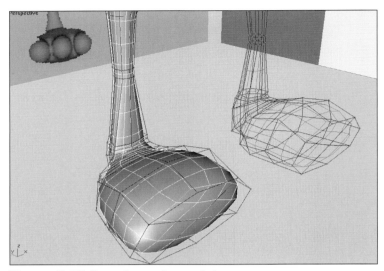

Figure 3.85 *Foot after vertices scaled*

Preparing to Extrude the Toes

1. Uncheck **Use Soft Selection** if it is still checked.
2. Switch to the **Edge** sub-object level.
3. Select the four edges at the end of the foot, and press the **<Delete>** key to delete them.

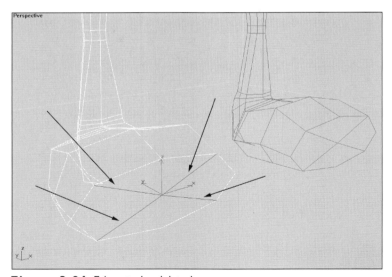

Figure 3.86 *Edges to be deleted*

4. Switch to the **Vertex** sub-object level.

5. As we did with the hand before extruding the fingers, adjust all the vertices so the edges at the end of the foot form a rectangular shape. See the figure that follows.

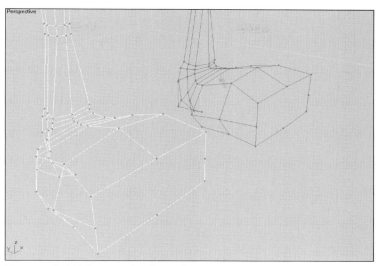

Figure 3.87 *Rectangular foot after vertices are scaled*

6. In the Bottom viewport, move the center vertices over as shown in the figures that follow.

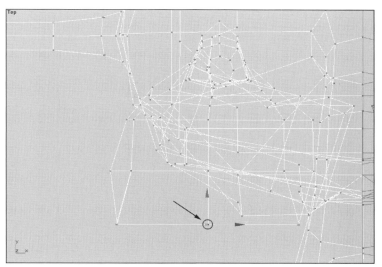

☀ **TIP** ☀

Here we are using the same techniques we used to create the three polygons for the finger extrusions.

Figure 3.88 *Top viewport, center vertex selected*

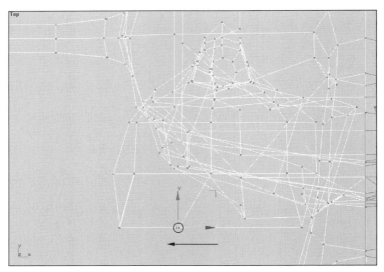

Figure 3.89 *Center vertex moved to the left*

Using the same technique that you used for the hand, create the appropriate cuts on the foot so that we will have 3 different polygons that we can extrude into toes. The steps for this procedure are:

- Divide the two edges
- Perform all the cuts to create the three polygons

We are almost ready to start extruding and outlining the toes. But one problem remains. If you look closely at the new polygons, you will see that the two outside edges actually have three vertices each.

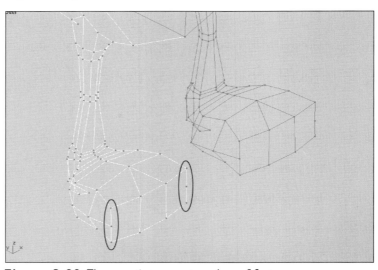

Figure 3.90 *Three vertices on outer edges of foot*

Eliminating Extra Vertices

We need to get rid of the middle vertex on each side so the toes can extrude.

1. Switch to the **Edge** sub-object level.
2. Select the edge connecting to the center vertex on each side of the foot, and press the **<Delete>** key to delete the selected edge.

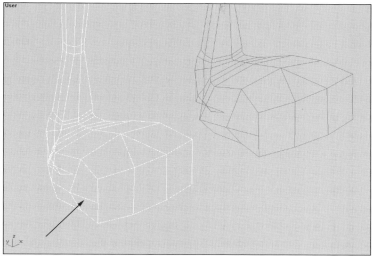

Figure 3.91 *Select this edge on both sides of the foot*

3. Switch to the **Vertex** sub-object level. You will see a floating (unused) vertex on each side of the foot.

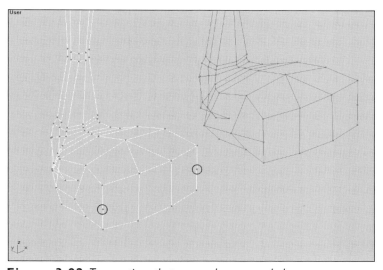

Figure 3.92 *Two vertices that are no longer needed*

4. Right-click **3D Snap** and make sure that only the **Vertex** option is checked on the Grid and Snap Settings dialog. Close the dialog and turn on **3D Snap**.

5. Select the left vertex and snap it down to the vertex directly beneath it.

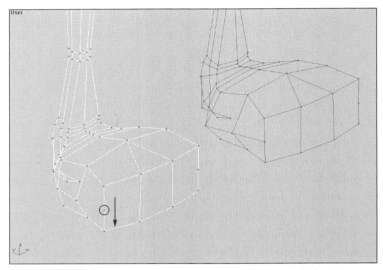

Figure 3.93 *Snapping the middle vertex to the vertex below it*

6. Use a marquee to select around the new bottom vertex.

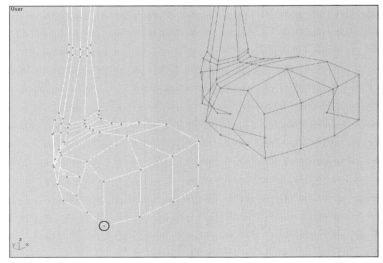

> ☼ **TIP** ☼
>
> *There are currently two vertices in the marked spot, so use a marquee to be sure you select them both.*

Figure 3.94 *Bottom two vertices selected*

7. Press the **Collapse** button to weld the two vertices into one.

8. Turn off **3D Snap**.

9. Repeat the previous steps to remove the corresponding vertices on the right side.

Making the Polygons Planar

1. Switch to the **Polygon** sub-object level.
2. Select each of the three polygons at the ends of the toes, and click **Make Planar** for each one.
4. Select all three polygons and set the extrusion type to **By Polygon**.
5. Extrude by approximately 10 units, and outline inward by -0.5 units.

   At this point, start paying close attention to the reference pictures to make sure you're matching them exactly with your extrusions and outlines.
6. Extrude by another 8 units, and outline outward by around 3 units.
7. Extrude by another 8 units.
8. Press **Collapse** to narrow the toe ends.

Alien Touch-Up

Let's do any necessary touch-up before mirroring and attaching the alien body.

1. Turn off the **Polygon** sub-object level.
2. Turn on **Use NURMS Subdivisions**.

3. Access the **Vertex** sub-object level.
4. Using the Front and Left viewports, carefully adjust all of the alien's vertices until you are satisfied with the way he looks.

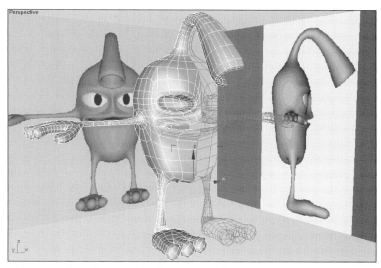

Figure 3.95 *Alien model after vertex adjustments*

5. When you are satisfied with the alien model, save the scene with the filename *Alien10.max*.

TUTORIAL 3.4 FINISHING THE MODEL

To finish the alien, we're going to weld the two halves together. Before doing this, it's a good idea to add the eyeball so we can make any changes necessary to accommodate it.

Creating the Eyeballs

1. On the **Create** panel, click **Sphere**.
2. In the Front viewport, create a sphere over the left eye socket.
3. Change the sphere's object color to white, and name the object **Left Eye**.
4. Using the Front, Top, and Left viewports, position the eye to match the location in the reference pictures.

5. Select the alien and go to the **Modify** Panel.

6. Switch to the **Vertex** sub-object level.
7. Adjust the vertices around the eye until the eyelids conform nicely to the eyeball. See the following figure for reference.

> ☼ **TIP** ☼
>
> You might find it necessary to adjust the eyeball's position from time to time while getting the eyelid and eyeball to line up. If this happens, you will need to turn off the Vertex sub-object level in order to select and move the eyeball.

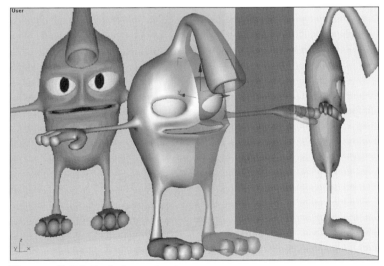

Figure 3.96 *Eyeballs are in place*

8. Turn off the **Vertex** sub-object level.
9. In the Front viewport, hold down the **<Shift>** key and move the eyeball to the other side of the head. On the Clone Options dialog, set the Clone type to **Copy** and the Name to **Right Eye**.

Attaching the Two Sides Together

1. Switch to the **Polygon** sub-object level and verify that the polygons on the left side of the alien are still selected.

2. Turn off the **Polygon** sub-object level so you are at the base level of the object.

3. Under the Surface Properties rollout, uncheck the **Use NURMS Subdivision** checkbox.

    This causes both sides to lose their smoothing and appear boxy for the time being.

4. Select the right side of the alien.

5. Right click on the modifier stack and convert this object to an **Editable Poly**.

    This will break the object's connection with the alien's left side, making it possible for us to attach it.

6. Press **<Alt-X>** to turn off the semi-transparency if it is still enabled for this side.

7. On the Geometry rollout, click **Attach** and click on the left side. Right-click to turn off **Attach**.

    Although the two alien halves are now one object, we have a slight problem. There are two sets of identical vertices running down the middle of the alien, one from each half. The vertices must be welded together to avoid an unsightly seam at the center. When the vertices are correctly welded, the seam up the center line will disappear.

    There are two techniques you can use to weld the vertices together. The technique you use depends on whether the two halves are lined up perfectly.

    If the vertices are lined up perfectly, you can weld them together all at once:

- Switch to the **Vertex** sub-object level.
- In the Front viewport, use a marquee to select all the vertices up the center of the alien, where the two halves meet.
- On the Edit Geometry rollout, set the weld threshold (the value next to Selected) to 0.2 and press the **Selected** button.
- If you receive a message telling you that no vertices were welded, try increasing the threshold to 0.3 and repeating the process.

If the vertices are not lined up perfectly, you will have to work with the weld threshold setting a little more.

- Switch to the **Vertex** sub-object level.
- In the Front viewport, use a marquee to select all the vertices up the center of the alien, where the two halves meet.
- On the Edit Geometry rollout, set the weld threshold (the value next to Selected) to 0.5 and press the **Selected** button.

- Inspect the body. Look for vertices that may not have been welded together. To check a particular vertex, use a marquee to select around it and see how many vertices get selected. You can look on the Selection rollout under the Copy and Paste buttons to see the number of selected vertices. If necessary, check **Ignore Backfacing** so you do not accidentally select vertices on the opposite side of the alien.
- If you have a selection that contains two vertices, set a high threshold value such as 2 or 3 and press the **Selected** button again. Continue this process until all of the vertices have been welded together.

When you have finished welding the two sides of the alien together, you're all done modeling the alien!

8. Save the scene with the filename *Alien Model.max*.

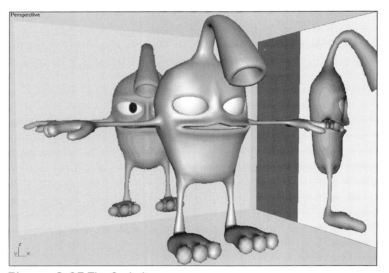

Figure 3.97 *The final alien*

SUMMARY

In this chapter, you learned how to:

- Use front and side reference pictures to model a character
- Sculpt character details such as eyes, mouth, hands and feet
- Work with a smoothed and boxy model at the same time with NURMS subdivision
- Subdivide vertices, edges and polygons in a variety of ways
- Use an instanced object as an aid in modeling

Many of these box modeling techniques, and more, are utilized in the next chapter. There you will model the remote control car, also known as the rover.

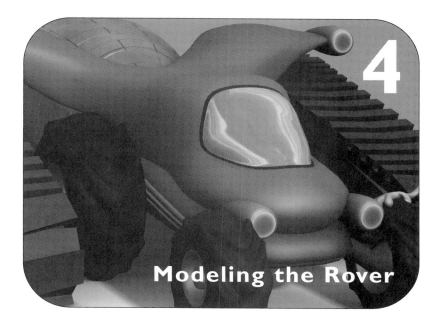

4
Modeling the Rover

In this chapter we are going to model the alien's remote control car, better known as the rover. The rover will be created with a combination of modeling techniques, some familiar from earlier chapters, and some completely new.

TECHNIQUES IN THIS CHAPTER
In this chapter, you will learn how to:
- Create lines and work with vertices.
- Create a framework of lines to form an object's shape.
- Put a solid surface over the framework of lines.
- Create individual polygons from scratch by drawing them on the screen.

In addition, you will use and build on many of the box modeling techniques you learned in earlier chapters.

TUTORIALS IN THIS CHAPTER

This chapter contains six tutorials for creating the rover car.

- *Tutorial 4.1 Creating the Spline Network* shows you how to create an intersecting network of splines that will later be surfaced.

- *Tutorial 4.2 Surfacing the Rover* uses the Surface modifier and other techniques for putting a solid surface on the rover body.

- *Tutorial 4.3 Creating the Wheels*, *Tutorial 4.4 Creating the Reactor*, and *Tutorial 4.5 Creating the Track Assembly* use both familiar and new box modeling techniques to create the additional parts of the rover.

- *Tutorial 4.6 Finishing the Rover* attaches the wheels to the rover and finishes up a few additional parts.

SPLINES

The rover body will be created by constructing a network of *splines*. A spline is a continuous line or curve. A shape is made of one or more splines (continuous lines or curves). A shape can have many splines.

In previous chapters, you worked with splines briefly. In this chapter, you will be creating numerous splines as a framework for the vehicle body.

Since you'll be working with splines a great deal in this chapter, let's start off by talking about how they work.

CREATING SPLINES

All the standard shapes in **3ds max** are made up of one or more splines. In this chapter, you'll be working almost exclusively with the Line shape. Along the way you'll convert it to an Editable Spline, which has many of the same features as Editable Mesh and Editable Poly objects.

You will also attach several splines together to create the framework for the rover.

VERTEX TYPES

Splines are made up of vertices and segments. A *segment* is a connecting line between two vertices.

A vertex can be set to any one of four types. The type determines the appearance and flexibility of the segments around it.

The following illustrations show the four types of vertices and how they work.

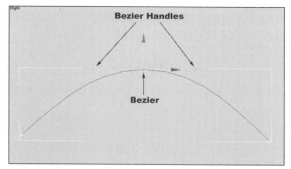

A *Bezier* vertex makes a smooth curve and has handles for manipulating the curve. Bezier vertices are named after an engineer who developed this method of representing smooth curves in computer drawings.

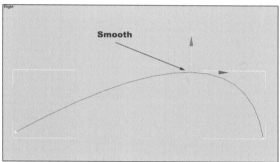

A *Smooth* vertex makes a smooth curve and has no handles.

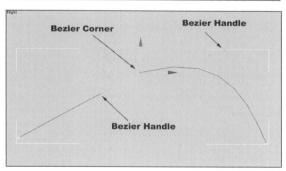

A *Bezier Corner* vertex has sharp angles and handles for manipulating the angle.

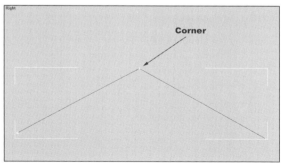

A *Corner* vertex makes a sharp angle, and has no handles.

Changing the Vertex Type

In order to change a vertex type, you must use the Vertex sub-object level. This sub-object level is available only for splines created with the Line tool, or splines converted to Editable Splines. Any spline can be converted to an Editable Spline by right-clicking the shape on the modifier stack and choosing Convert to Editable Spline from the pop-up menu.

Once you are at the Vertex sub-object level, you can right-click a vertex to display the Quad menu and select a new type for the vertex in the upper left quadrant.

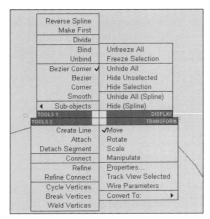

Figure 4.1 *Quad menu with vertex type selection*

In the same way, you can select and change several vertices at the same time.

You must click directly on one of the selected vertices to display vertex types on the Quad menu.

Before doing the first tutorial in this chapter, we recommend that you draw a few lines in **3ds max** and play around with the vertex types to get a feel for how they work.

Setting the Vertex Type When Drawing

When you draw a line, you have some control over the vertex types on the line.

- A quick click creates a Corner type vertex.
- Clicking and dragging while creating the line creates a Bezier vertex.

Some shapes are created automatically with specific vertex types. A circle, for example, is created automatically with four Bezier vertices.

THE SURFACE MODIFIER

When the splines are in place, the Surface modifier is applied to make a surface over the splines. The underlying spline framework can easily be modified to change the shape of the surface.

There are two approaches a modeler can use when creating a spline framework for use with the Surface modifier. One is to make all the vertices the Bezier or Smooth type, which creates a smooth, rounded surface when the Surface modifier is applied. Another is use to use only Corner vertices and create a boxy object with the Surface modifier which is then smoothed with other tools.

In this chapter, we'll be using the second approach. Using this method, we'll have a low-resolution framework that we can use to control higher resolution geometry. This approach will give us the ability to make subtle refinements to underlying geometry while simultaneously seeing the final smoothed result.

The Surface modifier is excellent for creating organic surfaces. Many modelers use it as their primary tool for complex objects such as character models.

THE ANATOMY OF THE ROVER

The rover consists of several parts. Each will be modeled individually in this chapter. As you go through the tutorials, refer to the illustration below as needed.

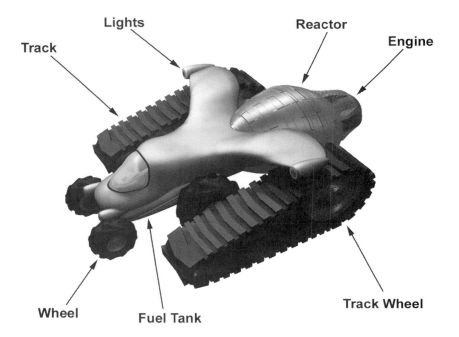

Figure 4.2 *The rover and its parts*

TUTORIAL 4.1 CREATING THE SPLINE NETWORK

In this tutorial you'll create the splines that will form the basis for the rover body. You'll create the splines for one half of the rover, then mirror other half later.

As always, we need a virtual studio for the rover. This virtual studio has three images for the top, side and front view.

Setting up the Virtual Studio

If you want to load our virtual studio and get started with modeling, load the file *Rover Virtual Studio.max* from the *Scenes/Virtual Studios* folder on the CD that comes with this book.

If you want to create your own virtual studio, basic instructions are given here. More detailed information on setting up a virtual studio can be found in *Chapter 2 Modeling the Remote Control*.

1. Set up three planes as shown in the following figure. Create and assign materials to the planes using the following reference image files from the *Maps/Rover Refs* folder on the CD:
   - *Rover_Top.jpg*
   - *Rover_Front.jpg*
   - *Rover_Side.jpg*

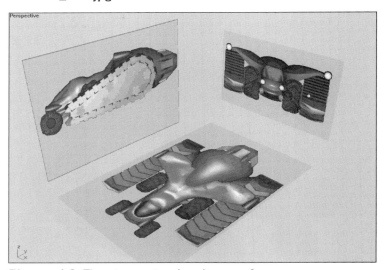

Figure 4.3 *Three-image virtual studio setup for rover*

2. Select all three planes. On the **Display** panel, under the Display Properties rollout, uncheck **Frozen in Gray**.
3. On the Freeze rollout, click **Freeze Selected**.

Now we're ready to start creating the rover.

Creating the Body Profile

We'll start off by creating the spline for the body profile.

1. Activate the Right viewport and click **Min/Max Toggle** at the bottom right of the screen to bring the viewport to the full screen size.

2. On the **Create** panel, click **Shapes**. Click **Line.**

3. Starting behind the windshield, draw a line with 19 vertices to outline the shape of the body. The line should end where the windshield begins. Use the following image as a reference for placing the vertices.

> **☆ TIP ☆**
>
> Since you are not able to see all of the body, you will need to make some educated guesses as to where the line should fall.

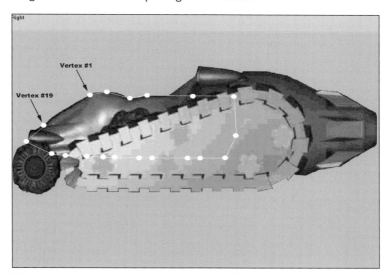

Figure 4.4 *Line to form the vehicle's profile*

4. Go to the **Modify** panel.

5. Click the **[+]** next to the Line in the stack view, and select the **Vertex** sub-object.

6. If any of the vertices need repositioning, use **Select and Move** to move them into place.

Shaping the Battery Door Opening

1. Counting from the first vertex placed, select vertices **10** and **11**.

2. In the Perspective viewport, move the two selected vertices out from the line on the X axis.

This will be the outer edge of the battery compartment.

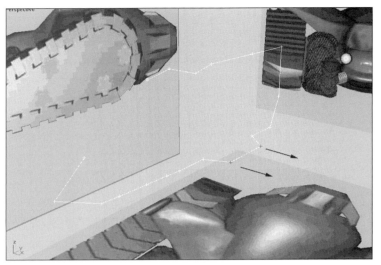

Figure 4.5 *Two selected vertices moved out on X axis*

3. In the Top viewport, move each vertex separately in its Y axis so it aligns with the vertex beside it.

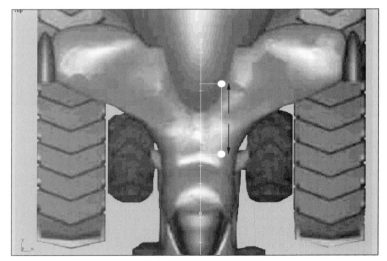

> ☀ **TIP** ☀
>
> *Move the vertices only on the Y axis as you have already taken care of the X axis placement in the previous step.*

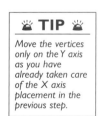

Figure 4.6 *Two selected vertices moved out on Y axis*

While creating the spline network, it will become increasingly important to align certain vertices with others. We will turn on vertex ticks (visible dots) to make the vertices visible at all times.

4. Right-click on the line and select **Properties** in the lower right quadrant of the Quad menu. On the Object Properties dialog, check the **Vertex Ticks** checkbox.

This causes all vertices on the line to be visible at all times.

Creating the Leading Edge Wing Splines

1. In the Top viewport, use the **Line** tool to draw a line on the leading edge of the wing (the vehicle's left wing). This line should have 6 vertices.

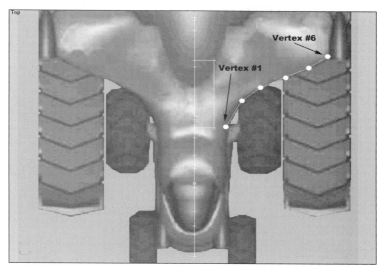

Figure 4.7 *Line matching leading edge of wing*

 2. On the **Modify** panel, access the **Vertex** sub-object level.

3. Using the following figure as a reference, align the vertices in the Front viewport with the top front of the wing.

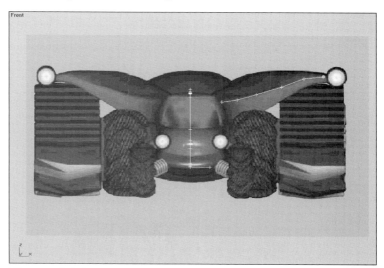

Figure 4.8 *Line matched up with reference picture*

 4. Turn off the **Vertex** sub-object level to return to the base **Line** object.

5. Turn on **Vertex Ticks**.
6. In the Front viewport, copy the line to the bottom of the wing.

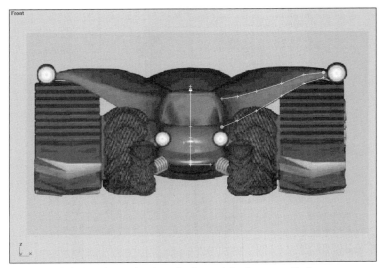

Figure 4.9 *Leading edge line duplicated for bottom of wing*

 7. Access the **Vertex** sub-object level and position the vertices so they align with the bottom of the wing.

Creating the Middle Wing Splines

1. In the Top viewport, use the **Line** tool to create another line with six vertices positioned at the center of the wing.

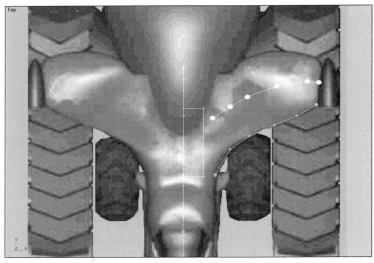

Figure 4.10 *Vertices positioned at center of wing*

2. In the Front viewport, position the vertices to match up with the reference image.

> **TIP**
> Move vertices up or down on the Y axis only.
>
> Note that the middle of the wing is raised higher than the front or back.
>
> You will need to do a little guesswork since you are unable to clearly see where the bottom is in any of the reference pictures.

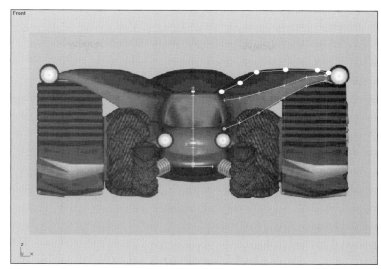

Figure 4.11 *Center vertices positioned with reference*

3. Enable **Vertex Ticks** for the line.
4. In the Front viewport, copy the line down, and position the vertices for the bottom of the wing.

Creating the Trailing Edge Wing Splines

1. In the Top viewport, draw a line with 7 vertices starting at the center of the rover and extending to the end of the wing. Align the line with the wing's trailing edge.

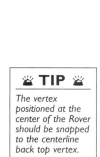

> **TIP**
> The vertex positioned at the center of the Rover should be snapped to the centerline back top vertex.

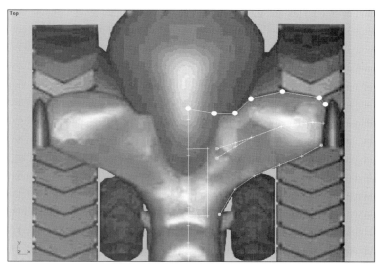

Figure 4.12 *The wing's trailing edge line*

2. In the Front viewport, move the vertices until they have been positioned in what you feel is the best location.

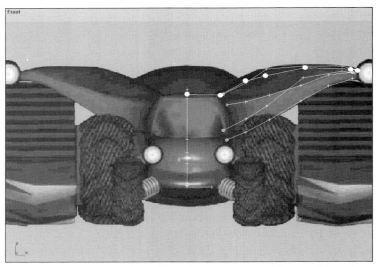

Figure 4.13 *The wing's trailing edge line positioned in the Front viewport*

3. Enable **Vertex Ticks**.

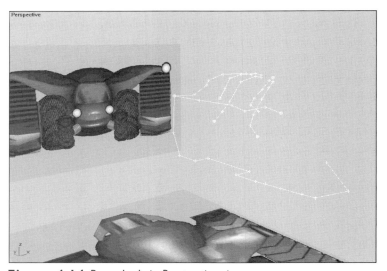

Figure 4.14 *Rover body in Perspective view*

Attaching the Lines

In order for the Surface modifier to work with the spline network later on, all the splines must be attached together.

1. Switch to the Perspective view.
2. Select any one of the lines that you have created.
3. In the Modify panel, click the **Attach** button on the Geometry rollout.

   As you move your mouse over the various lines, you will notice the cursor changes. This indicates that the cursor is over a spline which can be attached.
4. Click on each line until they have all been attached.
5. Right click to turn off **Attach**.

Creating the Wing Ribs

In the next set of steps, you'll be creating new lines by drawing from one existing vertex to another. When doing so, you will often see messages asking you whether you want to close the spline or weld vertices.

- If you are asked if you would like to close the spline, answer **Yes**. The Surface modifier works best with closed splines.

- If you are asked if you would like to weld the vertices, answer **No**. Generally, this happens when you have just clicked on an existing vertex with a spline running in a direction other than the spline you just clicked on. We will want to keep these vertices separate.

1. Right-click the **3D Snap** button and make sure that only the **Vertex** option is checked on the Grid and Snap Settings dialog. Close the dialog and turn on the **3D Snap** button.
2. Position the Perspective view so that you have a good view of all vertices at the end of the wing tip.

3. On the **Modify** panel, click the **Create Line** button.
4. Place your mouse over the last vertex on the trailing edge of the wing. When the mouse pointer changes to the snap cursor, click to start the new line. Continue to click in a counter clockwise direction until you have returned back to the starting vertex.

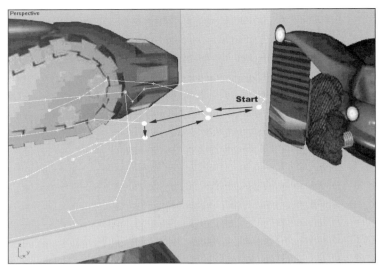

Figure 4.15 *The pattern for the new line*

5. Right click to complete the line. If you are asked if you want to close the spline or weld vertices, see the instructions on the previous page.

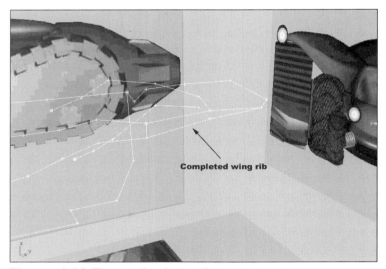

Figure 4.16 *The completed wing rib*

6. Using the technique just shown, create five more ribs. The **Create Line** button should still be active, so you can just go ahead and draw. Use the following figure as a reference.

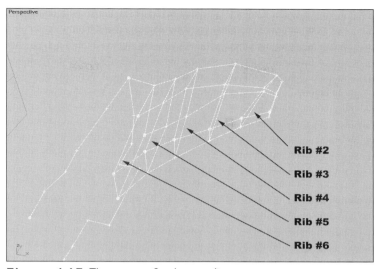

Figure 4.17 *The pattern for the new line*

Creating the Windshield Outline

1. Verify that the **Create Line** option is still on.

2. In the Top viewport, draw an outline around the rover's windshield. The line should have six vertices. Use the following figure if you need assistance.

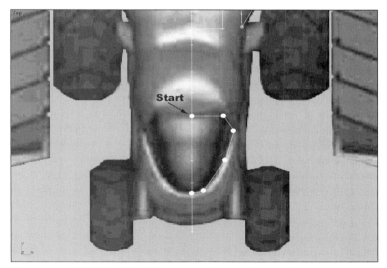

Figure 4.18 *Windshield line created in the Top viewport*

3. Right click twice to complete the line and exit the **Create Line** tool.

4. Turn off **3D Snap**.

5. In the Front viewport, position the vertices to match up with the reference image. Again, only move them up or down on the Y axis.

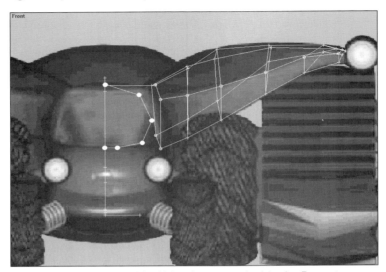

Figure 4.19 *The windshield has been matched In the Front viewport*

6. Turn on **3D Snap**.
7. Click the **Create Line** tool again.
8. In the Right viewport, draw another line that will cut through the middle of the rover going from the nose to the tail end. There should be a total of ten vertices. Use the reference following figure for placement. When finished creating the line, right click twice to exit the **Create Line** tool.

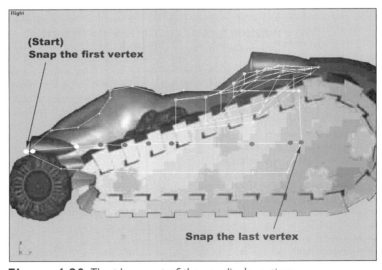

Figure 4.20 *The placement of the new line's vertices*

9. In the Top viewport, manipulate the vertices on the X axis so that the line matches the shape of the body.

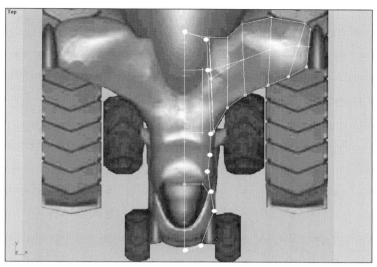

Figure 4.21 *The placement of the vertices from the Top viewport*

Connecting the Wings to the Body

1. Turn on **3D Snap**.

2. In the Right viewport, create two lines that connect the bottom two vertices at the beginning of the wing to the next two vertices on the line below them.

> ☀ **TIP** ☀
>
> *By now you should have the hang of turning on the Create Line tool, drawing the line and right-clicking once to end the line and once more turn it off. We're assuming you don't need specific instruction on this task any more, so it is not mentioned in the steps that follow.*

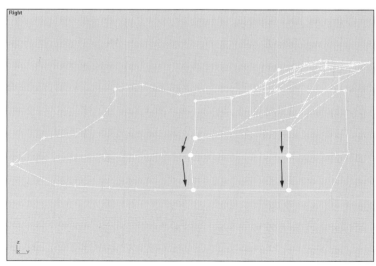

Figure 4.22 *Two lines connecting bottom wing through centerline to bottom*

3. In the Perspective view, connect the middle line vertex that was used in the last step to the back right upper corner of the rover. Use the following figure as a reference.

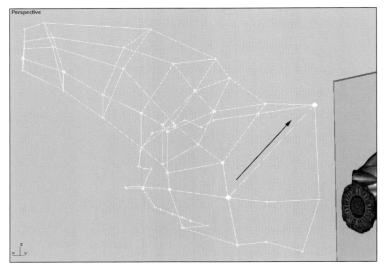

Figure 4.23 *New line snapped between the two back corner vertices*

4. In the Top viewport, complete the bottom corner with an L-shaped line.

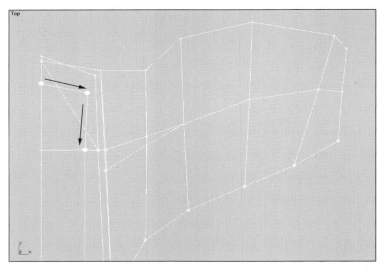

Figure 4.24 *The L-shaped line that completes the bottom corner of the rover*

5. Turn off **3D Snap**.
6. Select the corner vertex on the line that you just created.

7. In the Perspective or Right viewport, move the vertex into position so that it is even with the other back vertices.

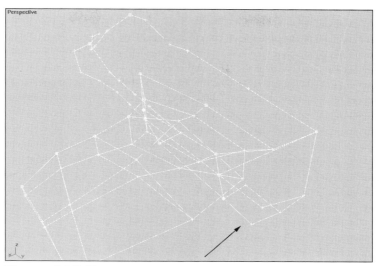

Figure 4.25 *The corner vertex moved out*

8. Turn on **3D Snap** again.
9. Connect a line between the bottom corner vertex and the corresponding vertex above it.

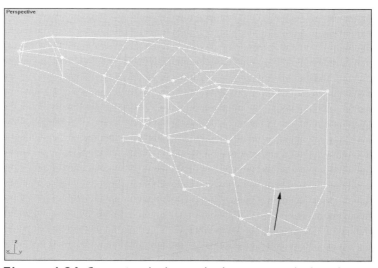

Figure 4.26 *Connecting the bottom back corner into the line above it*

Filling in the Front of the Rover

In the next series of steps you will create the bottom of the framework near the front of the rover's body.

1. In the Top viewport, create a backwards L-shaped line at the bottom of the cage.

    Be sure to snap only the first and last vertices, and to snap to the correct vertices,

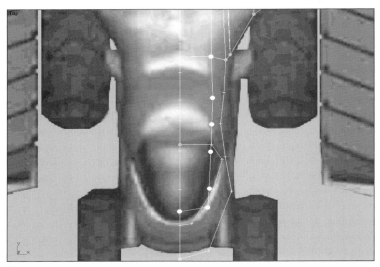

☼ TIP ☼

Refer to figure 4.31 to help you see where you're going with this line.

Figure 4.27 *Backwards L-shaped line*

2. In the Perspective or Right viewport, move the vertices on the Z axis so they are aligned with the bottom of the rover.

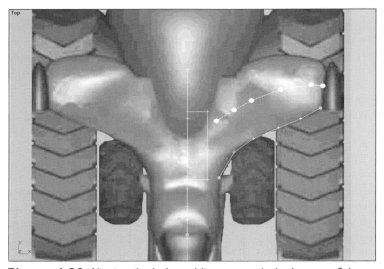

Figure 4.28 *Aligning the L-shaped line to match the bottom of the rover*

3. In the Right viewport, connect the midsection of the cockpit window to the bottom inner most vertex on the wing. Use the following figure as a reference.

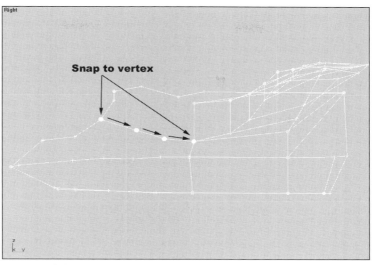

Figure 4.29 *Connecting the mid section of the cockpit window to the inner wing*

4. Create another line above the last line that was just created. Use a total of four vertices and make sure to snap the first and last ones to the current spline network.

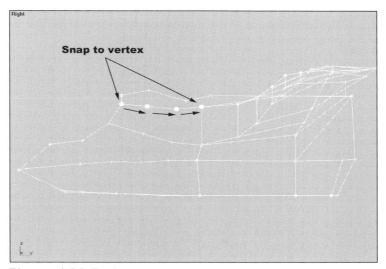

Figure 4.30 *Top line*

5. In the Top viewport, move the two inner vertices on the newly created line on the X axis so that they help create a continuous line that connects the cockpit window to the upper inner wing vertex.

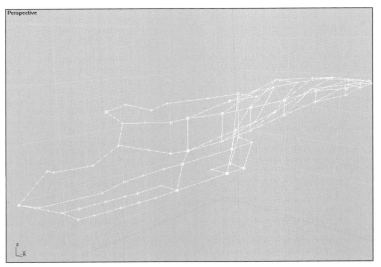

Figure 4.31 *Rover body starting to take shape*

Creating the Front Ribs

1. In the Perspective window, create the first of the front ribs that will connect all of the lines together. Make sure to snap to every vertex. Use the following figure as a reference for exact placement of vertices.

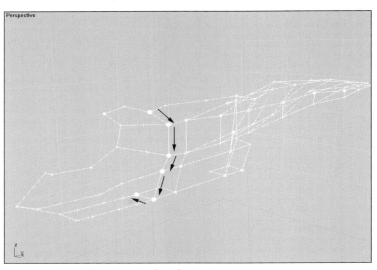

Figure 4.32 *Creating the first front rib*

2. Using the same approach as the last step, create the 2nd, 3rd, and 4th ribs.

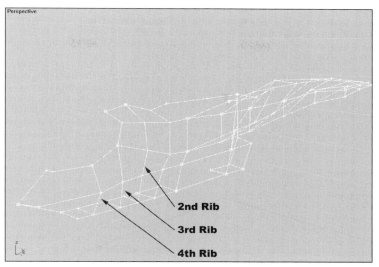

Figure 4.33 *Creating additional ribs*

3. Connect the top line to the inner top leading wing vertex. Again, make sure to snap to the existing vertices.

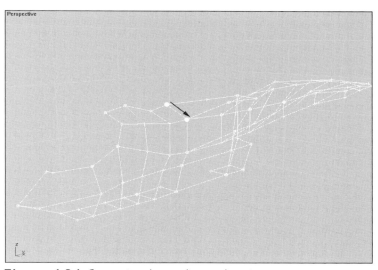

Figure 4.34 *Connecting the top line to the wing*

4. Use **Arc Rotate** to rotate the view around to the front of the rover and connect the three remaining vertices at the vehicles front.

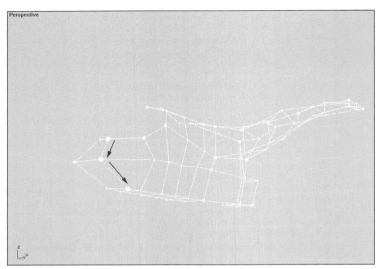

Figure 4.35 *Connecting the remaining front vertices*

5. Rotate the Perspective view so you can see the top of the rover. Create a line that will complete a connection to the vertex that starts the middle of the wing spline.

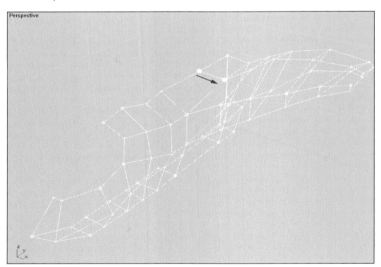

Figure 4.36 *Connecting the middle wing spline*

6. Use a marquee to select all of the vertices and change them to **Corner**. Even if you are sure that they are already set to **Corner**, set them anyway. This way, you will prevent any problems form occurring in the next few steps.

7. Turn off the **Vertex** sub-object level.

   The spline framework is now complete.

8. Save the scene with the filename **Rover01.max**.

TUTORIAL 4.2 SURFACING THE ROVER

Now that the spline framework is complete, we can apply the Surface modifier to it to make a solid 3D surface. The Surface modifier creates a patch surface over all areas surrounded by three or four connecting splines.

Applying the Surface Modifier

1. Click the down arrow next to **Modifier List** on the **Modify** panel and choose the **Surface** modifier.

   A surface is placed over all or some of the spline framework.

   If you are lucky, the surface will look smooth and solid. However, it is common to find that the surface has been created with the normals flipped the wrong way, as shown in the following figure.

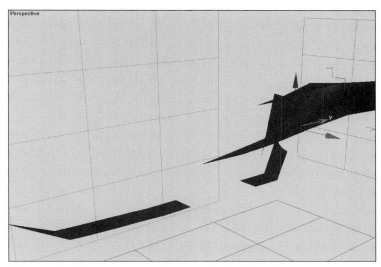

Figure 4.37 An example of normals being flipped the wrong way

2. If the new surface looks strange or appears to be inside-out, check the **Flip Normals** checkbox on the Parameters rollout.

The ends of the wings should remain open for the time being. You will cover them with the light housing later on.

   If the surface still does not look solid, then some of your vertices are not snapped correctly. Remember that the Surface modifier only surfaces areas that are surrounded by three or four segments. If a vertex is out of place in a particular area, a surface will not be placed over that spot.

   If your surface doesn't appear solid, remove the **Surface** modifier and see if you can find out where the vertices are not snapped. Turn on **3D Snap** and move the vertices into place, then apply the **Surface** modifier again. Repeat this operation until the surface appears solid when the **Surface** modifier is applied.

   Continue with these steps only when your surface looks solid.

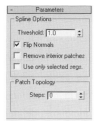

3. On the command panel, set **Steps** to 0.

   The Surface modifier automatically creates a patch surface. What we want is an Editable Poly.

4. Right-click in the modifier stack and select **Collapse All** from the pop-up menu. If a warning appears regarding collapsing the stack, click **Yes** to continue.

   The object becomes an Editable Patch when the modifier stack is collapsed.

5. Right click on the modifier stack and select **Convert to: Editable Poly** from the pop-up menu.

6. Check the **Use NURMS Subdivision** option on the Surface Properties rollout

7. Set **Iterations** to 2.

   This will cause the rover body to look very smooth.

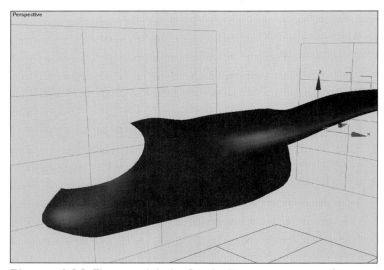

Figure 4.38 *The smooth look after the Iterations are set to 2*

8. Name the object **Rover Body.**

Reshaping the Body with a NURMS Cage

By setting the Surface modifier's Steps option to 0 and using NURMS subdivisions as the smoothing method, we've created a way to quickly modify the overall shape of the object using the simpler shape of the framework.

This technique is very similar to the polygon modeling with NURMS that we did when modeling the alien. This time we just created our basic shape out of splines rather than from a primitive object.

1. Select the rover object.

2. On the **Modify** panel, turn on the **Vertex** sub-object level.

   A wire cage appears around the geometry.

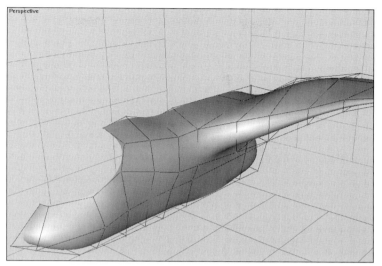

Figure 4.39 *Wire cage surrounding the geometry*

3. Move the vertices on the cage to make the object look similar to the following figure.

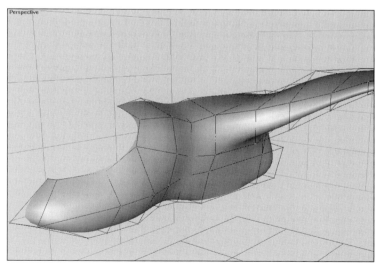

Figure 4.40 *The final geometry after the vertices have been manipulated*

Here are a few tips for working with the NURMS cage.

- Do not move any of the centerline vertices toward or away from the rover body. If you do, the second half of the body will not align properly when it is mirrored later on.

- The geometry tends to shrink a little when you use a NURMS cage. You will need to move some of the vertices outward so that the geometry will fill the shape properly. For example, in the Top viewport you can see that the wings have shrunken and no longer match the reference picture.
- Make the cockpit bubble-shaped.
- Make the nose narrow for some aerodynamics.
- Give the underside (just behind the cockpit) some aerodynamics too.
- Don't worry about the holes at the end of the wings. You will fill the holes later.

4. Select and move vertices to shape the rover. Do the best you can to match the reference pictures.

5. After you have finished adjusting the vertices, turn off the **Vertex** sub-object level.
6. Save the scene with the filename **Rover02.max**.

Mirroring the Other Half

Now that one half of the rover is perfect, we can mirror it to the other side.

1. With the rover body selected, click **Mirror Selected Objects** on the Main Toolbar.
2. In the Mirror dialog, set the mirror axis to **X** and clone selection to **Copy**. Adjust the offset until the new half of the rover is aligned with the first half. Click **OK**.
3. In the Top viewport, move the new half on the X axis so its center is aligned with the original half's center.

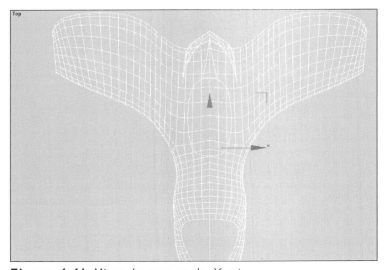

Figure 4.41 *Mirror the rover on the X axis*

4. Select the original half.

5. On the **Modify** panel, click the **Attach** button on the Edit Geometry rollout. Click the other half to attach it, then right-click to deactivate **Attach**.

6. Access the **Vertex** sub-object level.

7. In the Top viewport, use a marquee to select all of the vertices down the center of the geometry.

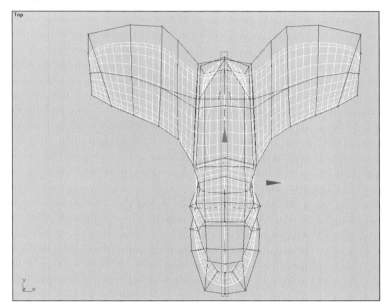

Figure 4.42 *Use a marquee to select the centerline vertices*

8. On the Edit Geometry rollout, locate the Weld section. Set the **Selected** field to 0.5.

9. Click the **Selected** button. All of the vertices overlapping one another will now be welded together into a single vertex.

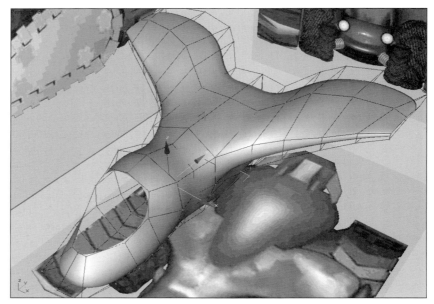

Figure 4.43 *All centerline vertices have been welded together*

Centering the Pivot Point

We want the object's pivot point to be at its center. We can accomplish this on the **Hierarchy** panel.

1. Go to the **Hierarchy Panel**.

2. Click the **Affect Pivot Only** button.
3. Click **Center to Object**.

   This will force the pivot point to the center of the object.
4. Click **Affect Pivot Only** button to turn it off.

Fine-Tuning the Rover Body

Now that the two halves have been joined into one, you will need to make one final pass at fine-tuning the vertices. The main focus should be on the center vertices that were welded in the previous steps. Move vertices as necessary to adjust the rover body.

Pay close attention to the front and back of the rover. Make sure that you have the desired shape before moving on.

Creating a Windshield

A modeling technique we have not used yet is the creation of polygons by drawing directly on the screen. With so many vertices already making up the object, we can simply snap to a series of vertices to create a new polygon. We'll use this method to create a windshield for the cockpit.

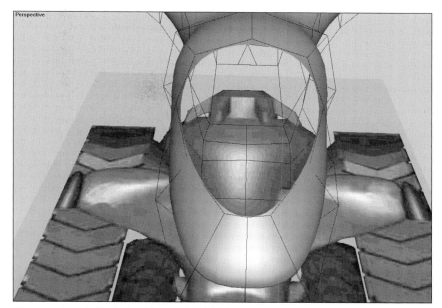

Figure 4.44 *Cockpit without windshield*

1. Switch to the **Polygon** sub-object level.
2. Turn on **3D Snap**.
3. Click the **Create** button on the Edit Geometry rollout.
4. In the Perspective viewport, start at the top and click on each vertex around the windshield area, moving in a counter-clockwise direction until you have gone all the way around the rim and returned to the starting vertex. Click the starting vertex to complete the polygon.

    A new polygon is created for the windshield.

5. Right-click to deactivate the **Create** tool.

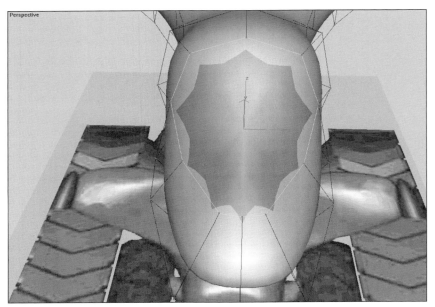

Figure 4.45 *The new windshield polygon*

6. Extrude the new polygon inward by -4.0 units, and outline it inward by -2.0 units.
7. Extrude outward by 4.0 units and outline inward by -3.8 units.
8. Click the **Make Planar** button to flatten out the polygon.

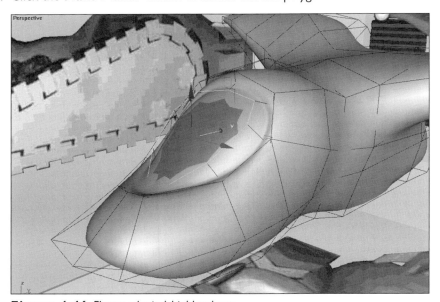

Figure 4.46 *Flattened windshield polygon*

9. Move, rotate and scale polygons or vertices to complete the windshield.

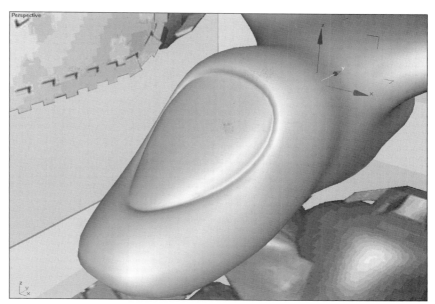

Figure 4.47 *The final windshield*

Making the Battery Compartment

1. In the Perspective view, use **Arc Rotate** to show the opening under the rover.

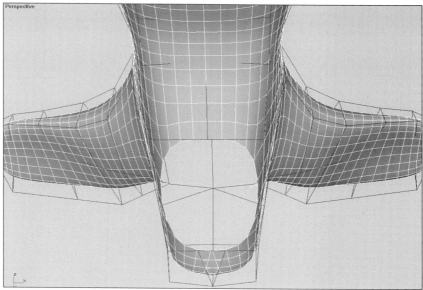

Figure 4.48 *Bottom of rover showing battery compartment opening*

2. Click the **Create** button on the Edit Geometry rollout. Using the same technique you used to create the windshield, draw a polygon that closes the battery cavity.

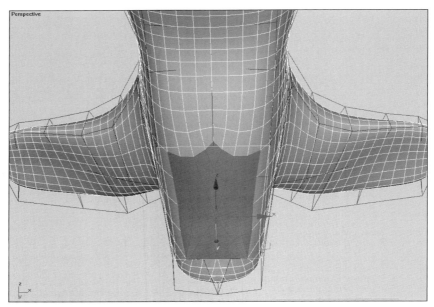

Figure 4.49 *New polygon covering the battery compartment*

3. Click the **Extrude** down arrow spinner once, and outline inward -1 unit.
4. Extrude inward approximately -17 units, and outline inward one click (-0.1).
5. Click the **Extrude** down arrow spinner once, and outline inward -8 units.

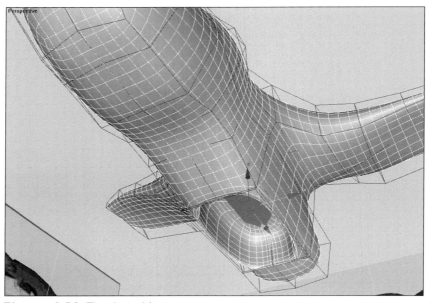

Figure 4.50 *The shaped battery compartment*

Creating the Light Housing

There are numerous methods that could have been used for creating the lights on the rover. We have chosen to use soft-selections as a modeling tool to turn a simple sphere into a light.

1. In the Front viewport, create a sphere that matches the size of the left wing's light in the reference image.
2. Name the sphere **Light01**.

   This ensures that the duplicate lights we create in a moment will automatically be named **Light02**, **Light03**, and **Light04**.

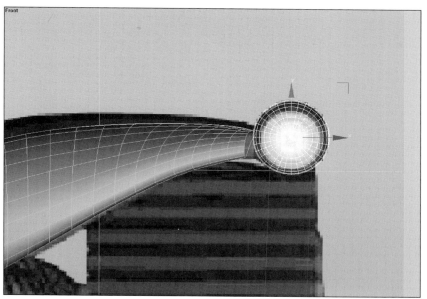

Figure 4.51 *Sphere matching the reference picture of wing light*

3. Convert the sphere to an Editable Poly.

Shaping the Light Housing

1. On the **Modify** panel, access the **Vertex** sub-object level.
2. In the Perspective view, select the single center vertex at the back of the sphere.
3. Expand the Soft Selection rollout and check the **Use Soft Selection** option. Set **Falloff** to about 14 units.

   This will cause vertices about halfway around the sphere to be soft selected.
4. In the Right viewport, move the selected vertex on the X axis until it lines up with the back of the light in the reference picture.

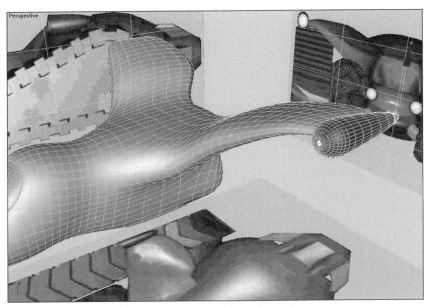

Figure 4.52 *The shape of the sphere, viewed in the Perspective viewport*

 5. In the Perspective viewport, use **Arc Rotate** to rotate around to the front of the new light housing, and select the single center vertex at its front.

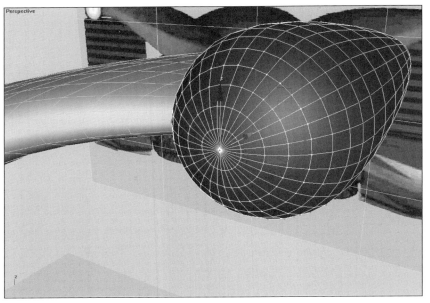

Figure 4.53 *The center front vertex selected*

6. Change **Falloff** to 12, and move the selection inward by 1 unit on the Y axis.
7. Uncheck the **Use Soft Selection** checkbox.

Finishing the Light and Housing

1. In the Right viewport, use a marquee to select the first five rows of vertices on the front side of the light, and move them inward.

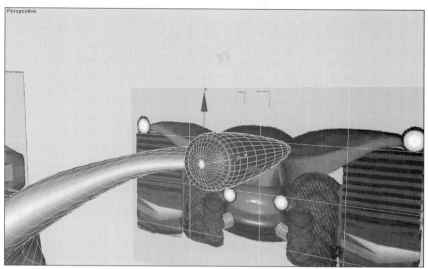

TIP
Moving these vertices will create a separation between the light-emitting area and its housing.

Figure 4.54 *The final shape of the light*

2. Turn off the **Vertex** sub-object level.
3. Copy the light and housing to the three other locations where lights appear in the reference pictures. Scale the lights at the front of the rover if necessary.

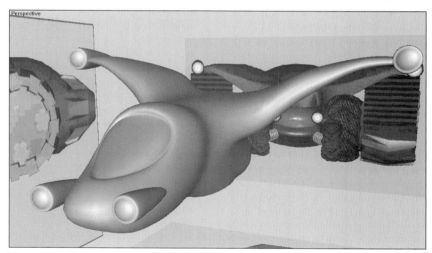

TIP
Position the wing lights so they cover the holes at the end of each wing.

Figure 4.55 *Lights copied and positioned*

The main body of the rover is now complete.

5. Save the scene with the filename **Rover03.max**.

TUTORIAL 4.3 CREATING THE WHEELS

Now that the rover body is complete, we can create the wheels.

Creating the Wheel Cylinder

1. In the Right viewport, create a cylinder with the following settings:

Name	Wheel01
Radius	24
Height	22
Height Segments:	3
Cap Segments	1
Side	18

2. In the Right and Front viewports, line up the cylinder over the reference picture's right front tire.

3. On the **Modify** panel, convert the cylinder to an Editable Poly.
4. On the Surface Properties rollout, check the **Use NURMS Subdivision** checkbox.
5. Set **Iterations** in both the Display and Render sections to 2.

Defining the Treads

1. Access the **Vertex** sub-object level.
2. In the Front viewport, use a marquee to select the two middle rows of vertices

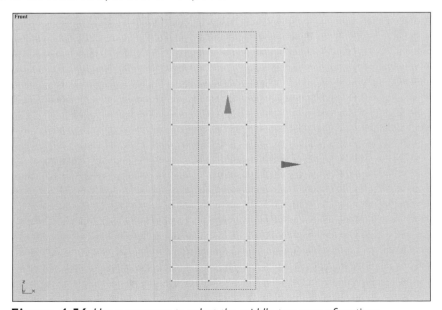

Figure 4.56 *Use a marquee to select the middle two rows of vertices*

Modeling the Rover

3. Use **Select and Non-uniform Scale** to scale the vertices on the X axis toward the center of the selection.

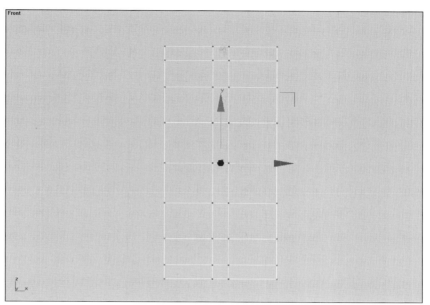

Figure 4.57 *Vertices scaled on X axis*

4. In the Perspective view, use **Select and Non-uniform Scale** to scale the selected rows of vertices up on the YZ Axis by a very small amount.

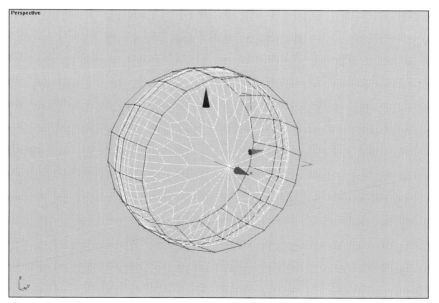

Scaling Y and Z to about 104% should be plenty.

Figure 4.58 *Non-uniform scale on the YZ axis*

5. Switch to the **Polygon** sub-object level.
6. Select the three polygons on every other row around the outside of the tire.

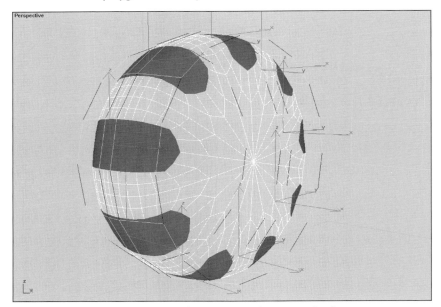

Figure 4.59 *Every other row of polygons on the outer wheel*

7. Extrude the selected polygons by approximately 2.5 units.

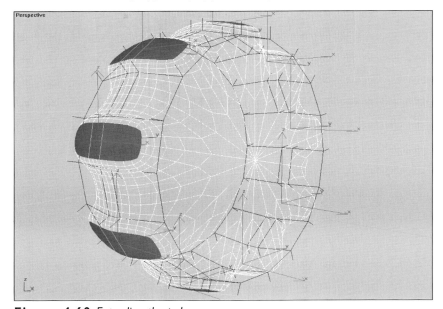

Figure 4.60 *Extruding the polygons*

8. Switch to the **Vertex** sub-object level. The same vertices we selected earlier are still selected.

9. In the Perspective viewport, rotate the selected vertices counterclockwise around the X axis by approximately 13 degrees.

   Rotating the selected vertices creates treads on the tire.

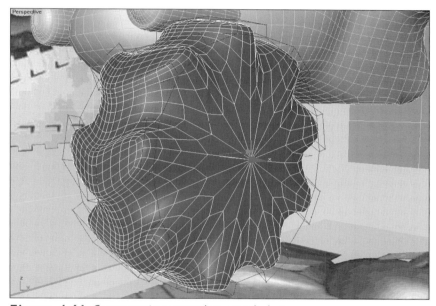

Figure 4.61 *Center vertices rotated counterclockwise*

Making the Tire Rounder

1. Switch to the **Polygon** sub-object level.
2. Select the two polygons on the inner and outer sides of the tire.

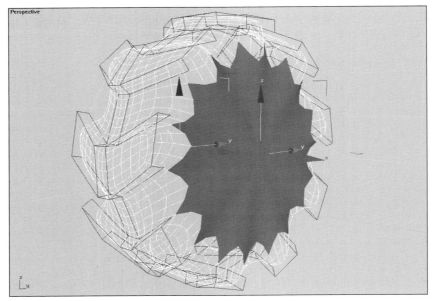

Figure 4.62 *Inner and outer polygons selected*

3. Extrude the selected polygons 0.7 units, and outline them -5.3 units.

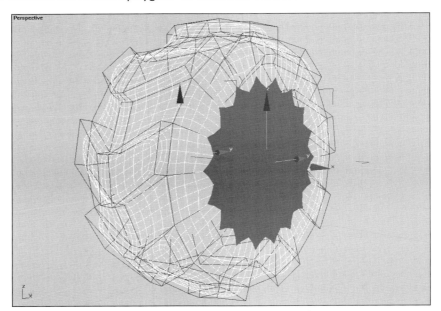

Figure 4.63 *The wheel after the first extrusion and outline*

4. Click **Select and Non-uniform Scale**.

5. Choose **Use Selection Center** from the **Use Pivot Point Center** flyout on the Main Toolbar.

   This will cause the selection center to be used as the center for scaling operations.

6. In the Front viewport, scale the selected polygons away from one another on the X axis to approximately 115%.

   This will give the tire some roundness.

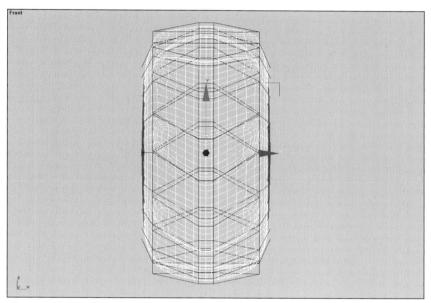

Figure 4.64 *The wheel after applying a non-uniform scale to the selected polygons*

Creating the Rim

1. Change the **Use Selection Center** button back to **Use Pivot Point Center**.
2. Extrude the selected polygons -0.7 units, and outline -4.0 units.
3. Extrude the selected polygons -0.5 units.
4. Extrude the selected polygons 0.1 units, and outline -0.9 units.
5. Extrude -10 units to give the depth of the tire rim.
6. Extrude 0.1 units to create a flat edge at the back of the rim.

Creating the Hub

Next you will create the hub at the center of the wheel.

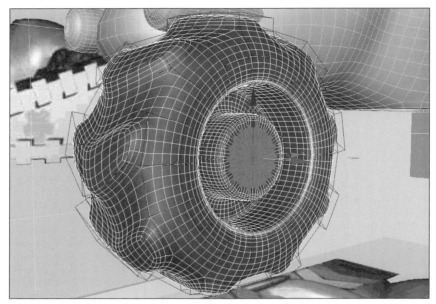

Figure 4.65 *The hub*

1. Select only the centermost polygon on the outside of the tire/rim (on the side away from the rover).
2. Extrude the selected polygon -0.1 units, and outline -5.0 units.
3. Extrude the selected polygon 1.6 units, then extrude again 6.5 units.
4. Extrude the selected polygon -0.1 units, and outline -1.1 units.

5. Turn off the **Polygon** sub-object level.

Duplicating the Wheel

Now that we have completed the first tire, you may notice that your system is starting to slow down. We can fix this by reducing the Display Iterations.

1. On the Surface Properties rollout, change the Display **Iterations** setting to 0.

    With this setting you will still be able to see the general shape of the tire, but there is no need to keep it at such a high resolution in viewports.

2. With the Front viewport active and the tire selected, use **Mirror Selected Objects** to mirror and copy the tire on the X axis. Position the tire to match the front tire on the opposite side in the reference pictures.

3. Copy the left front tire and position it where the left back tire belongs.
4. In the Front viewport, use **Select and Uniform Scale** to uniformly scale the tire up until the height matches the reference picture.
5. Use **Select and Non-uniform Scale** to scale the tire on the X axis until it matches the width of the tire in the reference picture.

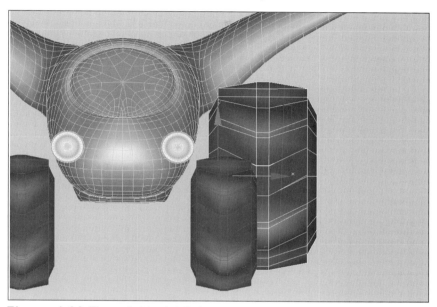

Figure 4.66 *The scaled back left tire*

Changing the Treads on the Back Tire

1. Switch to the **Vertex** sub-object level. All the center vertices are still selected.
2. Rotate the selected vertices clockwise to create a tread.

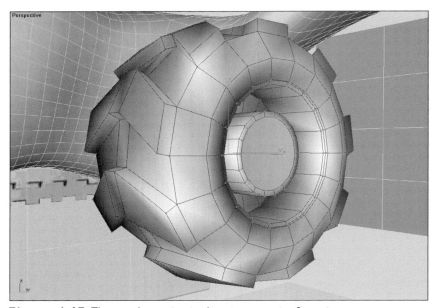

> ☸ **TIP** ☸
>
> *Make sure the treads go in the direction opposite the front tire treads.*

Figure 4.67 *The treads running in direction opposite front tires*

3. Turn off the **Vertex** sub-object level.
4. In the Front viewport, mirror the tire on the X axis and align it with the right back tire in the reference picture.

   The wheels are now complete.
5. Save the scene with the filename **Rover04.max**.

TUTORIAL 4.4 CREATING THE REACTOR

The rover is powered by a reactor on its back.

Creating the Reactor Sphere

1. Create a large sphere in the Front viewport. Match the sphere as closely as possible to the reactor in the picture reference. Name the sphere **Reactor-Engine**.

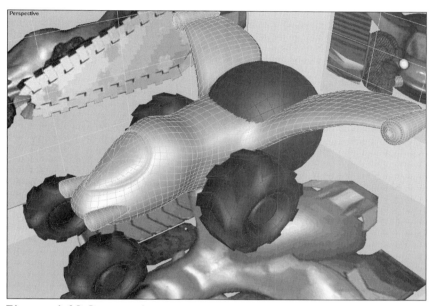

Figure 4.68 *Reactor sphere*

2. In the Top viewport, move the new sphere back so that only about 1/8 of it is sticking into the rover body geometry.

3. If necessary, increase the sphere's **Radius** so it matches the reactor in the reference picture.

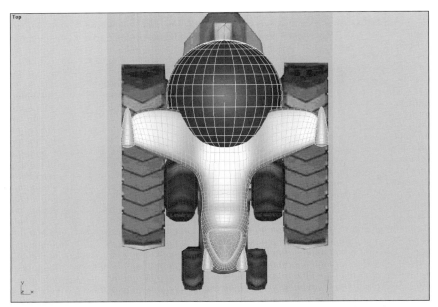

Figure 4.69 *Top view of rover with reactor sphere positioned and sized*

4. In the Front viewport, non-uniformly scale the sphere on the Y axis so that it fits the reference picture.

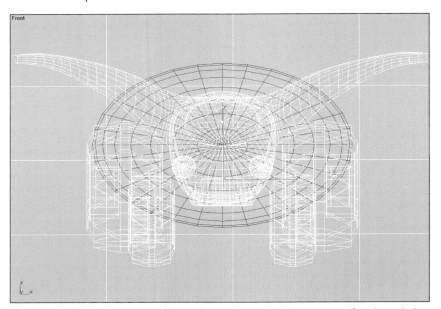

Figure 4.70 *A front view of the sphere after it has been non-uniformly scaled to match the reference*

5. Convert the sphere to an Editable Poly.

Shaping the Reactor

1. Press **<Alt-X>** to make the reactor semi-transparent.

2. Access the **Vertex** sub-object level.
3. In the Top viewport, use a marquee to select the first 3 or 4 rows of vertices on the sphere that are closest to the rover body.
4. On the Soft Selection rollout, check the **Use Soft Selection** checkbox and adjust the **Falloff** so you affect vertices far back enough to properly shape the reactor.
5. In the Top viewport, non-uniformly scale the vertices inward on the X axis until the geometry matches the reference picture.

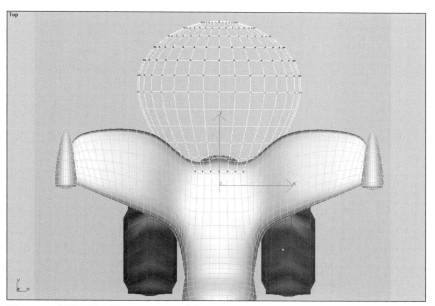

Figure 4.71 *The front portion of the sphere has been scaled in along the X axis*

6. Uncheck the **Use Soft Selection** checkbox.

Preparing for the Engine Nozzle

1. Switch to the **Edge** sub-object level.
2. Use a marquee to select the three rows of polygons at the back of the reactor.

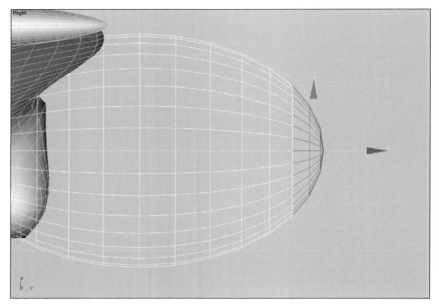

Figure 4.72 *The back three rows of edges are selected*

3. Press **<Delete>** to delete the selected edges.
4. Press **<Alt-X>** to turn off semi-transparency.

Extruding the Engine Nozzle

1. Switch to the **Polygon** sub-object level.
2. Select the polygon at the back of the reactor sphere.

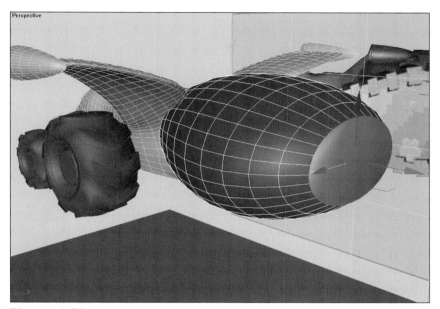

Figure 4.73 *Back polygon on reactor selected*

3. Extrude the polygon inward by approximately -25 units.

4. Use **Select and Uniform Scale** to scale the polygon down to about 50% of its original size.
5. Extrude the polygon outward approximately 32.5 units, and outline the selected polygon 7.5 units.

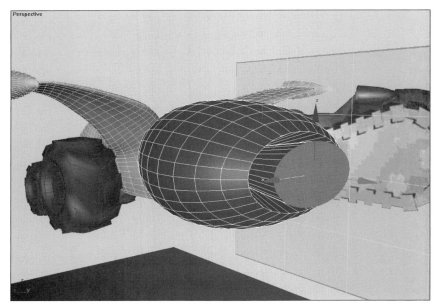

Figure 4.74 *The engine is starting to take shape*

6. Non-uniformly scale the selected polygon on the X axis to make the polygon into a circle rather than an oval.

7. In the Right viewport, extrude the polygon until it matches the reference picture where the engine nozzle is ready to slope downward.

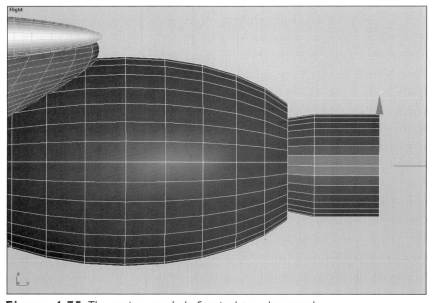

Figure 4.75 *The engine nozzle before it slopes downward*

8. Extrude again until the new polygon matches the end of the engine in the reference picture, and outline downward so that the engine matches the slope.

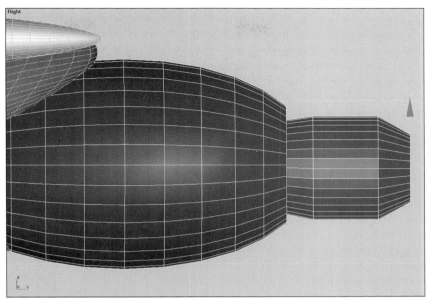

Figure 4.76 *The engine nozzle after sloping the end*

9. Use the techniques that you've become familiar with to hollow out the inside of the engine. This can be completed with a few extrusions and outlines.

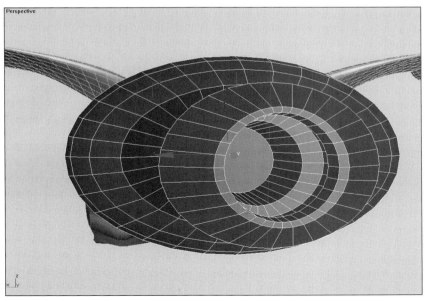

Figure 4.77 *Be creative in creating the inside of the engine nozzle*

Creating the Nozzle Fins

In this series of steps, you will be extruding selected polygons to make the nozzle fins. In selecting polygons around the engine, there is a very specific pattern that you will need to follow:

- Select the two polygons on the top of the engine. Be sure to select polygons only on the flat portion -- do not select polygons on any of the sloping areas.
- Going in a counter-clockwise pattern, skip the next 3 polygons then select 2, skip the next 4 polygons and select 2, skip the next 3 polygons and then select the polygons at the bottom of the engine.
- Continue with skipping the next 3 polygons and then selecting 2, skip the next 4 polygons and select 2. If you were to skip the next 3 polygons you would be back to the top.

1. Select polygons on the engine as described above.

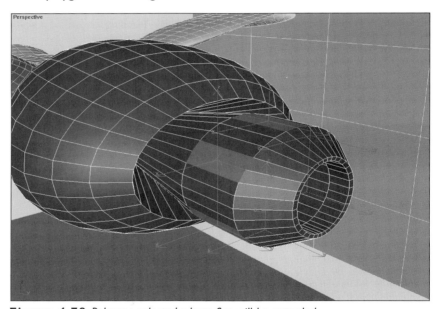

Figure 4.78 *Polygons selected where fins will be extruded*

2. Extrude the selected polygons approximately 7.5 units, and outline down by about -2 units.

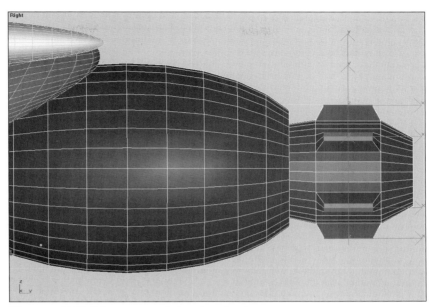

Figure 4.79 *Polygons extruded and outlined*

3. Move the selected polygons toward the reactor on the X axis.

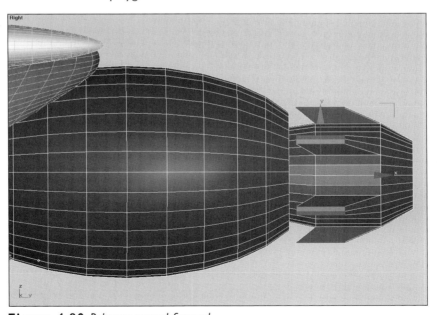

Figure 4.80 *Polygons moved forward*

4. Extrude another 4.5 units.

5. Move the new polygons so they are halfway inside the engine.

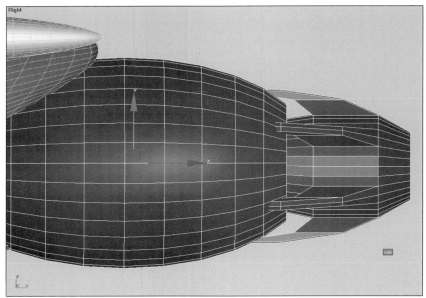

Figure 4.81 *Polygons moved partially into the reactor*

This completes the reactor engine.

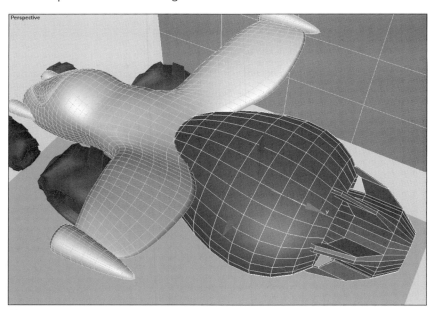

Figure 4.82 *Perspective view of the final engine/reactor*

6. Save the scene with the filename **Rover05.max**.

Modeling the Rover 217

TUTORIAL 4.5 CREATING THE TRACK ASSEMBLY

As part of its design, the rover also has two track assemblies for powering the car over rough terrain and obstacles. Each track goes around two additional wheels near the back of the rover.

Here you will build the track wheels and the support. The track support goes around the wheels and supports the treads.

Creating the Track Wheel

1. Turn off the **Polygon** sub-object level if it is still on.

> **TIP**
> To hide objects, select them and go to the Display panel, and click Hide Selected.

2. To assist in building the track wheels and belts, hide all objects except the reference planes.

3. In the Right viewport, create a cylinder with the following settings:

Name	TrackWheel01
Radius	37.5
Height	40
Height Segments	1
Cap Segments	1
Sides	18

4. Position the cylinder so that it is aligned within the back portion of the track in the reference pictures in both the Right and Front viewports.

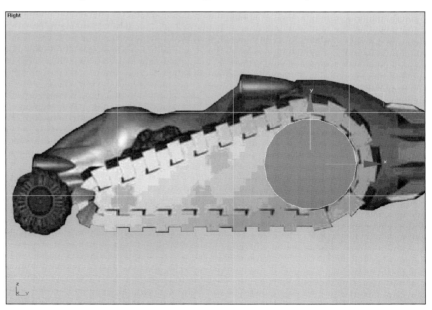

Figure 4.83 *The positioning of the cylinder*

5. Convert the cylinder to an Editable Poly.

Shaping the Track Wheels

 1. Access the **Polygon** sub-object level.

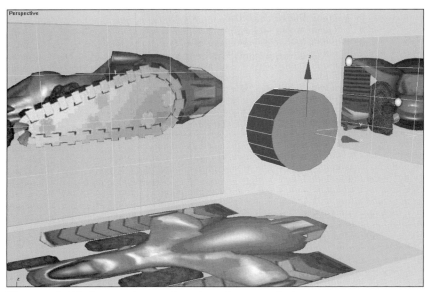

Figure 4.84 *Outer polygon selected on the new cylinder*

2. Select the cap polygon facing away from the rover.
3. Perform a series of extrusions and outlines to make the wheel look similar to the following figure.

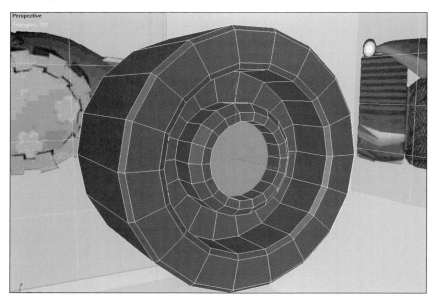

Figure 4.85 *The final shape of the track wheel*

Modeling the Rover   219

4. Use a marquee to select all the polygons on the track wheel.
5. On the Surface Properties rollout, click the **Auto Smooth** button.
6. Exit the **Polygon** sub-object level.
7. Copy, position and scale the back wheel to the front. Check the front wheel's width in the Front viewport, and scale the width separately on the X axis if necessary.

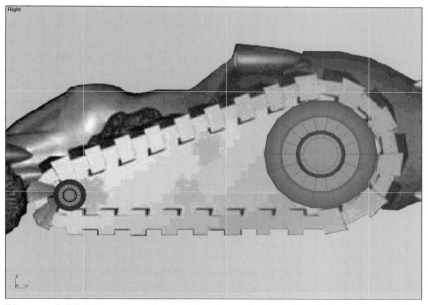

Figure 4.86 *The new track wheel has been duplicated, scaled, and positioned*

Creating the Track Support

Next you will create the track support, an object that goes around the wheels and supports the treads.

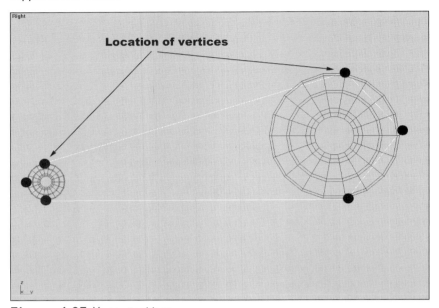

Figure 4.87 *Vertex positions*

1. In the Right viewport, use the Line tool to trace the inside of the track. Use a total of 6 vertices, and make sure your last vertex is placed at the same location as the first vertex. When asked if you would like to close the spline, answer **Yes**.

 2. In **Vertex** sub-object mode, fine tune the locations of the vertices.

3. Select all of the vertices, right-click on any one of them, and convert them all to the **Bezier** type.

4. Adjust the Bezier handles until the shape accurately matches the reference picture and flows around the front and rear track wheel tires.

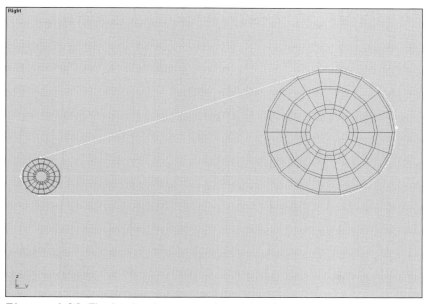

Figure 4.88 *The line has been properly positioned*

> ☼ **TIP** ☼
>
> *Converting the track support outline to an Editable Poly turns it into a 3D object and causes one side of it to become shaded.*

5. Right-click in the modifier stack and convert the line to an Editable Poly.
6. Name the line **Track Support**.
7. In the Front viewport, move the track object to align with the outside of the track wheels.

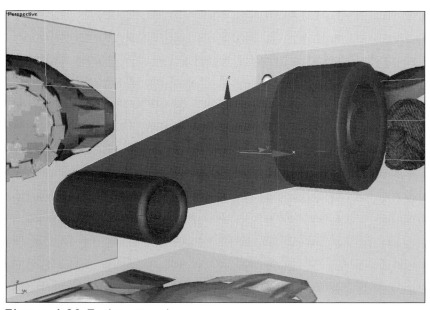

Figure 4.89 *Track positioned*

Extruding the Track Support

1. Access the **Polygon** sub-object level.
2. Select the polygon making up the entire track support.

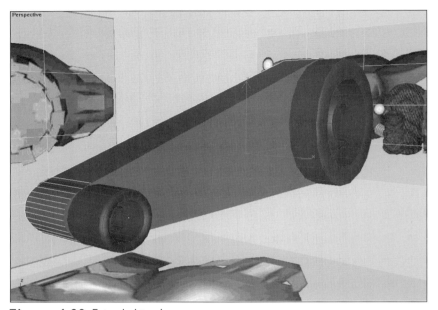

Figure 4.90 *Extruded track*

3. Extrude the polygon until the track support is a little wider than half the thickness of the track wheels.

4. Switch to the **Border** sub-object level.

   This sub-object, available only with an Editable Poly, selects only open edges.

5. Use a marquee to select the entire object. This will select the open border of the extruded polygon, which is currently not shaded.
6. Press the **Cap** button, located under the Edit Geometry rollout. This will create a surface on the open side of the track support.

Adjusting the Track Support

The tracks must be connected to the rover body in some way. A polygon extruded from the track support will serve this purpose.

1. Switch to the **Polygon** sub-object level.
2. Select the polygon that will be facing toward the rover.
3. Extrude the selected polygon 4 units, and outline -4 units.

4. In the Perspective view, non-uniformly scale the selected polygon down on the Y axis and on the Z axis.

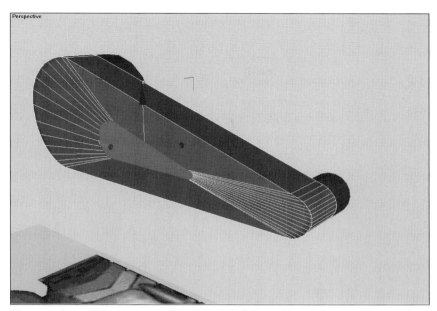

Figure 4.91 *Polygon non-uniformly scaled down on Y and Z axes*

5. Move the selected polygon back on the Y axis and up on the Z axis to complete the support. Use the following figure as a reference.

Figure 4.92 *New polygon moved to proper position*

6. Extrude the selected polygon approximately 52 units.
7. Exit the **Polygon** sub-object level.

Positioning the Track Assembly

Now we need to move the track objects into place and make sure they are positioned correctly in relationship to the other rover objects.

1. Unhide all objects in the scene.

2. If necessary, select and move the track support and wheels to sit appropriately with the other rover objects. The newly extruded polygon on the track support should sit just inside the reactor.

3. Select the track support alone and access the **Polygon** sub-object level.

4. Move the selected polygon up on the Z axis and then back on the Y axis until the geometry looks good.

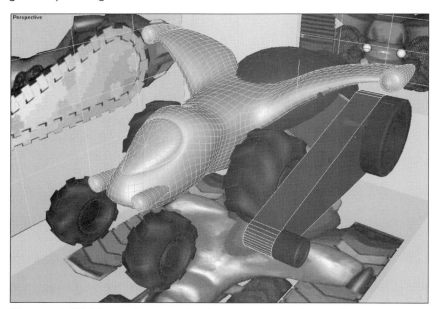

Figure 4.93 *The track support has been attached to the rover's reactor*

5. Exit the **Polygon** sub-object level.

The track itself (treads) require special attention because they will be expected to move. For this reason, the treads will be built in a later chapter when we prepare to animate the rover. In the meantime, we will duplicate the existing track assembly to the other side.

Mirroring the Track Assembly

1. Select the track support and the two track wheels.

2. In the Front viewport, use **Mirror Selected Objects** to mirror and copy the objects on the X axis.

3. Move the new pieces to the other side and use the Front viewport to properly align them.

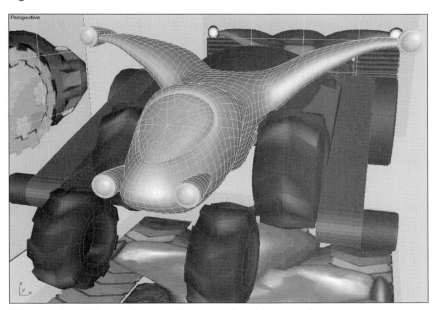

Figure 4.94 *The track assembly mirrored and positioned*

This completes the track assembly for the time being. As mentioned earlier, you will create the treads in a later chapter as they need to be animatable and thus require special attention.

4. Save the scene with the filename **Rover06.max**.

TUTORIAL 4.6 FINISHING THE ROVER

To complete the rover, we need to attach the rover wheels to its body, and create a fuel tank and a battery compartment door.

Connecting the Wheels to the Rover

Each wheel will be connected to the rover via a simple strut, as shown in the following figure.

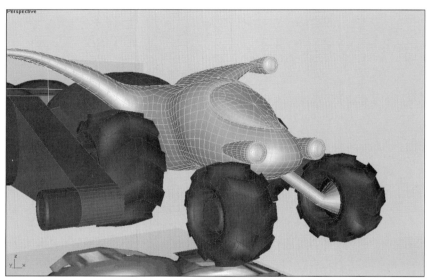

Figure 4.95 *Wheel strut*

1. In the Right viewport, create a cylinder with the following settings:

Name	Strut01
Radius	10
Heigh	17
Height Segments	2
Cap Segments	1
Sides	18

2. Position the cylinder at the center of the front left wheel.

3. In the Perspective viewport, move the cylinder on the X axis so it is up against the inner rim of the wheel's tire.

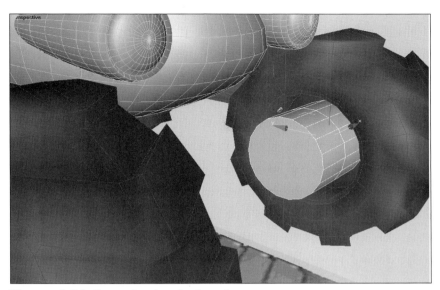

Figure 4.96 *Connecting the wheels to the rover*

4. On the **Modify** panel, convert the cylinder to an Editable Poly.
5. Access the **Vertex** sub-object level.
6. In the Front viewport, uniformly scale the first two rows of vertices (from the left) down to approximately 60% of their original size.

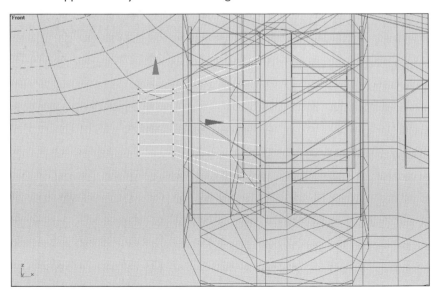

Figure 4.97 *First two rows scaled down to 60% of original size*

7. Use a marquee to select only the first row of vertices, and move them into rover body. This will flatten the cylinder. Rotate the vertices until the volume returns to the cylinder.

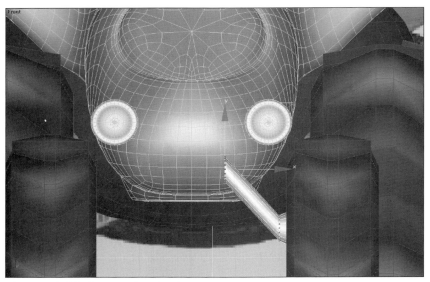

Figure 4.98 *Shaping the wheel strut geometry*

8. In the Perspective view, move the selected vertices back on the Y axis until they are embedded into the front of the rover body.

9. Rotate the first and second rows of vertices until you are happy with the way it looks.

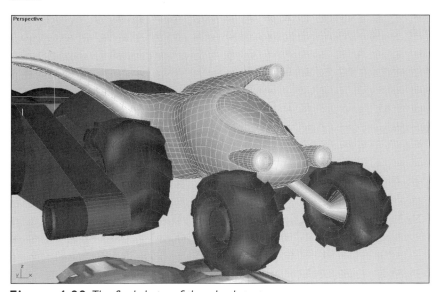

Figure 4.99 *The final shape of the wheel strut*

Creating the Strut Spring

Each front strut will be surrounded by a spring for shock absorption.

1. On the **Create** panel, choose **Dynamics Objects** from the pulldown list.
2. Click **Spring**.
3. In the Top viewport, create a spring around the same size as the strut. Name the object **Spring Strut01**.

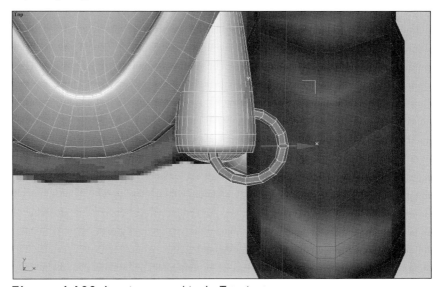

Figure 4.100 *A spring created in the Top viewport*

4. Go to the **Modify** panel.
5. In the Perspective viewport, move and rotate the spring until it is positioned over the strut. You will also need to adjust the **Height**, **Diameter** and **Turns** values on the **Modify** panel to get the spring to look right.

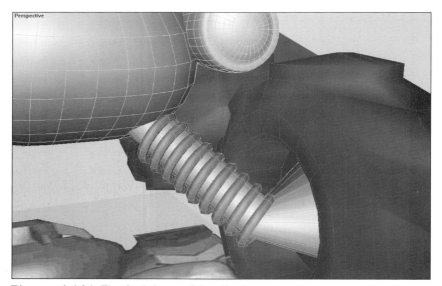

Figure 4.101 *The final shape of the wheel support system*

Mirroring and Copying the Struts

1. Mirror, copy and position the strut and spring to the other side of the rover for the right wheel.
2. Copy and move one of the struts (without the spring) to the center of one of the large tires. Scale the new strut and work with its vertices to shape it.
3. Mirror and copy the strut to the wheel on the other side.

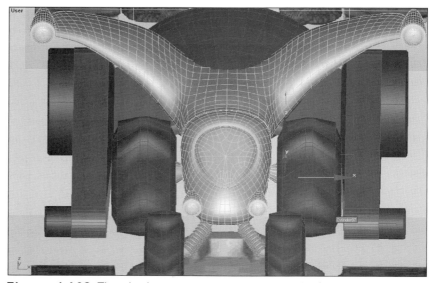

Figure 4.102 *The wheel support systems connecting all of the wheels to the rover*

Creating the Fuel Tank

TIP

On the Create panel, you will need to choose Standard Primitives from the pulldown list in order to find the Box option.

1. In the Front viewport, create a **Box** to represent the fuel tank. Set its dimensions so it fits beneath the front of the rover body. These are the dimensions we used, though yours may differ.

Name	Fuel Tank
Length	30
Width	52
Height	110
Length Segs	1
Width Segs	2
Height Segs	1

2. Move the box so it starts close to the nose of the rover and extends all the way to the start of the battery compartment. Check all viewports to make sure the positioning is correct before continuing.

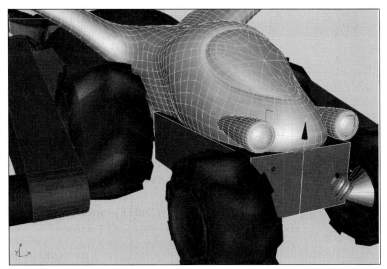

Figure 4.103 *The position of the fuel tank*

3. Convert the box to an Editable Poly.

4. Access the **Vertex** sub-object level.

5. In the Top viewport, use a marquee to select the six vertices at the back of the box (closest to the rover).

6. Use **Select and Non-uniform Scale** to scale the vertices on the X axis until they come close to the tires, but don't actually meet them.

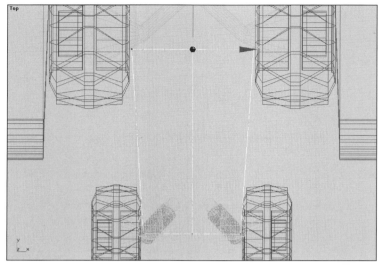

Figure 4.104 *Non-uniform scale the back of the fuel tank*

  7. Turn off the **Vertex** sub-object level.

Creating New Polygons on the Fuel Tank

We need to create some new polygons so we can shape the tank. Next, you will slit the edge so there will be a new set of polygons on each side of the tank.

  1. Switch to the **Edge** sub-object level.

2. In the Perspective view, select the back center vertical edge that is closest to the battery compartment.

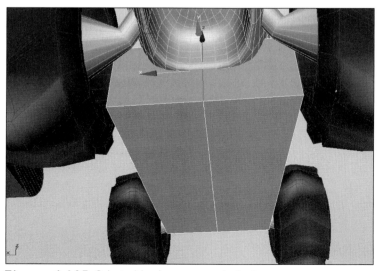

Figure 4.105 *Selected back center vertical edge*

3. Chamfer the selected edge by approximately 22.5 units.

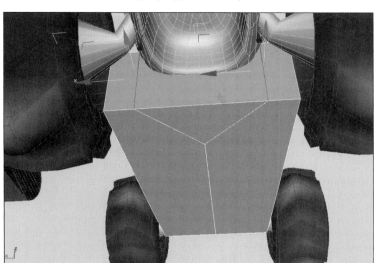

Figure 4.106 *Chamfered edge*

Shaping the Fuel Tank

Now we can start shaping the fuel tank.

1. Switch to the **Polygon** sub-object level.
2. Select the two new outer polygons on the side of the box closest to the battery compartment.

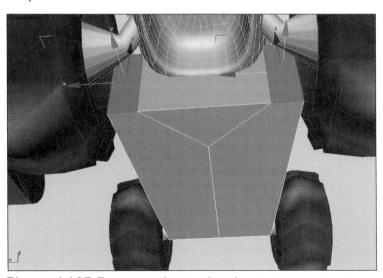

Figure 4.107 *Two outer polygons selected*

3. Extrude the polygons by about 26 units.

   The extrusion should end around the halfway point of the battery compartment.

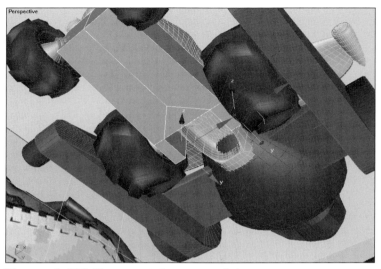

Figure 4.108 *Two outer polygons extruded*

 4. Use **Select and Non-uniform Scale** to scale the selected polygons on the Z axis.

5. In the Right viewport, move the selected polygons up on the Y axis so that the top of the geometry travels in a straight line.

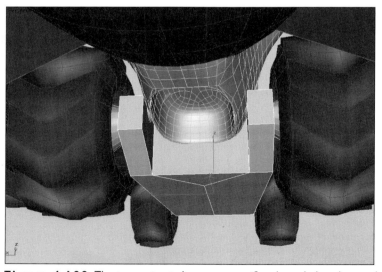

Figure 4.109 *The two outer polygons non-uniformly scaled and moved up*

 6. Exit the **Polygon** sub-object level.

Finishing the Fuel Tank

1. With the fuel tank selected, check the **Use NURMS Subdivision** checkbox.
2. Set the Display **Iterations** to 2.
3. Check **Iterations** in the Render section, and set the Render **Iterations** to 2.

4. Access the **Vertex** sub-object level.
5. Move the fuel tank's vertices until you are happy with the look of the fuel tank.
6. When you have finished adjusting the fuel tank, change the Display **Iterations** back to 1.

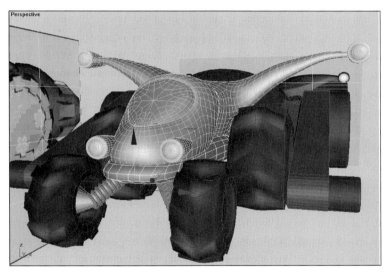

Figure 4.110 *The final fuel tank*

As with the wheels, this practice will allow us to keep faster interaction in viewports while rendering the higher level of detail.

Creating the Batteries and Battery Door

Keep the battery as simple as possible Remember, in production you are always on a tight schedule!

1. Using the techniques you've learned throughout these modeling exercises, create a cylindrical battery and position it in the battery compartment. Name the object **Battery**.
2. Activate the Top viewport and press the **<B>** key to change it to the Bottom viewport.
3. In the Bottom viewport, use the **Line** tool to create the outline of the door that will enclose the battery compartment. Give the line the name **Door**.
4. Apply the **Extrude** modifier to the line to give it a little thickness. Set the **Amount** to a small number such as 2 or 3.

5. On the **Modify** panel, right-click the **Line** and choose *Collapse All* from the pop-up menu to collapse the modifier stack.

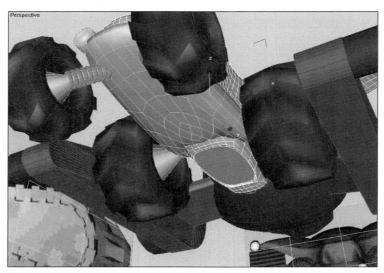

Figure 4.111 *The battery door*

The door should be able to slide back into the rover body and become completely hidden. You may need to move the door a little farther into the battery cavity so it cannot be seen at all when it slides back.

The rover and all its parts are now complete!

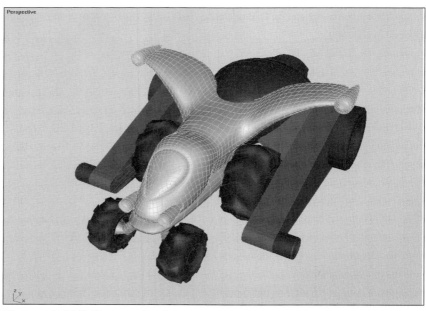

Figure 4.112 *The completed rover*

6. Unfreeze and delete the virtual studio reference planes.

    In **Chapter 6 Technical Setup for the Rover**, you will be creating the tracks with treads for use with the rover animation setup. For this reason, the model you have just created should be named **Rover - No Tracks.max**.

7. Save the scene with the filename **Rover - No Tracks.max**.

SUMMARY

In this chapter, you learned how to:

- Create lines and work with different types of vertices
- Create an intersecting framework of lines to form an object's shape
- Use the Surface modifier to put a solid surface over a framework of lines
- Create individual polygons by drawing them on the screen

All the major modeling tasks for our commercial are complete. Next you will learn how to use MAXScript, which is essential for our animation setups.

Introduction to MAXScript

The purpose of this chapter is to introduce you to the wonderful and powerful world of MAXScript. MAXScript is a *scripting language* built into **3ds max**. A scripting language a special programming language that uses lines of text commands to give instructions and perform tasks. MAXScript uses commands specifically for controlling user interface elements, creating and editing objects, and other **3ds max** functions.

In this chapter you will write scripts from scratch and learn how to use them to improve your productivity. This will provide a foundation on which you can build so you can one day become a true MAXScript expert.

TECHNIQUES IN THIS CHAPTER

This chapter provides an introduction to MAXScript, while later chapters will elaborate on specific uses of MAXScript for our project.

In this chapter, you will learn how to:

- Write a simple script
- Create objects and edit their parameters with MAXScript
- Create custom user interface elements such as rollouts and buttons
- Check for errors in MAXScript programs

TUTORIALS IN THIS CHAPTER

The tutorials in this chapter are designed to build from the simplest of techniques to the point where you are writing fully functional scripts that perform real tasks.

- *Tutorial 5.1 Basic Scripting Commands* shows you how to use MAXScript in its simplest form to create a primitive object.

- *Tutorial 5.2 Variables* take you through the steps of changing an object's parameters in the MAXScript Listener window.

- *Tutorial 5.3 A First Script* uses the MAXScript editor to write a complete script from scratch with what you learned in the first two tutorials.

- *Tutorial 5.4 Functions* and *Tutorial 5.5 Arguments* builds on previous tutorials to add more functionality to the script and show you how to use these more advanced programming features.

- *Tutorial 5.6 Custom User Interface* takes you through the steps of building a custom rollout or floater containing buttons and sliders.

By the time you do all these tutorials, you will be able to go on to the next chapter and start building the control elements that will be used for our commercial scene.

UNDERSTANDING MAXSCRIPT

MAXScript is designed to be so simple that non-programmers can easily learn to use it, yet powerful enough to accommodate the needs of experienced programmers.

Knowing how to write basic scripts will open the door for many advanced uses of **3ds max**. You can develop scripts that can automate complex and laborious tasks, enhance the user interface, and expand **3ds max** feature sets. Scripts can be made to execute when certain events occur, such as when the time slider is moved. The bottom line is that MAXScript can help improve your productivity immensely.

One additional advantage to learning MAXScript is that the skills you learn can be applied to other programming languages you learn later on, such as HTML or C++. All these programming languages use similar commands and require the same logical skills used with MAXScript.

SCRIPTING STEPS

Writing and using a MAXScript consists of several steps.

1. Write the script.
2. Test the script, either one line at a time or as a body of commands.
3. Save the script.
4. Run the script to perform the tasks indicated by the script's commands.

You will get to perform all these tasks over the course of this chapter.

ACCESSING THE MAXSCRIPT OPTIONS

There are three areas in **3ds max** where you can access MAXScript options, labeled in the diagram that follows.

1. MAXScript menu
2. Utilities panel
3. Mini Listener, a miniature version of the MAXScript Listener window

The fourth area shown is the MAXScript Listener window, which is shown as open and ready to use.

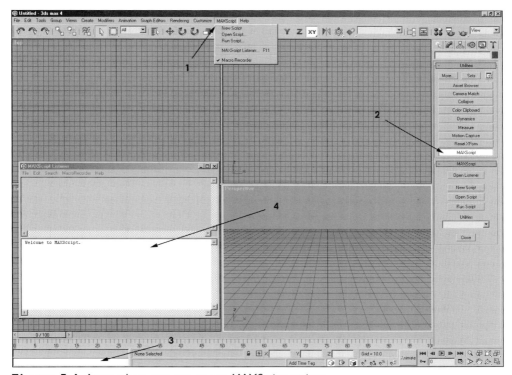

Figure 5.1 *Areas where you can access MAXScript options*

Both the MAXScript menu and the Utilities panel provide a way to:

- Open the MAXScript Editor to write and save a script
- Open a saved script
- Execute (run) a script
- Open the MAXScript Listener

WRITING A SCRIPT

There are three areas of **3ds max** where scripts can be written and tested.

The Mini-Listener at the bottom left of the screen provides an area where one line of a script can be typed and executed. You will use this area at first to get accustomed to using MAXScript commands.

You can also write scripts with the MAXScript Listener window. This window allows you to type in several lines of the script. As each line is typed, it is executed immediately, and any result is displayed in the window. The Listener is a good place to test all or part of a script, especially when you're first learning MAXScript.

Finally, the MAXScript editor is a window where you can type in and save scripts for execution all at once. This is where you put your scripts when you are ready to run them for real.

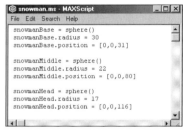

Figure 5.2 *MAXScript Editor*

You can also use any text editor such as Windows Notepad or Wordpad to write scripts. If you use a word processing program such as Microsoft Word, be sure to save the file as text only.

The word *code* is sometimes used to refer to the lines or commands of a script.

TUTORIAL 5.1 BASIC SCRIPTING COMMANDS

The Mini Listener is a text field at the lower left of the screen that accepts a single line of commands. The Mini-Listener provides a quick and easy way to type one script line to see how it works.

Using the Mini-Listener is an excellent way to start learning MAXScript.

Entering a Command

1. Reset **3ds-max.**
2. Locate the Mini-Listener at the bottom left of the screen.
3. Type the following into the white area of the Mini-Listener:

   sphere( )

   Press **<Enter>** after you have typed in the line.

TIP
If the sphere doesn't appear, check your spelling. Also ensure there is no space between the open and close parentheses.

Figure 5.3 *Command typed into Mini Listener*

You have just created a sphere in the viewports. Now that wasn't that hard, was it?

The use of the parentheses with nothing inside them indicates that you want to make a sphere with the default parameters.

Changing Object Parameters

Now that a sphere has been created, you can change its editable properties such as its Radius parameter.

Accessing object parameters requires the use of a programming technique known as *dot notation*. This simply means indicating the parameter by placing a dot between the object name and its parameter.

When referring to an object by name in MAXScript, the name is always preceded by a dollar sign ($).

Let's use this information to adjust the radius of the sphere.

1. Locate the name of the sphere you just created.

   Most likely, the object name is Sphere01. If it has a different name, just substitute the object name where you see **sphere01** in the next step.

2. In the Mini Listener type the following:

   $sphere01.radius = 50

   Press **<Enter>** after you have typed in the line.

   The sphere's radius has changed to 50.

Note that MAXScript is not case sensitive. Although the sphere's name has a capital S in the scene, you could use either a lower or upper case S at the beginning of the object name when referring to it in MAXScript. The same is true for the Radius parameter.

3. Delete the sphere.

THE MAXSCRIPT LISTENER WINDOW

The MAXScript Listener window is an another area where you can type in commands and try them out. The purpose of the MAXScript Listener is to give you a place to try out MAXScript commands before typing them into a final script file.

You can open the MAXScript Listener window by right clicking the Mini Listener area and selecting *Open Listener Window* from the pop-up menu.

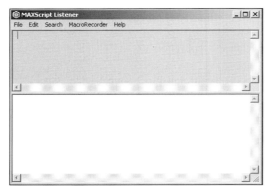

Figure 5.4 *MAXScript Listener window*

The Listener window functions as an interactive interpreter. As you type one line of commands into the pink area at the top and press <**Enter**>, the commands are executed immediately, and any result is displayed in the white area at the bottom. Your typed commands are black, output text is blue, and error messages are red.

You can also type commands into the white area to display the commands and results together.

MAXSCRIPT SYNTAX

MAXScript programs must be written with a specific *syntax* (arrangement of letters and symbols). When MAXScript is interpreting your script, it must understand exactly what you mean, and it is very picky — there is no forgiveness if you type something wrong, even by just a little bit.

MAXScript will let you know when you have typed something it doesn't understand by displaying a red error message in the Listener.

VARIABLES

There will be times when you need to store values for later use. You can store these values in *variables*. A variable is a simply a letter or series of letters that are assigned a value. You assign the value to the variable, and you can use it later on with other commands.

Variables are easy to create. Simply specify the letter or letters that make up the variable name, then follow it by an equals sign and a value. For example, the following statement will create the variable x and assign the value 15 to it:

x = 15

With the variable x holding the number 15, you can put the variable into use.

The following statement shows one method of using a variable:

$sphere01.radius = x

When this line is typed, MAXScript knows to assign the value being held in the **x** variable to the sphere's radius, making the radius 15.

Variables can hold more than just numeric values. They can hold letters, colors, parameters, and even object names.

Variable names should not contain spaces or punctuation, but they can contain numbers. When creating variable names, keep in mind that not every name can be used to represent a variable as some are reserved by MAXScript for specific uses. Try to come up with unique names that describe what you are doing.

TUTORIAL 5.2 VARIABLES

In this tutorial, you will learn how to assign an object name to a variable, then change the object's parameters using the variable.

Using an Object Name as a Variable

1. Reset **3ds max**.

2. From the **MAXScript** menu, choose **MAXScript Listener**.

   The MAXScript Listener appears.

   The Listener is divided into two panes The top pane (pink) is called the Macro Recorder pane. The lower pane (white) is called the Output Pane. For now we will work only with the Output Pane so you can see the results under each command as it is typed and executed.

   In this tutorial and in those that follow, it is assumed that you will press **<Enter>** after each line of code.

3. In the Output Pane (white area), type the following:

   myBox = box( )

   A box is created in the viewport, and its name is assigned to the variable **myBox**.

TIP

If you do not see two panes in your Listener window, move your cursor just below the file menu until it turns to two horizontal lines. You can then click and drag the separator down to expose the top pane.

4. In the Output Pane, type the following commands. Be sure to press **<Enter>** after each one.

   myBox.length = 55
   myBox.width = 55
   myBox.height = 10

   Each time you type a command and press **<Enter>**, the command is executed immediately. The value for each parameter is printed to the Output pane beneath the command that you just executed, and the box's size adjusts appropriately.

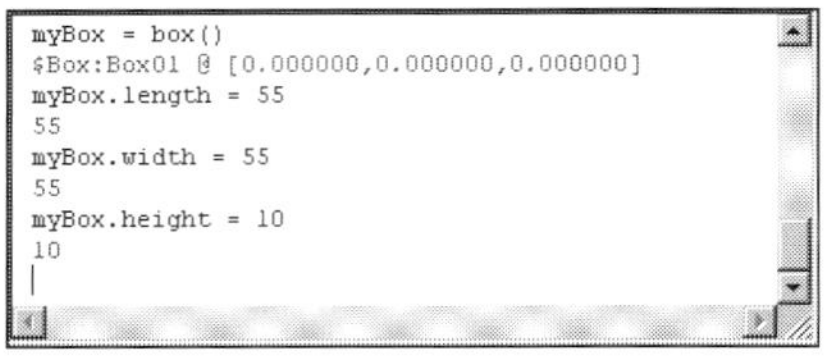

   ```
   myBox = box()
   $Box:Box01 @ [0.000000,0.000000,0.000000]
   myBox.length = 55
   55
   myBox.width = 55
   55
   myBox.height = 10
   10
   ```

   Figure 5.5 *The MAXScript Listener window with myBox commands*

5. Delete the box.

Listing Parameters

If we've done our job right, by now you are itching to find out how you can see a listing of all the parameters you can change with MAXScript. A built-in MAXScript command called **showProperties** can be used for this purpose. This command provides a listing of all parameters that can be accessed for a particular type of object.

1. In the Output Pane, type the following commands:

   myBall = sphere( )
   showProperties myBall

   This lists all the available parameters for a sphere in the Output Pane.

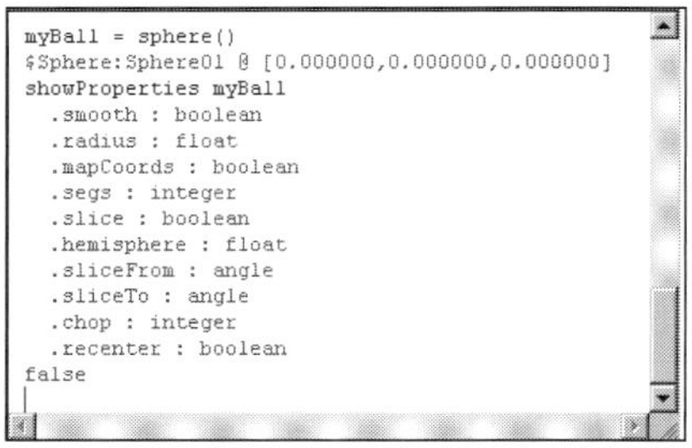

   ```
   myBall = sphere()
   $Sphere:Sphere01 @ [0.000000,0.000000,0.000000]
   showProperties myBall
     .smooth : boolean
     .radius : float
     .mapCoords : boolean
     .segs : integer
     .slice : boolean
     .hemisphere : float
     .sliceFrom : angle
     .sliceTo : angle
     .chop : integer
     .recenter : boolean
   false
   ```

   Figure 5.6 *Results from the showProperties command*

   The listing will show you the parameter type and the type of data that it requires. For example, the **smooth** parameter requires a boolean, which is a true or false value.

2. In the Output Pane, type:

   myBall.smooth = False

   In the Perspective viewport, you can see that the sphere now appears as faceted, as it is no longer smoothed.

3. To change the sphere back to smooth, type:

   myBall.smooth = True

   The sphere has returned to its previous smoothed state.

4. Delete the sphere.

   Here you have learned that each parameter expects a certain type of data. Below is a listing of common data types and the type of data they expect.

   - **Booleans** - True or False. You can also use On or Off.
   - **Float** - Any number with or without decimal points.
   - **Integer** - Any whole number, meaning no decimal points. Negative whole numbers are okay as long as they would work for the parameter on the command panel.
   - **String**: - Any combination of letters and numbers, with double quotes around the text.

5. Create a few more objects using MAXScript in the Listener window.

   - Use the technique shown in step 1, where the object is created and assigned to a variable at the same time.
   - Use the showProperties command to get a listing of the objects parameters.
   - Set a few parameters using MAXScript commands.

   This is a great way to become familiar with the use of objects and their parameters in MAXScript.

ARRAYS

As you've seen, variables are basically containers that store a single piece of information. An *array* is a series of variables stored under a single name. In other words, an array is nothing more than a large variable that is divided up into compartments, and each of these compartments can store one piece of information. These compartments are called *elements*. Each element is assigned a number, and the information in the compartment is accessed via this number.

One way to understand an array is to think of an apartment building. The apartment building has a single address such as 127 Main Street, and each apartment (element) has its own number such as Apt. 12 to indicate the individual compartment. Arrays work the same way.

Arrays are one of the most useful ways of handling large blocks of data in MAXScript. We will be using arrays in the next chapter to make the animation process quick and painless.

Creating an Array

To create an array, you first assign it an overall name. This is accomplished by specifying a variable name followed by the equals sign, the pound sign (#) and open and closed parentheses. For example, an array called **MyFirstArray** can be created with the following command:

MyFirstArray = #( )

This command creates an *open array*. An open array means there is not a predetermined number of elements — we can simply add elements as we go along.

Adding Elements to an Array

After creating the array, you need to add elements to it. Elements are added by placing brackets after the array name with the element number, and assigning a value to the element.

You can add anything you want as elements — integers, strings, true/false values and float numbers. The following are examples of commands that would add elements to the array created above.

MyFirstArray[1] = "Peter"   (adds a string)
MyFirstArray[2] = 10   (adds an integer)
MyFirstArray[3] = $sphere01   (adds an object name)

The last line shown above would work only if there were actually an object named Sphere01 in the scene.

A note to experienced programmers: In MAXScript, the first element in the array is assigned the number 1, not 0 as in other programming languages.

Adding Multiple Elements

You can add multiple elements to an array by specifying them within the parentheses when you first create the array. For example, the following line would create an array with three elements, each containing a string:

MySecondArray = #("Jelly","Glazed","Plain")

You could just as easily create an array with different kinds of data:

MyThirdArray = #(23,True,"Plain")

You could later change the value of any element in the array by assigning it a new value:

MyThirdArray[3] = "Jelly"

The Position Parameter

Each object has a position parameter that specifies where the object sits in 3D space. The position parameter is an array with three elements that is automatically set up by MAXScript each time you create an object.

The three elements define the object's position along the X, Y and Z axes in world space respectively.

For example, to set the position for Sphere01 in the scene at the XYZ position of 20,30,50, you could use the following command:

>   **$sphere01.position = [20,30,50]**

You will get to use this parameter in the next tutorial.

THE SCRIPT EDITOR

Up until now you have been typing all of your MAXScript commands in the Mini-Listener or in the MAXScript Listener and executing the commands one at a time. Now it's time to graduate to writing real scripts in the MAXScript editor.

The MAXScript editor is a simple text editor that looks very similar to the Windows Notepad. It has the added advantage of a feature that can evaluate the script (test the commands) right from within the editor. The commands are tested one line at a time, starting at the beginning and going through to the end.

You might wonder why we stress the testing aspect of scriptwriting so much. If you are new to programming, it might not seem as though there would be many mistakes in a script, or that errors would be difficult to find and fix. As experienced programmers, we can tell you that it is very easy to mistype a word or leave out a letter that will make the entire script fail, and that it is sometimes difficult to find the error. This is why we are teaching you testing as we go along, so you will get in the habit of testing commands before you attempt to write a long script. The more familiar you are with the commands, the easier it will be for you to detect and correct mistakes.

We've already practiced with a sphere's parameters, so we can go ahead and use them in a real script, which we'll type into the editor.

TUTORIAL 5.3 A FIRST SCRIPT

Here you'll write a simple script that will create a snowman with three spheres. Leave the MAXScript Listener window open while you do this tutorial as you will need it when checking for errors.

Creating the Snowman Script

1. Delete any objects currently in the scene.
2. From the **MAXScript** menu, choose **New Script** to open the MAXScript editor and start a new script.

☆ **TIP** ☆

You can also open a new script in the MAXScript Editor on the Utilities Panel by clicking MAXScript, then New Script.

The MAXScript editor appears.

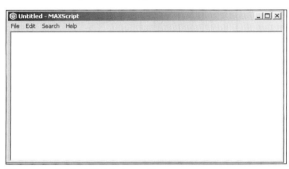

Figure 5.7 *MAXScript editor*

3. Type the following commands into the Editor.

   snowmanBase = sphere( )
   snowmanBase.radius = 30
   snowmanBase.position = [0,0,31]
   snowmanMiddle = sphere( )
   snowmanMiddle.radius = 22
   snowmanMiddle.position = [0,0,80]
   snowmanHead = sphere( )
   snowmanHead.radius = 17
   snowmanHead.position = [0,0,116]

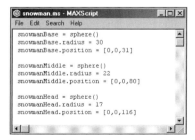

Figure 5.8 *Command typed into MAXScript editor*

4. Save the script by choosing **Save** from the **File** menu on the MAXScript editor. Name the file **Snowman.ms**.

   By default, **3ds max** saves all scripts in the **Scripts** folder.

   Next, we want to evaluate the script and see if all the commands make sense to the MAXScript interpreter.

Testing for Errors

The advantage of using the MAXScript editor over Windows Notepad or other text editors is that you can check the viability of the script right from within the window. In order to use this feature effectively, you should have the Listener window open on your screen.

1. If the MAXScript Listener is not currently open, open it by choosing **MAXScript Listener** from the **MAXScript** menu.
2. On the MAXScript editor, choose **File/Evaluate All** from the menu.

   If the script has been typed correctly, a basic snowman will appear on your screen.

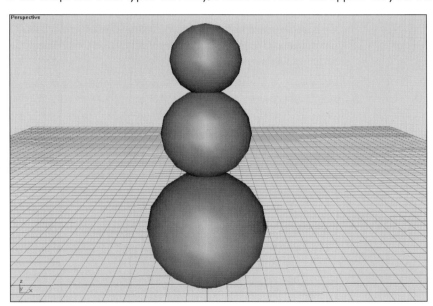

Figure 5.9 *Results from snowman script*

The Listener displays only blue text and the snowman appears, then the script is fine. If red error messages appear in the Listener window, then the script has an error that needs to be fixed.

3. If the script has errors, compare it carefully with the script on the previous page, looking for mistyped letters or punctuation. Make sure there is no space between parentheses, and no space before or after the dot between the object name and parameter.

Choose **File/Evaluate** to test the script as many times as necessary until no red text appears in the Listener and the snowman appears. When the script is correct, you might have several more spheres than necessary, so delete the spheres on the screen and choose **File/Evaluate** once more to create them. Then save the script again by choosing **File/Save** from the menu.

Alternate Ways to Set Parameters

Note that you can create an object with no parameters then set them later, or you can create an object and set its parameters when you create it.

So far you have been creating the object, then setting the parameters in separate lines of code:

```
snowmanBase = sphere( )
snowmanBase.radius = 10
snowmanBase.segments = 20
```

You could also create the same object with one line of code:

```
snowmanBase = sphere (radius:10 segments:20)
```

By entering parameters inside the parentheses you are able to set the sphere's parameters as it is created. Note that you do not need the parentheses in the second example.

We won't be using this method of creating objects just yet, but you should be aware of it for future examples.

Adding Comments

The more code you add, the more complicated it looks. It is a good idea to get into the habit of adding comments in your scripts. *Comments* are lines in the code that are ignored by MAXScript when it is executing your script, but can be read as plain English by anyone trying to figure out how the script works.

Comments provide you and others with a road map to various parts of the script. You may have the entire picture in your head when you write the script, but six weeks from now (or even tomorrow morning!) you may find that you can't remember what's where and why you put it there. Then you end up spending just as much time as you did when you first created it trying to relearn what you had done. (Not that this has ever happened to us or anything.) With good use of comments, this nightmare can be avoided.

To create a comment, you simply place two dashes at the beginning of the line, then type your text. MAXScript will ignore anything on a line that comes after two dashes.

Our snowman script is very simple, and is not likely to confuse experienced scripters if it remains uncommented. However, it's a good practice script for creating comments.

TIP

If you plan to make a script available to the professional community, you should include as comments your name, the date the script was created and any other relevant information in addition to comments about the commands themselves.

The following lines show the script with sample comments included.

```
-- Snowman Script, created by Ford Prefect
-- Creates a simple snowman with three spheres

-- Create the snowman base, set its radius, and position it
snowmanBase = sphere( )
snowmanBase.radius = 30
snowmanBase.position = [0,0,31]

-- Create the snowman middle
snowmanMiddle = sphere( )
snowmanMiddle.radius = 22
snowmanMiddle.position = [0,0,80]

-- Create the snowman head
snowmanHead = sphere( )
snowmanHead.radius = 17
snowmanHead.position = [0,0,116]
```

1. Add comments to the snowman script explaining what everything does.
2. Evaluate the script after you have added the comments to ensure that everything still executes correctly.
3. Save the script as **Snowman01.ms**.

FUNCTIONS

Now that you have seen how to write a basic script, let's see how we can organize our code a little better.

You could say that the snowman script has three distinct sections, one for each sphere. We could organize each of these three sections into a *function*. A function is a group of commands that do a specific task.

If you anticipate doing the same task over and over again, you can use a function instead of including the same code several times in the script. When you want the script to perform the function, you simply call (request) the function to perform the task.

The simplest syntax for a function is:

```
function myFunction =
(
)
```

You must first tell MAXScript that you are declaring a function. You do this by starting the line with the word **function**, or you can use **fn** for short. Follow this word by the name of the function, then the equals sign. The commands inside the function are bounded by an open and a close parenthesis.

The open and close parentheses are used frequently in MAXScript as a way of *blocking* the code, which means placing it in blocks or chunks. This tells MAXScript when a specific portion begins and ends.

TUTORIAL 5.4 FUNCTIONS

In this tutorial, you will edit the snowman script to use functions.

Here is an example of the snowman script with each task set up as a function.

Indenting the code inside a function makes it easier to see what's a function and what's not.

```
function createBase =
(
    snowmanBase = sphere( )
    snowmanBase.radius = 30
    snowmanBase.position = [0,0,31]
)
function createMiddle =
(
    snowmanMiddle = sphere( )
    snowmanMiddle.radius = 22
    snowmanMiddle.position = [0,0,80]
)
function createHead =
(
    snowmanHead = sphere( )
    snowmanHead.radius = 17
    snowmanHead.position = [0,0,116]
)
```

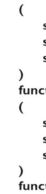

It is a common programming practice to start function and variable names with a lower case letter, then use an uppercase letter in the middle of the function or variable name to make it easier to read.

1. Delete all objects in the scene.
2. Update the code in the script editor to use the functions shown above.
3. Save and evaluate the script.

   What happened? Where is the snowman?

   Functions must be called before they will execute. Simply placing them in the program is not enough. To better understand this, visualize the following scenario.

   Imagine that you have a teenage son that is currently playing a video game. Your son knows how to wash dishes, as this is a function that has been previously defined. He is not going to get up and start washing dishes on his own.

   The only way he will perform the function is if he is told to wash the dishes. This is the part where you call the function. After he completes the function, he returns to playing his video game without being told to do so. This is what happens when a function is complete.

In creating functions in the snowman script, we have defined three functions, but we never told MAXScript to run any of them. To actually make the code execute, we must call it. Calling a function is a matter of typing the function name followed by open and close parentheses.

4. In the MAXScript Listener, type:

    createHead( )

    You will see that the head for the snowman has just appeared in the viewport. This is because we called a function (group of commands) that has already been evaluated and stored, waiting idle until a script calls it. Try typing in the other two functions.

    So how can we integrate the function calls into a script? All you would need to do is call the function from within the script.

5. To the snowman script, add the following three lines after the last function:

    createBase( )
    createMiddle( )
    createHead( )

6. Save and evaluate the script.

    This time a snowman is created.

You can call the functions only after you have defined them, so these three lines must be placed after the functions.

ARGUMENTS

While we have broken the code up into groups that perform specific tasks, the code is still not as efficient as it could be. A sphere is created, its radius is adjusted, and finally the sphere is moved. We have three different procedures doing basically the same thing.

The code could be made far more efficient if we were to have only one function that would create and position a sphere based on information we sent it. To do this, we need to pass the *arguments* to the functions.

An argument is not a disagreement, as you might think from its ordinary English language definition. An argument is a value that is passed from the caller to the function. A real life example would be calling a friend to meet you for lunch. When you call, you don't just say, "Meet me for lunch," you also specify the time and place to meet. The time and place would be arguments passed along with the call for lunch.

Here, we'll create a function that creates a sphere and requires two variables to be defined:

Remember that you can use either function or fn to start a function.

fn makeSphere
(
    newSphere = sphere( )
    newSphere.radius = x
    newSphere.position = y
)

In this case, the radius of the sphere is set to x and its position is set to y. But in order for the function to know what x and y are, they have to be passed to the function as arguments. Even if x and y are defined elsewhere in the script, the function won't be able to use them unless their values are passed to the function when it is called.

```
fn makeSphere x y =
(
    newSphere = sphere( )
    newSphere.radius = x
    newSphere.position = y
)
```

Note the x y = after the function name, indicating that x and y are to be passed to the function as arguments. The equals sign is always placed after argument variables passed to a function.

When you call the function later in the script, you need to specify the values to be passed to the function. The following call would create a sphere with a radius of 14 and an XYZ position of 10,20,30:

```
makeSphere 14 [10,20,30]
```

Now let's take a look at how we can use arguments to rewrite the snowman script.

TUTORIAL 5.5 ARGUMENTS

In this tutorial you will rewrite the snowman script to have one function and a few calls.

1. From the **MAXScript** menu, choose **New Script** to open the MAXScript editor.
2. Type the following code into the MAXscript editor: Comments are included to make the code clearer to you.

```
-- Creates a sphere with radius and position passed as arguments
fn createSnowman theRadius thePosition =
(
snowman = sphere( )
snowman.radius = theRadius
snowman.position = thePosition
)
-- Calls the function to create the sphere three times
-- with different radius and position each time
createSnowman 30 [0,0,31]
createSnowman 22 [0,0,80]
createSnowman 17 [0,0,116]
```

3. Save and evaluate the script.

The above script creates a function called createSnowman. This function accepts two arguments: theRadius and thePosition. The function then uses these two variables to define the radius of the sphere and its position. The function is then called three times, each time sending a different radius and position value. This is a far more efficient approach to writing the snowman script. We do not have to repeat similar code over and over again.

PROGRAM FLOW

Up to this point, every program we have written has been executed linearly, one line after another. There will be times when you want a script to make decisions based on certain conditions, then alter the flow of the program accordingly. This is nonlinear code.

To further clarify what we mean by linear and nonlinear code, the following examples give an analogy for linear vs. nonlinear actions using the task of getting up in the morning to get ready for work. The first example is linear, with no changes or deviations along the way.

> **Get out of bed**
> **Get dressed**
> **Sit at kitchen table**
> **Eat breakfast**
> **Put on coat**
> **Go to work**

How dull life would be if it were the same routine every day! In this second example, we'll make some decisions to control the flow of events.

> **Get out of bed**
> **Get dressed**
> **Sit at kitchen table**
> **Pick up morning paper**
> **Evaluate: The front-page headlines are interesting. True or False?**
> > **If true,**
> > > **Read first article**
> > > **Discuss it with your spouse**
> > **If false,**
> > > **Turn on TV**
> > > **Find favorite news channel**
> **Eat breakfast**
> **Evaluate: It looks like rain. True or False?**
> > **If true,**
> > > **Put on raincoat**
> > > **Take umbrella**
> > **If false,**
> > > **Put on sweater**
> **Go to work**

In the first linear example, you simply do one thing after another. The second example starts linearly, but then you make a few decisions that cause the program's outcome to vary depending on conditions.

THE "IF" STATEMENT

The easiest way to control a program's flow is to use a statement beginning with the word **if**, known as an *if statement*. The **if** statement performs a test and executes one set of code or another depending on whether the statement is true or false.

Here is an example of an **if** statement in action. Comments are included so you can see what is going on.

```
a = $sphere01.radius
b = $sphere02.radius
if  a > b then                          -- test if A is larger than B
(
    print "Sphere01 is larger"          -- if A is larger than B,
)                                       -- print this in the Listener
else
(
    print "Sphere02 is larger"          -- otherwise, print this in the Listener
)
```

In the example, the variables **a** and **b** are created, and each is assigned the same value as a specific sphere's radius. A test is then performed to see if the value of variable **a** is greater than the value of variable **b**. If the result of the test is true, meaning that **a** is greater than **b**, then the **if** block is executed. If it is false, the **else** block is executed.

LOOPS

Loops allow you to do repetitive tasks without having to write the same or similar code over and over again.

There are several commands that create loops. For our scripts, we will be using two types of loops called *for loops* and *while loops*, named after the commands that start them off.

As an example of where looping would come in handy, suppose you have 30 objects in your scene and you want to change the same 10 parameters on each of them. You could type 300 very similar lines of code, or you set up a loop with 10 lines, one to change each of the 10 parameters. The loop would execute 30 times, once for each object. You would simply set up the loop to use a different object name each time it executes the loop.

For Loops

A **for** loop works with a variable to determine how many times to execute. Programmers often use the variables **i** and **j** to count the number of times the **for** loop is executed. A loop could start with a statement like this:

```
for i=1 to 10 do
```

This says to do the loop 10 times. Each time, the variable **i** is incremented. You can use the variable **i** to change other parameters in the body of the loop if you like, but you don't have to.

Here is a simple example of how to use a **for** loop. The commands you haven't learned yet are explained after the example.

```
myBox=box( )                          -- create a box
for i=1 to 10 do                      -- do this loop 10 times
(
    newbox=copy myBox pos:[i*30,0,0]  -- copy and offset the box
    j = i as string                   -- make j a string, not integer
    newbox.name = "Newbox" + j        -- make new name for box
)
```

The first line beginning with **newbox** creates a copy of the box, and offsets the new box's position on the X axis by an amount equal to **i** times 30.

There is one completely new command here that you haven't seen yet:

j = i as string

This command converts the integer **i** to a string so it can be appended to the name **Newbox** in the next line.

If you like, enter the example above in MAXScript editor and see what it does. You should end up with a series of boxes, each with a name from **Newbox1** to **Newbox10**. There aren't any specific tutorials on the use of loops in this chapter, so feel free to experiment until you get the idea. You will use loops in the chapters that follow.

Another kind of **for** loop is usually used with an array. With this type of loop, you simply hand the loop an array and tell it to run through them all. The syntax goes as follows:

for [variable] **in** [array] **do** [specified action]

The loop goes through all the elements in the array, and for each one, it does the specified action.

For example, suppose you have an array of object names that you have created with the following commands:

```
bodyArray = #( )            -- create the array
bodyArray[1]=$Torso         -- put object names in the array
bodyArray[2]=$RightArm
bodyArray[3]=$LeftArm
bodyArray[4]=$RightLeg
bodyArray[5]=$LeftLeg
```

Now you want to print all the object names in the Listener. You could do it with one command, as follows:

for bodyPart in bodyArray do print bodyPart.name

The variable **bodyPart** is a name we made up just for this **for** loop. For each element in the array **bodyArray**, the **for** loop will put the element in the **bodyPart** variable, then print the object name.

If you like, go ahead and try this out in the Listener, setting up the array with actual object names in your scene.

While Loops

Let's take a look at another way of doing a loop construction. This time we will use the **while** loop.

A **while** loop works by testing an expression and continuing to loop as long as the expression is true. The syntax is as follows:

> **while** [test is true] **do** [specified action]

For example:

> **while altitude>0 do print "airborne"**

You can also flip it around to put the **do** before the **while**, and MAXScript will still know what you're talking about.

> **do** [specified action] **while** [test is true]

For example:

> **do print "airborne" while altitude>0**

Both these loops will continuously execute the **print** command as long as the **altitude** variable is greater than 0. The only difference is that in the first example, the **print** command won't be executed at all if the first test of the **altitude** variable finds it to be below 0. In the second example, the **print** command will be executed at least once no matter what.

USER INTERFACE DESIGN

A *user interface* comprises the visual elements of a program, such as buttons, rollouts, menus and so on. MAXScript allows you to create custom user interface elements that can float around the screen or be included in menus, panels and toolbars.

We have already stated that we are going to use MAXScript to control the animation in our scene with the alien and remote control car. All control will go through a custom user interface that you will design and create with MAXScript.

Throughout this book we are going to create several different user interface elements, each time using a different approach. In this section we are going to create a new utility which will have its own parameters and rollouts on the Utilities panel.

TUTORIAL 5.6 CUSTOM USER INTERFACE

Our first script will create a button for making a sphere. We know there's already a button that does this in **3ds max**, but creating this simple feature will give you great practice in creating buttons.

When you are asked to type in code, you probably won't understand what you're typing until after you've typed and evaluated it. That's okay. The quickest way to learn is to follow the instructions and observe the results.

Creating a Button

1. Reset **3ds max** and close the MAXScript editor and Listener if necessary.
2. On the Utilities panel, click **MAXScript**, and click **New Script**.
   This opens the MAXScript editor.
3. Type in the following code:

   Utility mySphere "Sphere Generator"
   (
       button btnCreate "Sphere"
)

4. Save the script with the name *SphereGenerator.ms*.

5. Evaluate the script.
   The Utilities pulldown menu now contains an option called **Sphere Generator**.
6. Click on the **Utilities** pulldown menu and select **Sphere Generator.**

A new rollout called Sphere Generator appears below the MAXScript rollout. It contains two buttons, one labeled Sphere and the other labeled Close.

Let's take a look at the relationship between what you typed into the editor and what we're seeing.

Utility - Tells MAXScript that this is a script for the Utilities panel

mySphere - The name of the utility script. This name could be anything you like. This name will not appear on the user interface, and is only for your internal use. It's always a good idea to keep the name as descriptive as possible. Please note that spaces *are* not allowed in the name.

"Sphere Generator" - The text between the quotes will be the name of the selection on the pulldown menu, and the label on the rollout itself. It is wise to use a descriptive name. Note that spaces *are* allowed in this name.

button - Tells MAXScript that it is going to create a button that can be clicked. There are a number of arguments that the user may pass to the button command. These arguments are optional and if not supplied, default values are assumed. More about this later.

btnCreate - The name that we have assigned to the button. Naming our buttons along with all other controls is a very important step. We will need to use this name later on to determine when the button is clicked.

"Sphere" - The text that appears on the button itself. Try to make this as descriptive as possible, because in the end, this is what the user will see.

When the script is evaluated, MAXScript automatically generates the rollout and the Close button.

You can compare interface component creation to a standard function call, just like the **createSnowman** function you created earlier.

Adding Functionality to the Button

Now that we have created a basic user interface, we need to add some sort of functionality to the button so it does something when clicked. To do this we will write a specific type of function called an *event function*.

An event occurs when a control's major function gets triggered. Examples would be: a button getting pushed, a checkbox getting checked, or a slider's value changing. The structure for the button event function is as follows:

```
on btnCreate pressed do
(
    [actions]
)
```

From our earlier code, MAXScript knows that **btnCreate** is the name of a button. It also recognizes that this special block of code will only get triggered if the button referenced by **btnCreate** is pressed. All the commands that need to occur when the button is pressed need to be placed between the opening and closing parentheses.

1. Press the **Close** button in the Sphere Generator rollout.

2. In the MAXScript editor, update the code to the following:

```
Utility mySphere "Sphere Generator"
(
    button btnCreate "Sphere"
    on btnCreate pressed do
    (
        b = sphere()
    )
)
```

3. Save and evaluate the script.

4. Reselect **Sphere Generator** from the Utilities pulldown menu.

5. Press the **Sphere** button.

This time a sphere is generated when the button is pressed. If you were to press the button again, another sphere would be created.

Each time a sphere is created, it is stored in the variable **b**. This will allow us easy access to the object that was just created. But beware, each time a new sphere is created, it is that sphere that exists in **b**. Any spheres created earlier cannot be accessed from the script.

Adding a Slider

To make our script a little more functional, we are going to add a slider control that will allow us to interactively control the sphere's radius.

1. Delete any spheres that have been created in your scene.
2. In the Script Editor, add the following code right after the **button btnCreate "Sphere"** line:

   Slider radiusSLDR "Radius:" range:[0,200,25]

3. Save and evaluate the script.

Figure 5.10 *The updated Sphere Generator rollout*

The Sphere Generator rollout now contains a Radius slider. Currently, the slider does nothing. The slider command only tells MAXScript to create a slider. Just as we saw with the button command, there is no functionality until we write the event procedure.

The slider command line is explained as follows:

Slider - Creates the slider itself.

radiusSLDR - The name of the slider control, used internally.

"Radius:" - The label that displays just above the slider.

range:[0,200,25] - Gives the slider as to its minimum, maximum, and default values respectively.

Adding Functionality to the Slider

Now let's add the event code for the slider and make it fully functional.

1. Update the script to the following:

   Utility mySphere "Sphere Generator"
   (
       button btnCreate "Sphere"
       slider radiusSLDR "Radius:" range:[0,200, 25]
       on btnCreate pressed do
       (
           b = sphere()
)
       on radiusSLDR changed val do
       (
           b.radius = val
)
)

   The slider event function includes an internal MAXScript variable, **val**, which holds the current value on the slider.

2. Save and evaluate the script.
3. Choose **Sphere Generator** from the pulldown to reinitialize the script.
4. Click on the **Sphere** button to create the sphere.
5. Move the slider. Whoops! An error message appears.

Figure 5.11 *Error generated when adjusting the slider*

This message appears because the variable **b** is not being passed between functions. Remember earlier, when we mentioned that a variable has to be passed to a function in order to be seen, even if it's defined elsewhere in the script? Well, we've just seen proof of this.

Declaring a Global Variable

Rather than passing the variable back and forth between functions, we can declare the variable **b** to be a *global* variable. A global variable is seen by all functions in the script at all times. Making **b** a global variable will make it easier to use in the script.

1. Insert the following line of code as the first line in the script:

   global b

2. Save and evaluate the script.
3. Choose **Sphere Generator** from the pulldown to reinitialize the script.
4. Click the **Sphere** button.
5. Adjust the radius slider. The sphere's radius will interactively change.

   Even though the script appears to be functioning fine, there is still a problem. Can you identify what the problem is? If you can't think of it right away, take a few minutes to look at how it runs. Think about any potential pitfalls that may be hiding in the dark. After you've come up with an idea, continue reading. We'll reveal the problem in the next section.

Adding Error Checking

We still have one major problem: What if the Radius slider is changed before a sphere is created? We will end up receiving the same error as we did earlier when the variable **b** was not visible. You cannot simply count on users not to make this error.

It is very simple to prevent the current problem. We basically need to set up a few lines to provide some level of error checking. In this case, we need to make sure that the variable **b** is defined before allowing the user to change the slider. To do this, we will use an **if** statement.

1. Update the code to the following:

   ```
   global b
   Utility mySphere "Sphere Generator"
   (
       button btnCreate "Sphere"
       slider radiusSLDR "Radius:" range:[0,200, 25]
       on btnCreate pressed do
       (
           b = sphere()
       )
       on radiusSLDR changed val do
       (
           try
           b.radius = val
           catch
           messageBox "A sphere must be created first"
       )
   )
   ```

2. Save and evaluate the script.
3. Reset **3ds max.**
4. Choose **Sphere Generator** from the pulldown to reinitialize the script.

5. Try adjusting the **Radius** slider, and you will receive a message.

Figure 5.12 *The message box*

6. Click the **Sphere** button.
7. Adjust the **Radius** slider again. This time, it works.

   In the previous tutorial, we set up an error handling routine. This routine will prevent the program for prematurely ending. We have done this by using a **try/catch** command combination. The script will try to execute the code following the **try** statement and if something fails, it will execute the **catch** code instead of displaying an error message.

   The **messageBox** function generates a window that can be used to provide some sort of feedback to the user. In this case we have let the user know that a sphere is required before the Radius slider will work.

   The **try** and **catch** functions are great for quick error handling. However, you should replace **try** and **catch** with another error-checking method once you finalize your code. These tools can very easily cause problems for less experienced programmers when they are trying to troubleshoot your code. It is also considered more elegant to simply disable a tool if it shouldn't be used.

   A better way to handle errors is to use the **enabled** parameter built in to every UI element. When **enabled** set to **false**, the UI element cannot be clicked or changed. You then change the parameter to **true** when it's okay to use the UI element. We will use this method to replace **try** and **catch** in our code.

> ☼ **TIP** ☼
>
> *Note that the enabled parameter is set to false at first, then is set to true after the sphere is created.*

8. Change your code to the following:

```
global b
Utility mySphere "Sphere Generator"
(
    button btnCreate "Sphere"
    slider radiusSLDR "Radius:" range:[0,200, 25] enabled:false
    on btnCreate pressed do
    (
        b = sphere()
radiusSLDR.enabled=true
    )
    on radiusSLDR changed val do
    (
        b.radius = val
    )
)
```

Creating Rollouts

So far we have created both simple linear scripts and non-linear utilities to appear on the Utilities panel. You can also arrange your UI elements into rollouts and floating windows.

Rollouts optimize your workflow by allowing you to organize related UI element groups into categories so that they can be collapsed and expanded when needed. In this way, the UI will change to meet the user's needs. Rollouts also provide a way for you to arrange your tools, and your code, in a logical manner.

This series of steps will walk you through the creation of a collapsible rollout.

1. Reset **3ds max**.

2. On the **Utilities** panel, click **MAXScript**, then **Open Listener**. You will use the Listener window later in this tutorial.
3. Click the **MAXScript** button, and click **New Script**.
4. In the MAXScript editor, type the following code:

   ```
   Utility rolloutTest "Rollout Example"
   (
       rollout rll_myFirstRollout "Main Rollout"
       (
           button btnHello "Hello World"
           on btnHello pressed do
           (
               print "Hello World"
           )
       )
       on rolloutTest open do
       (
           addrollout rll_myFirstRollout
       )
       on rolloutTest close do
       (
           removerollout rll_myFirstRollout
       )
   )
   ```

5. Evaluate the script.

6. Click on the **Utilities** pulldown menu under the MAXScript rollout, select **Rollout Example**.

   Two new rollouts called Rollout Example and Main Rollout appear on the panel. The second rollout contains a button labeled Hello World.

7. Click the **Hello World** button.

   The text **Hello World** is printed in the Listener.

   You must pay close attention to how the parentheses play their roles in the previous example. The first line of code sets the rollout called Rollout Example, then all of the code following the first line of code is enclosed inside opening and closing parentheses.

   Next is the rollout command for the Main Rollout rollout. Note that the command, through the use of parentheses, contains both the button and the button's event code. This illustrates how you make different buttons and controls lie on different rollouts.

   The rollout command takes two parameters: a name, and a caption. These are determine under which name the rollout object is stored and what text the user sees on the rollout border.

   The next two sets of statements are event handlers. The command **on rolloutTest open do** is executed when the script is first initialized, while **on rolloutTest close do** is executed when the Close button on the Rollout Example rollout is clicked. Yes, even rollouts can have event calls!

   The **addRollout** and **removeRollout** commands are used to actually add and remove the rollout from the Utilities panel. In this case, the commands require only one parameter, the rollout's name. In a minute we will see how we can supply the **addRollout** command with two parameters, the rollout's name and the name of the floating window to which we are adding the rollout.

Creating a Floater

Floaters are simply floating windows that contain rollouts. A floater can have UI controls only if it has rollouts.

Floating windows can be very helpful when the user needs to switch back and forth between different panels, or if the user works in expert mode where all of standard UI elements are hidden.

The syntax for creating a floater is slightly different than the syntax used to create a utility. This series of steps will walk you through creating a simple floater that changes an object's name.

1. Reset **3ds max**.
2. On the **Utilities** panel, click the **MAXScript** button, and click **New Script**.

3. In the MAXScript editor, type the following code:

   myWin = newrolloutfloater "Renamer" 200 200
   rollout rll_main "Main Rollout"
   (
       editText etRename "Name:" text:"<type name here>"
       on etRename entered newText do
       (
           if $==undefined then messagebox "Please select an object"
           else $.name = etRename.text
)
)
   addrollout rll_main myWin

4. Evaluate the script.

   A new floater is created.

   Figure 5.13 *The new floater*

5. To see the new floater in action, create a sphere in any viewport, and right-click when you are done. With the new sphere still selected, type a new name for the sphere in the floater dialog and press **<Enter>**.

6. Check the name of the sphere on the command panel. It should be the same as the one you just entered on the floater.

   Note that you might need to click on the sphere again for the name to update on the command panel.

7. Close the editor.

   The script that you just executed begins with the assigning of **newrolloutfloater** to a variable. The command **newrolloutfloater**, which creates the floater, takes three arguments, name, width and height.

   There is no opening parenthesis after this line to block in the code that would belong to it. Instead, all of the data goes into the various rollouts that will later be added to the floater.

   Assigning entire floaters to variables is an approach that later simplifies the code. When you begin adding rollouts to the floater, you can simply refer to the variable name that contains the floater. This is just one more demonstration as to how valuable variables can be.

The command **on etRename entered'newText do** line triggers an event when a user enters information into the edit text field **etRename** and presses the **<Enter>** key on the keyboard. The information is stored in a variable called **newText**, and the blocked code runs.

Once inside the blocked code, a check is made to ensure that something is selected. If not, a message box is displayed, letting the user know that an object needs to be selected first. If there is an object selected, the name is simply changed to the name that is contained in the variable **newText**.

This is only meant to be an example of how you can create a simple tool that will exist in a floating window and provide some sort of functionality. In ***Chapter 9*** we will be developing a very sophisticated floater that will allow you to quickly animate the alien character.

TIDY PROGRAMMING PRACTICES

As programming languages go, MAXScript is a very forgiving language. There are syntactical rules that must be obeyed, but in comparison to other programming languages, MAXScript is quite easy to write.

This has both pros and cons. It's great that you can quickly create scripts to meet your needs. However, this can easily lead you into messy programming practices that will make your scripts hard to understand and edit. Try to use tidy structured programming approaches whenever possible. Tidy programming practices include:

- Naming variables intelligently
- Adding comments liberally, especially for hard-to-understand blocks of code
- Whenever possible, pass arguments between functions rather than using global variables or try-and-catch functions
- Use indentations to make the script easy to follow visually

These are good programming practices no matter what language you use.

SUMMARY

In this chapter, you learned how to write scripts from scratch. Now you can do the following with scripts:

- Create and edit objects
- Write functions
- Pass arguments between functions
- Use loops to automate tasks
- Create custom rollouts, buttons and floaters

Now that you've been introduced to the basics of MAXScript, let's see if we can put it to good use in the chapters that follow.

The Job

You've just landed a job to create
a 30-second commercial for long-life batteries.
The commercial features an alien child
and his remote control rover.
Each component must be modeled from scratch,
and a live action background must be composited
with the 3D animation.

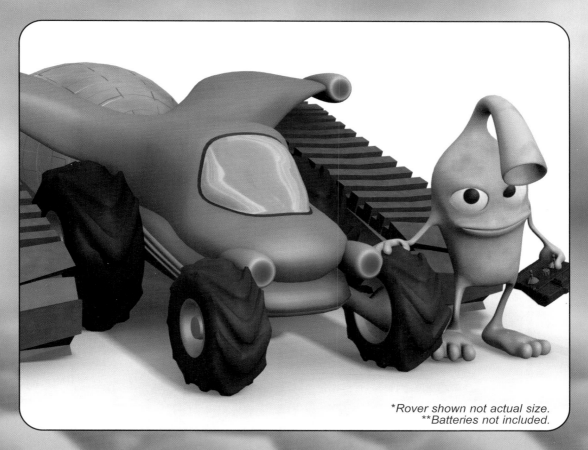

*Rover shown not actual size.
**Batteries not included.

You've got storyboards and a schedule,
and you're ready to go.

The Remote Control

The remote control is created using reference pictures and step-by-step instructions.
The box modeling techniques you learn while creating this object will be used to model the alien and the rover later on.

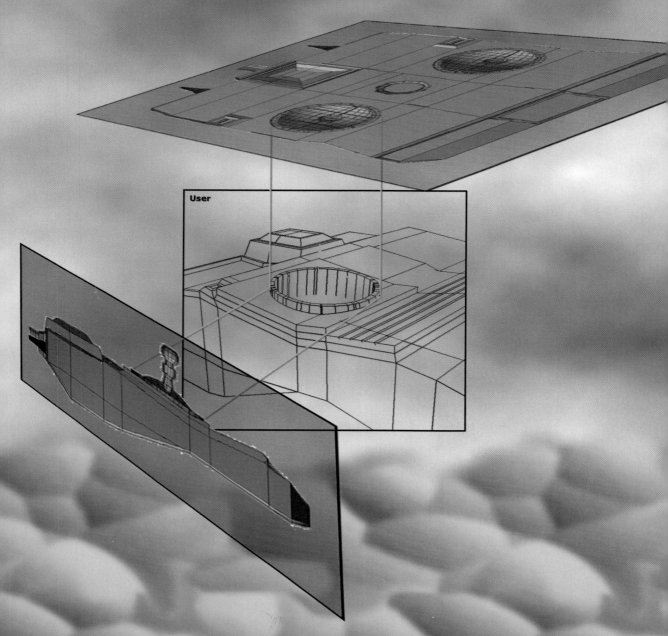

Extruding the Sides

1. Press **<Alt-X>** to make the remote control semi-transparent.
2. Switch to the **Polygon** sub-object level.
3. Select the polygon at the bottom of the indented area.
4. Extrude the polygon upward by about 4.5 units so it is just below the highest edge of the remote body when looking in the Left viewport.
5. Use the **Bevel** spinner to uniformly scale the new polygon inward by 1.0.
6. Click the **Select and Non-Uniform Scale** on the Main Toolbar. In the Perspective or Top viewport, scale the polygon inward on its Y Axis by 50%.
7. Click the **Extrude** spinner up arrow once.
8. Use the Bevel spinner to bevel the new polygon by -1.5 and create a top lip for the screw hole.
9. Use the **Extrude** spinner again to extrude the face down into the screw body by -4.0 units.

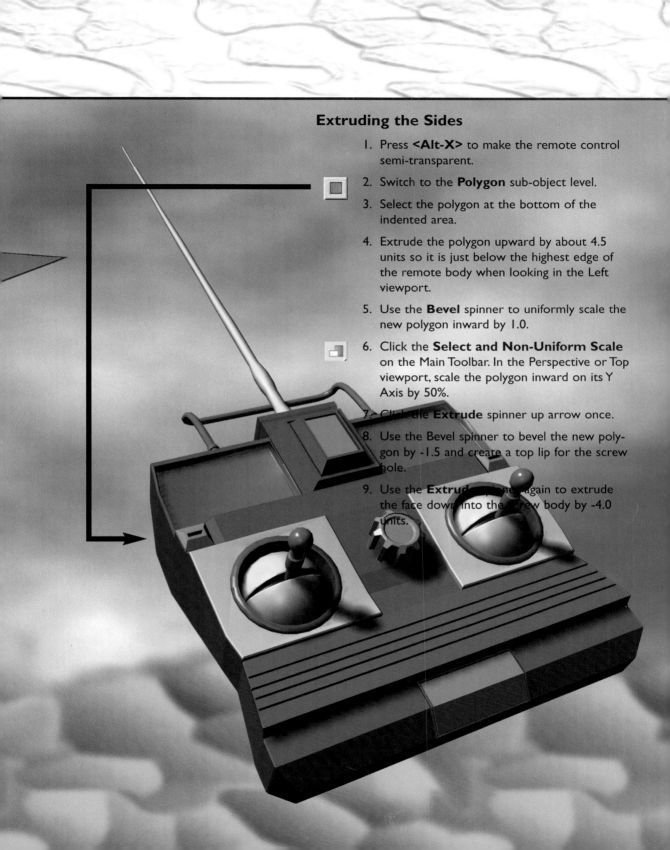

The Alien

The next step is to model the alien using advanced box modeling techniques. Starting with a humble box, we extrude, divide, cut and extrude again to mold the little guy into shape.

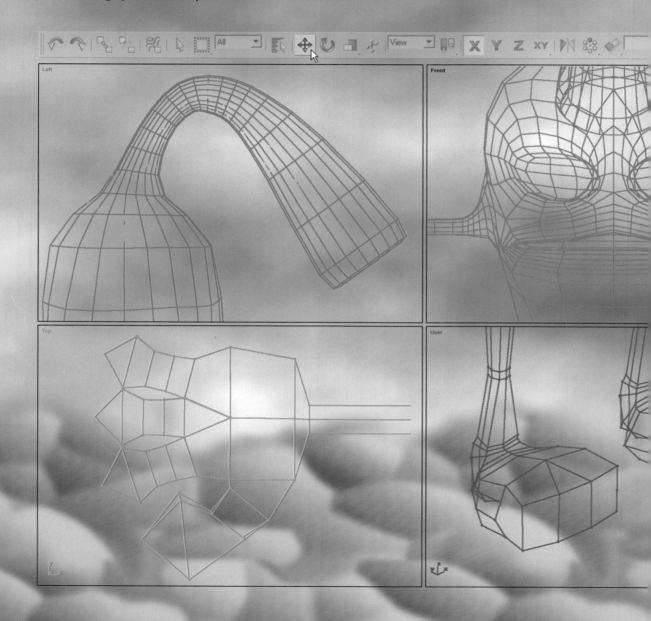

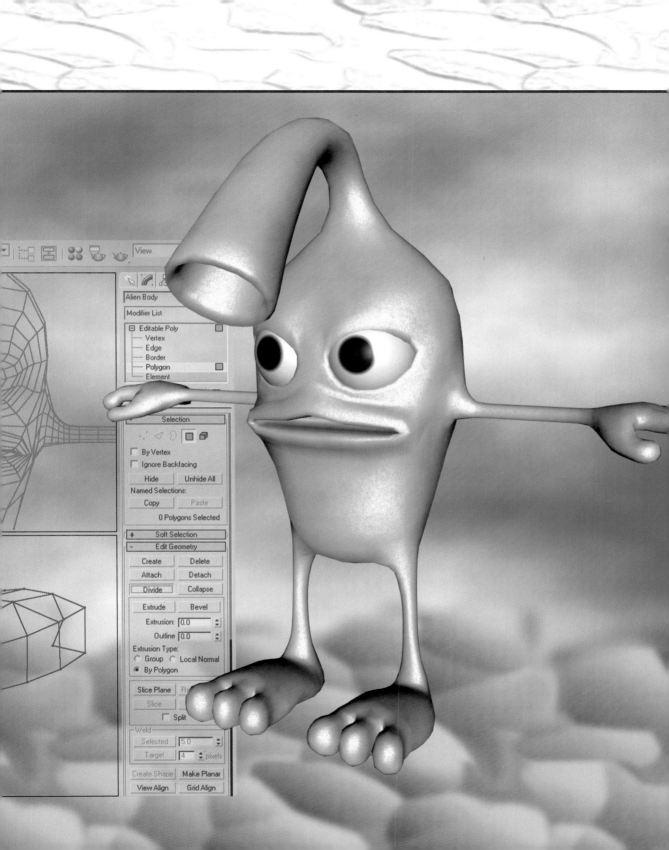

The Rover

The rover is built with a spline framework and surface tools plus additional box modeling techniques.

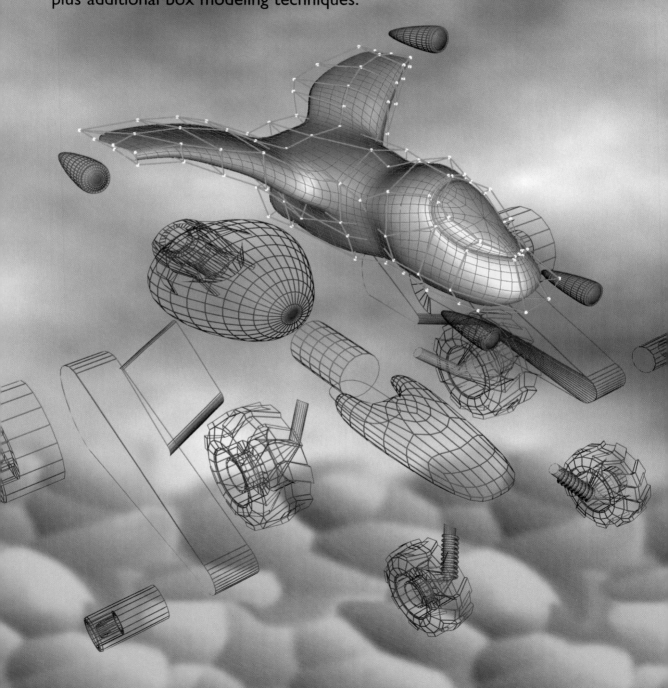

The wheels and tracks are built with special methods that make them easy to animate later on.

Rover Animation Controls

A few hand-written MAXScripts
and the motion capture controller
let you drive the rover around the screen in real time.

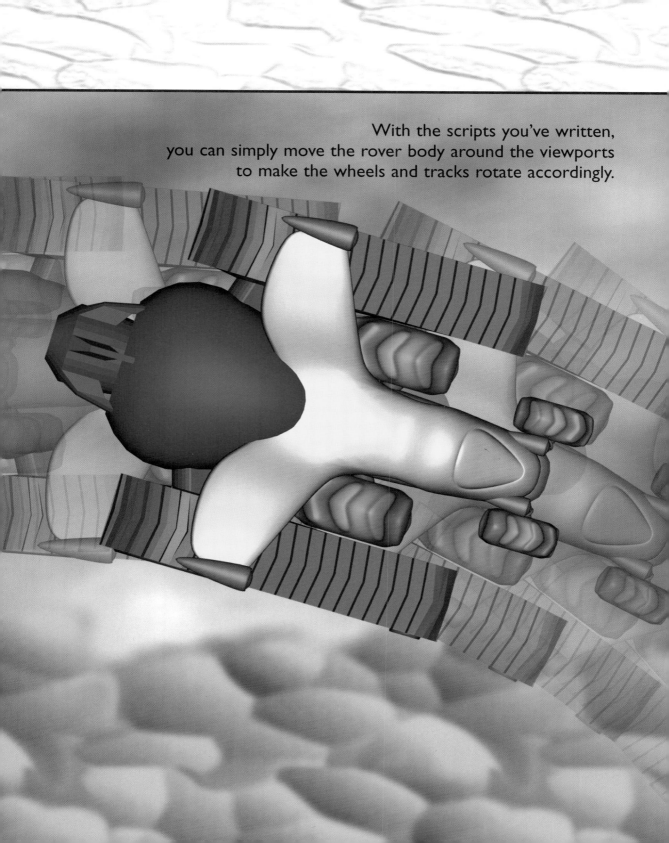

With the scripts you've written, you can simply move the rover body around the viewports to make the wheels and tracks rotate accordingly.

Alien Animation Controls

A skeleton of linked bones and a clever hand-built user interface make it easy to select and animate the alien's elbows, hands, knees and feet.

A series of morph targets created from the original alien model give you a range of facial expressions, which you also control with your custom user interface.

Live Background Footage

Camera motion from live background footage
shot at a local rock quarry
is tracked with the Camera Tracker utility
for seamless compositing with the alien animation.

In Combustion,
unwanted elements are painted out
and the landscape is colored and treated
for an alien appearance.

Compositing

You put it all together in Combustion.

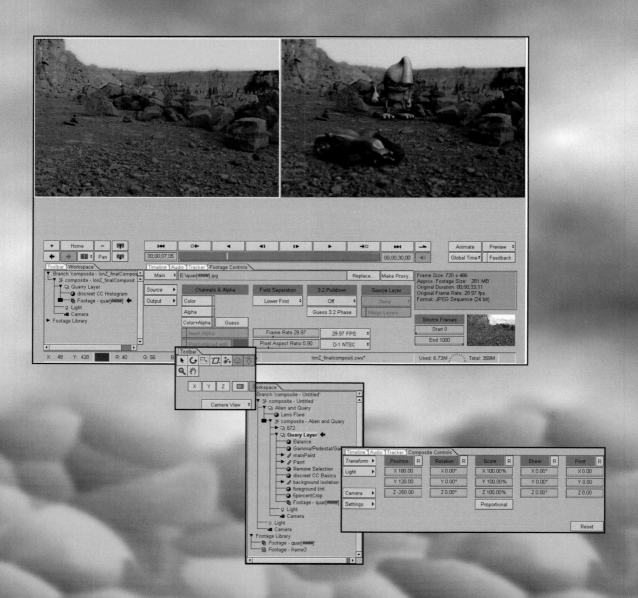

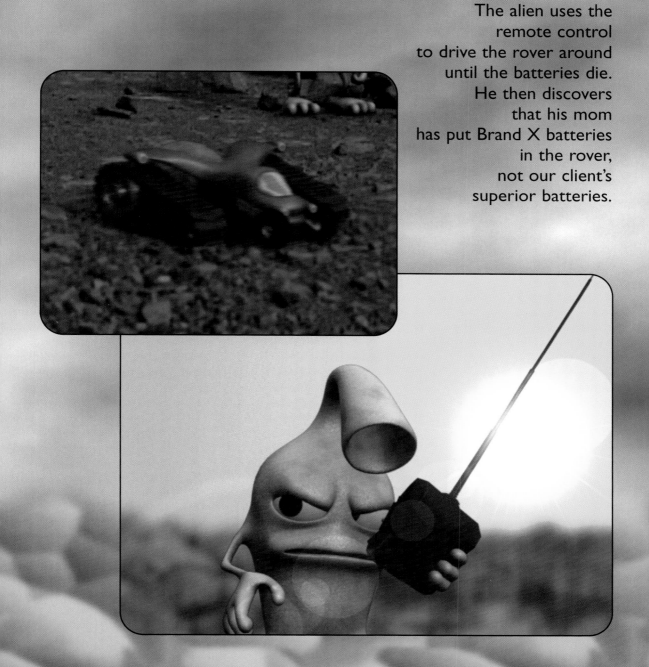

The alien uses the remote control to drive the rover around until the batteries die. He then discovers that his mom has put Brand X batteries in the rover, not our client's superior batteries.

The Final Animation

The project is complete,
the client is ecstatic,
and best of all,
you're a pro animator.

Technical Setup for the Rover

In this chapter we're going to look at some of the approaches you can take toward simplifying the rover animation process with scripting. The techniques will be applied to the rover, but can be used on many types of scenes.

TECHNIQUES IN THIS CHAPTER

In this chapter, you will learn how to:

- Make animatable tracks for the rover
- Use motion capture to move the rover interactively with the keyboard and mouse
- Create custom parameters for objects and use them to control other parameters
- Write scripts to turn the rover wheels and tracks appropriately when the rover is moved
- Write a custom user interface to control the motion capture process

Some of the techniques in this chapter do get a little complex. Just hang in there! The techniques will become clear as you go through the tutorials.

TUTORIALS IN THIS CHAPTER

In this chapter, you will complete the modeling of the rover tracks, and use scripts to set up a motion capture system for animating the rover.

- **Tutorial 6.1 Track Setup** will complete the modeling of the track treads and provide information about them for scripting the track motion.

- **Tutorial 6.2 Motion Capture** for the Rover will show you how to set up motion capture controllers that will capture keyboard presses and mouse motion.

- **Tutorial 6.3 Rover Control Scripts** goes over all the scripts necessary to make the wheels and tracks turn appropriately when the rover is moved.

- **Tutorial 6.4 Rover Control User Interface** sets up a simple user interface for the motion capture control system.

SCRIPTING

In this chapter, you'll be creating a lot of scripts. Scripts are small programs written within **3ds max** with MAXScript, its built-in programming language.

If you're not familiar with scripting, you should do the tutorials in *Chapter 5 Introduction to MAXScript*, which will introduce you to the basics.

In this chapter, we'll get into some of the more advanced uses of scripting to control the rover motion automatically. If you're not familiar with scripting you can still do the tutorials in this chapter, but you might not understand what we're doing. If you have a grasp of scripting and its basic commands, you should be able to follow the logic of the scripts in this chapter with no problem.

THE ROVER TRACKS

The tracks (treads) for the rover were not created earlier because a special procedure has to be used to make them.

Consider how the tracks will move. You can't simply rotate them because they are not round.

To animate the track we're going to place the treads along a path, then animate the positions of the treads along the path. This will give the appearance of the tracks rotating as they would in real life.

Having said that, we'll get right onto our next tutorial for completing the rover tracks.

TUTORIAL 6.1 TRACK SETUP

Preparing the Rover

1. Open the file **Rover - No Tracks.max** you created in **Chapter 4**, or open the file **Rover - No Tracks.max** from the **Scenes/Rover** folder on the CD.

2. On the **Create** panel, click **Helpers**. Click **Dummy**. Click and drag in any viewport to create a dummy object on the screen. The size of the dummy doesn't matter - just make it roughly the same size as the rover. Name the dummy object **mainControlNode**.

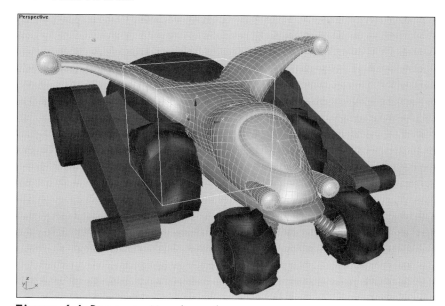

> **TIP**
>
> To quickly link all the objects to the dummy, select the objects, click Select and Link, then press the <H> key on the keyboard and select the dummy.

Figure 6.1 *Dummy positioned over the rover*

3. Move the dummy object near the center of the rover model. Be sure to check all your viewports to make sure the dummy is definitely over the rover before continuing.

4. Use **Select and Link** on the Main Toolbar to link all the parts of the rover to the **mainControlNode**.

5. Move the **mainControlNode** forward and backward and verify that all objects are moving with it.

6. Undo any moves to the **mainControlNode**.

7. Hide the **mainControlNode**.

8. Save the scene with the filename **RoverControl01.max**.

Isolating the Track Object

Next, we will isolate the track object. This will automatically hide all other objects and make it easier to see what we're doing.

1. In the Perspective viewport, select the rover's right track support.
2. **<Ctrl>**-right-click to access the Quad menu, and choose *Isolate*.

You have now entered a special mode where everything has been temporarily hidden except for the track support geometry. The scene will remain this way until you click the Exit Isolation button on the floating Isolated dialog. This is a quick and easy way to see just what you want to see in a scene.

Creating the Track Path Shape

One of the first things we want to do is measure the track circumference so we can build the treads, and later determine how much the treads should move when the rover moves.

This will be accomplished with a series of steps. The first is to create a shape that follows the track. You could draw this shape from scratch, but it's easy to create one that fits the tracks exactly with the **Border** sub-object.

A border is an edge on an Editable Poly without a polygon on either side of it. We will use an open border to create the shape of the track. To get an open border, we'll have to delete a polygon. After we create the track shape, we'll put the polygon back.

1. On the **Modify** Panel, access the **Polygon** sub-object level.
2. Change the Perspective view to *Wireframe* mode.
3. Select the outermost polygon.
4. Switch the Perspective view back to *Smooth + Highlights*.

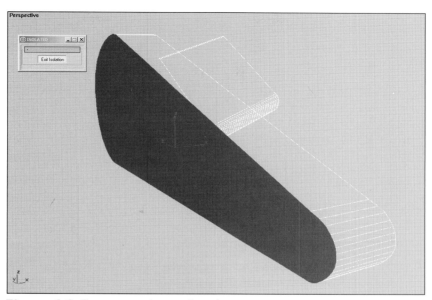

Figure 6.2 *Outermost polygon selected*

5. Press the **<Delete>** key to delete the polygon.
6. Switch to the **Border** sub-object level.
7. Select one of the edges around the new open hole that we created a moment earlier. You will see that a border is selected.

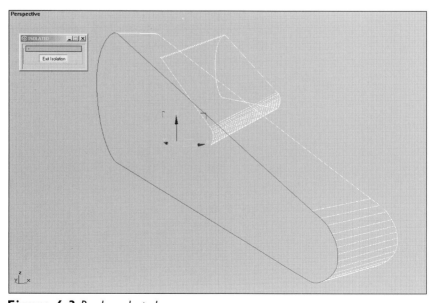

Figure 6.3 *Border selected*

We will now use the selected border to create a spline shape that will later be used as a path for the tracks.

8. Click **Create Shape** on the Edit Geometry rollout.

   The Create Shape dialog appears.

Figure 6.4 *Creating the new shape*

9. Name the new shape **Track Path**, and make sure the Shape Type is set to **Smooth**. Click **OK** to create the shape.

Putting the Polygon Back

1. To put the deleted polygon back, simply click the **Cap** button on the Edit Geometry rollout.

   This caps any open holes in the object.

2. Turn off the **Border** sub-object level to return to the base level of the object.
3. Click **Exit Isolation** on the Isolated dialog to bring all the objects back into the scene.

Creating a Track Measurement Box

In the following steps we are going to create a box and set it to follow the new shape we just created. In doing this, we will be able to get an estimate of the length of the track, and also provide a base for making the treads.

1. Create a box in the Front viewport with the following parameters:

Length	455
Width	56
Height	26
Length Segs	38
Width Segs	1
Height Segs	1

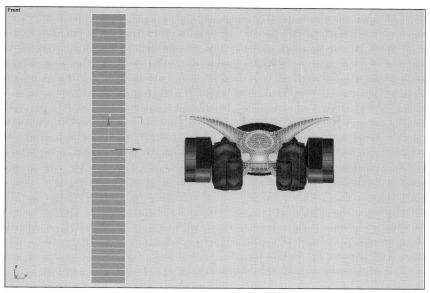

Figure 6.5 *New box in Front viewport*

2. Press the **<H>** key to open the Select By Name dialog, and select the **Track Path**.

3. On the **Hierarchy** panel, click the **Affect Pivot Only** button.

4. In the Pivot section, click **Reset Pivot**.

   This is required for the box to deform to the correct location on this path. If this step is skipped, your tracks will not align properly.

5. Click **Affect Pivot Only** once more to turn it off.

Deforming the Track Box

1. Select the box that you created a few minutes ago.

> **TIP**
> The * symbol indicates that this is a world space modifier. A world space modifier operates in world space rather than the object's local space.

2. With the new box selected, go to the **Modify** panel and apply a ***PathDeform** modifier to the box.

3. On the Parameters rollout, click **Pick Path**, and press the **<H>** key on the keyboard. Select **Track Path** and click the **Pick** button.

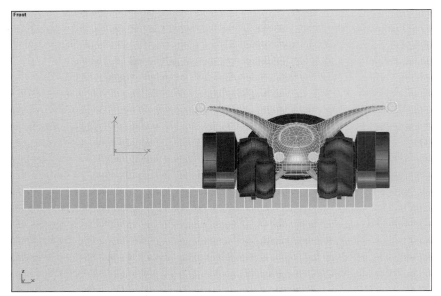

Figure 6.6 Don't panic if your box does not follow the path properly at first

4. Try a different **Pick Deform Axis** until the box appears similar to the shape of the path. Refer to the next figure for reference.

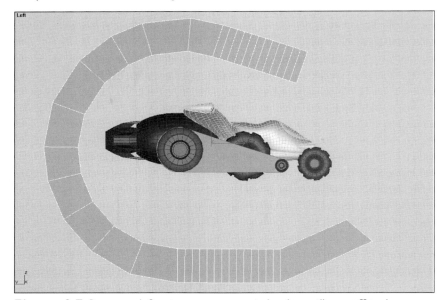

Figure 6.7 Box now deforming on correct axis but box still way off path

5. Click the **Move to Path** button.

The box should now actually be on the track path.

Figure 6.8 *Box deforming on track path*

Don't worry if the geometry appears inside out. This means that your normals are flipped. Since we are using this geometry as a sort of makeshift ruler for determining the length of our final tracks, we need not worry about fixing it.

Adjusting the Track Length

Now that the box is deforming properly on the track path, we can adjust some of the box parameters until we get the desired shape of our treads.

1. Select the box and go to the **Modify** panel.
2. Adjust the box parameters until the box deforms all the way around the path, no more and no less.

We adjusted our box parameters to the following:

Length	497
Width	52
Height	15
Length Segs	46

TIP

We increased the Length Segs to make the box deformation a little smoother.

Figure 6.9 Box deforming on track path

3. Highlight the ***Path Deform Binding** modifier on the modifier stack and click **Remove modifier from the stack** to delete it.

   The box will become fully extended again, and we can now use it as a measuring tool in determining the length of our tracks.

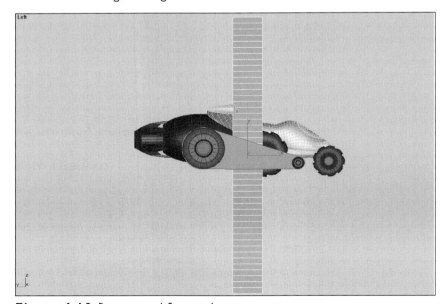

> ☀ **TIP** ☀
>
> *The length of the box will become very important later when writing the script. Record this number on a piece of paper so you can use it later.*

Figure 6.10 Box removed from path

Preparing to Build the Tracks

Because of the way the PathDeform modifier works, we will end up with distorted geometry as it wraps around the front curve of the tracks support geometry. This is due to the sharpness of the curve. The only way to combat this is to have a denser track. Instead of adding additional geometry to the box we created, we will double its length and then scale it back down once it is applied to the path.

1. Verify that the box is still selected.

   The box is no longer facing the direction it was earlier. This was caused by clicking the Move To Path button.

2. In the Left or Right viewport, make a copy of the box to the left side of the rover. Name the copy **Tracks01**.

   We will modify the copy and keep the original geometry as a reference.

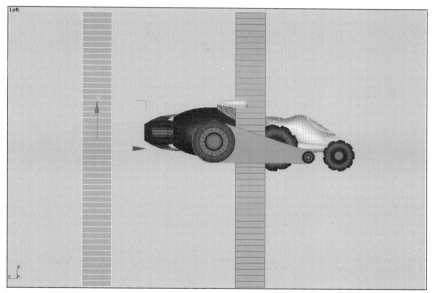

Figure 6.11 *Newly duplicated box that will be used as final track*

3. Make sure the new box is selected. On the **Modify** panel, set the **Width Segs** to 2.

   We are going to use the middle vertices to create an arrow shape in the tracks.

4. Double the current **Length** value.

   Ours has a current setting of 497 so we will set ours to 994. The new box is now twice the length of the original one.

5. So the box detail stays consistent with the original box's, set the **Length Segs** to twice its current value.

   Our current value is 46, so we will use a setting 92.

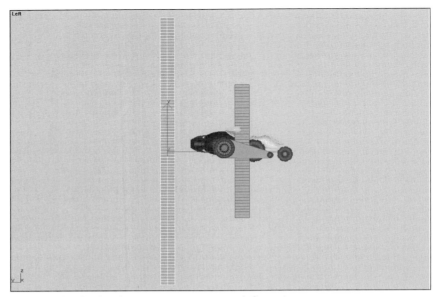

Figure 6.12 *New box just prior to vertex deformation*

6. Using the Left viewport as a reference, set the current **Height** parameter to a number slightly less than half the original value.

   We had a setting of 15 and have changed it to 6.

Shaping the Treads

1. Convert the **Box** to an **Editable Poly.**
2. Access the **Vertex** sub-object level.
3. In the Left viewport, use a marquee to select all the middle vertices.
4. In the Left viewport, move the selected vertices down on the Y axis enough to give an arrow shape in the tracks. Do not make the angles too sharp. The sharper they are, the more distortion you will have later on.

Technical Setup for the Rover 283

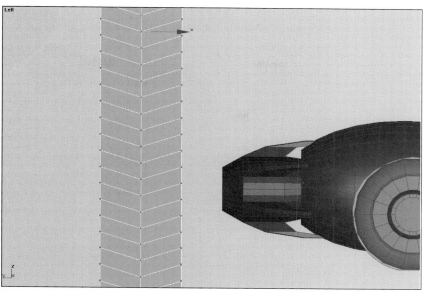

Figure 6.13 *Middle vertices moved downward*

In the Perspective view, you can see that the box's pivot point is located in the middle and on one side. The faces that you are about to select and extrude are located on the opposite side of the pivot.

 5. Switch to the **Polygon** sub-object level.

 6. In the Perspective view, select every other row of polygons opposite the pivot.

The side with the pivot will end up being against the track support.

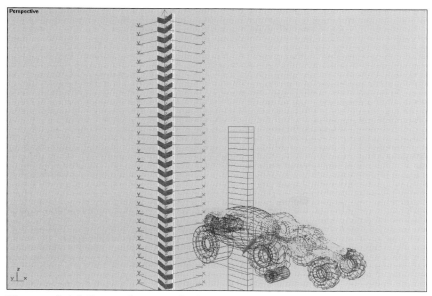

Figure 6.14 *Every other row of polygons selected on side opposite pivot point*

7. Using the Front viewport as a reference, extrude the selected polygons outward to a point just before the end of the original box that you created.

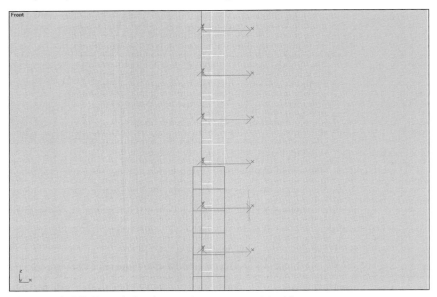

Figure 6.15 *Extruded polygons do not pass original box*

8. Turn off the **Polygon** sub-object level.
9. Delete the original box.

Wrapping the Treads Around the Support

1. On the **Modify** panel, apply a *****PathDeform** modifier to the **Tracks01** box.
2. On the Parameters rollout, click **Pick Path**, and press the **<H>** key on the keyboard. Select **Track Path** and click the **Pick** button.
3. Change the **Path Deform Axis** so that the track is deforming the way it should. We used the Y axis.
4. Click **Move to Path.**
5. If the track appears to be turned inside out, apply a **Normal** modifier and check the **Flip Normals** checkbox.
6. On the **Modify** panel, click on the *****Path Deform Binding** entry.
7. Test the track rotation by adjusting the **Percent** spinner under the Parameters rollout.

   This is the parameter we will be changing later on to make the tracks rotate. The treads will distort as they go around the front of the track.
8. If you do not like the way the treads are deforming, set **Stretch** to 0.5 and try the **Percent** spinner again.

> ☼ **TIP** ☼
>
> *If the treads end up inside the track support, try changing the Rotation parameter to 180.*

This removes much of the distortion but puts more treads on the track.

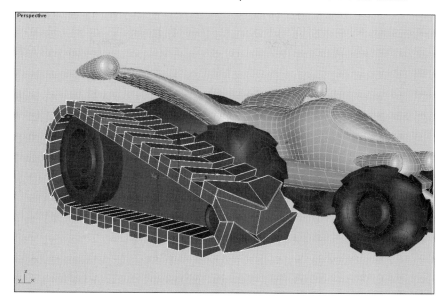

 TIP
Whether to use Stretch is based on personal preference, so we will leave it up to you. Just be sure to note whether you use it or not, as this will be important when writing the scripts later on.

Figure 6.16 *Stretch set to 1.0 leaving distortion on front*

 TIP
If you want to move the track, select both the track and the path and move them at the same time. Otherwise, the track will deform when moved.

Figure 6.17 *Stretch set to 0.5 - Less distortion, twice as many treads*

9. Fine-tune the position of the track so that it looks good on the supporting objects. You might find it necessary to move the two wheels that are inside the track.

Duplicating the Track

1. Select both the **Track** geometry and the **Track Path** shape.
2. Copy the two over to the other track support.
3. Position until everything looks good. Again, you may have to move the two inner wheels as well.

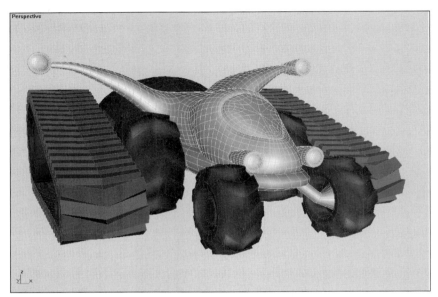

Figure 6.18 *Both tracks created and aligned*

4. Save the scene with the filename *RoverControl02.max*.

Linking the Tracks to the Rover

So the tracks will move properly with the rest of the rover, we need to link both the track paths and tracks to the mainControlNode.

1. Unhide the **mainControlNode.**
2. Press **<H>** to open the Select by Name dialog.
3. Select **Track Path**, **Track Path01**, **Tracks01** and **Tracks02**.

4. Use **Select and Link** to link the selection to the **mainControlNode**.

MOTION CAPTURE

When most beginners in the field of animation hear the term *motion capture*, they immediately picture an actor wearing a special suit with sensors at each joint that communicate to a computer. As the actor dances, kicks, tackles, runs or moves in any way, the computer captures every movement, which is in turn placed on a 3D character to make it come to life. This is definitely a form of motion capture, but it is only one of many uses.

Basically, motion capture (*mocap* for short) is the science of capturing a real-world motion via some sort of input device and translating it into keyframe data. Sure, a body suit can act as a giant motion capture input device, but what about other types of devices such as the mouse you use every day?

3ds max has a built in motion capture system that provides incredible animation power with little setup. Supported devices include: joysticks, mouse, keyboard, or even a MIDI device.

Our goal for this chapter is to set up a system for the rover so we can actually drive the car around the scene in real time. This will provide us with a very quick way of testing different types of motions.

You might think this is a silly approach to animating the rover. After all, you could perhaps keyframe the entire animation in the time it will take to write the scripts in this chapter. But what if you were all done with the animation and the client decided they didn't like it? In a real production job, this happens all the time. With this script, we can record different motions while the client watches. Once we get one they like, we can move on and without looking back. Besides, motion capture is a lot of fun to play with, but we won't necessarily tell the client about that part.

CONTROLLERS

In **3ds max**, the motion capture feature is a type of *controller*. Controllers provide an alternate means for setting an object's animation aside from the usual moving and rotating of objects directly in viewports. For example, you could set up LookAt controller to make one object's local Z axis always point at another object. This is very handy for skeet shooting and other weapon-related animation.

In this case, we will replace the default controllers with motion capture controllers, which will make the object move only in response to keyboard presses and mouse movement.

MOCAP/CONTROLLER SETUP FOR THE ROVER

The remote control car has two large tank-like tracks that are the driving force of the vehicle. You might think the best way to move the rover is the way it would work in real life, by making the tracks and wheels turn. But in 3D, the opposite approach is actually best.

We have determined that the best approach is to control the car's body in real time with motion capture, and use a script for the wheels and tracks so they automatically rotate correctly when the body moves. For example, when you move the body forward by a certain number of units, the tracks and wheels will rotate by the appropriate number of degrees.

This approach might seem backward, but if you try to go about it the opposite way you will find that our approach is much easier.

DESIGNING THE CONTROL SYSTEM FOR THE ROVER

Theoretically, the use of the motion capture control setup is straightforward:

- Press the **<Up>** arrow on the keyboard to move the vehicle forward, and press the **<Down>** arrow to move it backward.
- Press the **<Left>** arrow to turn to the left, and the **<Right>** arrow to turn right.

Using a Dummy Object

This all sounds great so far, but unfortunately, there is a downside. The limitation that we have with motion capture is that the moment you let off of a keyboard key used to control the motion capture, the object being controlled snaps back to its starting position. For example, if you press the up key, the remote control car would move forward. But the moment that you let off of the key, the car would snap back to its starting position. This is not very useful, and in fact is downright frustrating, so we're going to use a slightly different approach.

Problems like this are very common when you start getting into a technical directors position and you are assigned the task of finding new and easier ways to accomplish complex tasks. In this case, we are going to use a clever work-around to fix our problem. Instead of directly controlling the car, we are going to use a dummy object. Dummies are special types of objects that do not render. They are great to use in control systems and linked hierarchies.

In this case, our dummy will be controlled by the up, down, left, and right keyboard keys. Of course the moment a key is released the dummy snaps back to its starting position. So what we'll do is write a MAXScript that will preserve the car's position and rotation when the dummy snaps back.

Controlling Acceleration

Moving the car at a constant speed all the time would make the animation very dull. We need some sort of throttle to control the animation. We will then be able to take the throttle and multiply its value by the dummy's position and determine the amount of distance we will need to move the car forward to make the car accelerate.

For the throttle, we are going to use a slider manipulator with the mouse as the motion capture input device.

TUTORIAL 6.2 MOTION CAPTURE FOR THE ROVER

Continue with the file that you have been working on in this chapter.

Setting Up the Dummy Object

1. On the **Create** panel, click **Helpers**. Click **Dummy**.

2. In the Top viewport, create a **Dummy** helper object. Name the new dummy **RC_Controller**.

3. On the Move Transform Type-In dialog, set the RC_Controller's Absolute:World positions to: 0,0,0. This will move it to the 0,0,0 point on the construction grid.

> ☠ **TIP** ☠
>
> *This is a very important step. It will be much easier to determine how far the dummy has moved if we know its origin was 0,0,0.*

Figure 6.19 *Move Transform Type-In dialog values set to 0,0,0*

Assigning a Position Controller

Before we begin connecting up the motion capture controllers, it is very important to make sure that you clearly understand your objectives. At this time we are only configuring the movement and rotation of RC_Controller.

With the decisions we made earlier about what key will do what, we need to look at how these motions and rotations will translate into 3D space using the XYZ axes:

- When the **<Up>** arrow is pressed, RC_Controller will move in a positive direction on Y axis by 20 units.
- When the **<Down>** arrow is pressed, RC_Controller will in a negative direction on the Y axis by 20 units.
- When the **<Left>** arrow is pressed, RC_Controller will rotate 2 degrees around the Z axis.
- When the **<Right>** arrow is pressed, RC_Controller will rotate -2 degrees around the Z axis.

With these objectives clearly in mind, it will be easy to set up the motion capture controllers.

We will be assigning controllers to different aspects of the dummy object, namely its position and rotation. In **3ds max**, these aspects are called *tracks*. When we talk about tracks in these tutorials, don't confuse them with the treads on the car that we made earlier. Here, the term *track* means the transformation aspect of the object.

1. With the RC-Controller object selected, go to the **Motion** panel expand the Assign Controller rollout.

This is the rollout where we can assign new controllers.

Currently the Position track has a Bezier Position controller assigned to it. This is the default controller for the Position track on every object created in **3ds max**. For our motion capture session to work as we have planned, we will need two controllers on the Position track. To assign more than one controller you must use a special type of controller called Position List.

2. Highlight the **Position** track.

3. Click the **Assign Controller** button. On the Assign Position Controller dialog, choose the **Position List** controller.

This controller will allow us to layer various controllers.

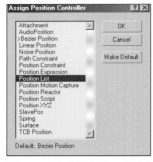

Figure 6.20 *Assign Position Controller dialog*

The Position track has changed. It now shows a Position List controller assigned to it, and a [+] symbol to the left of the track.

4. Click to expand the track. Inside the List controller you will find two tracks: One is set to the Bezier Position controller and the other one is labeled Available. Let us start by replacing the Bezier Position controller. Select this track.

5. Press the **Assign Controller** button again. On the dialog, select the **Position Motion Capture** controller and click **OK**. The Motion Capture properties window appears. Close this window for now; we will adjust its parameters in a minute.

6. Select the track labeled as **Available.**

7. Press the **Assign Controller** button. Select the **Position Motion Capture** controller again, click **OK**, and close the **Motion Capture** properties window.

Note that how you now have two different Position Motion Capture controllers and there is now a new track labeled Available. Every time you add a controller to this track, you will get a new one called Available which will allow you to select yet another controller. This is how we are able to layer controllers with the List controllers.

Assigning Devices and Bindings

A moment ago we closed each of the Motion Capture property dialogs. These dialogs are where we will configure all the motion capture parameters.

1. To open the properties window for the first Position Motion Capture controller, highlight the track on the Assign Controller rollout and right-click. From the pop-up menu, select *Properties*.

   The Position Motion Capture dialog will allow us to control the X, Y, and Z positions for RC_Controller. In the motion capture control system that we are setting up, we will only control the Y position of the Dummy. This will allow it to be pushed forward and backward.

   Remember, there is no reason to use the X or Z positions because we do not want the car to slide left or right or take off and fly!

2. Press the **Y Position Button** in the Device Bindings group.

Figure 6.21 *Motion capture properties window with no devices assigned yet*

3. On the Choose Device dialog, select **Keyboard Input Device** and click **OK**.

Figure 6.22 *Selecting Input device for controlling object*

With the keyboard input device assigned, we need to define which key on the keyboard will provide the input information.

4. Click on the **<Unassigned>** pulldown menu and select **Up Arrow** from the menu.

   This will be the controller responsible for moving the dummy object forward.

5. Under the Parameter Scaling group, set **Range** to 20.

The Range determines how far forward the dummy will move. We have set it up to move 20 units.

There are a few other parameters that can be used to determine the characteristics of the dummy's forward motion when the **<Up>** arrow is pressed. For this motion capture setup, the default settings for the remainder of the parameters will work just fine.

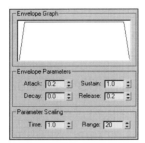

Figure 6.23 *Parameters for keyboard input device*

6. Close the Motion Capture properties dialog.

Setting up a Second Position Motion Capture System

Now we are going to set up the second Position Motion Capture controller. With the first controller set to drive the dummy forward, we will use the second controller to move it backwards. The steps will be similar to the previous setup so I will keep the direction short.

1. Select the second position motion capture controller in the Assign Controller rollout.
2. Right click on the track and select properties from the menu.
3. Click the **Y Position** button in the Motion Capture properties window.
4. Select **Keyboard Input Device.**
5. From the **Unassigned** pulldown menu, select **Down Arrow**.
6. In the Parameter Scaling group, set the **Range** to -20.

   We do this because we need the dummy to go backwards 20 units when the down arrow key is pressed.

7. Close the Motion Capture properties window.

Assigning a Rotation Controller

For the rotation of the dummy object we will use the Rotation track. We will need to assign two controllers, the same as we did with the position track. The left arrow will be used to rotate the dummy to the left while the right arrow will be used to rotate it to the right. These steps are very similar to the position controller steps, so this section will act more as an overview instead of an in-depth process.

Currently, the Rotation controller has a TCB Rotation controller assigned to it.

1. Highlight the **Rotation** track.

2. Click the **Assign Controller** button. On the Assign Position Controller dialog, choose the **Rotation List** controller.

3. Expand the Rotation track by clicking the **[+]** next to it.

   There are two tracks, just as there were for the Position track. The first track is the TCB Rotation controller while the second track is Available. We are going to replace the default TCB Rotation controller.

4. Highlight the **TCB Rotation** controller.

5. Click **Assign Controller** and select the **Rotation Motion Capture** controller.

   This time we will go ahead and setup the motion capture settings on the properties window.

6. Click **Z Rotation**.

7. Select **Keyboard Input Device** on the dialog.

8. From the **Unassigned** pulldown menu, select **Left Arrow**.

9. Under Parameter Scaling, set the **Range** to 2.

   We are finished setting up this motion, so you can close the Motion Capture properties window.

Setting up the Second Rotation Controller

1. Highlight the **Available** track on the Assign Controllers rollout.

2. Click **Assign Controller** and select the **Rotation Motion Capture** controller again.

3. Set up this controller similar to the previous one, but make it so the right arrow key causes the dummy to rotate -2 degrees around the Z axis.

4. Close the Motion Capture properties window.

   The keyboard controls are now set up.

5. Save the scene with the filename *RoverControl03.max*.

Creating a Throttle Slider

For the throttle system, we will create a slider manipulator and then tie its value, through motion capture, to the vertical movement of the mouse. The motion of the mouse will multiply the current motion to create acceleration.

The first step is to create the slider object.

1. On the **Create** panel, click **Helpers**. Choose *Manipulators* from the pulldown menu.

2. Click **Slider**.

3. Click once near the top center of the Front viewport to place a slider manipulator in the scene.

4. On the Name and Color rollout, change the name from Slider01 to **Throttle**.

5. On the **Modify** panel, set the following parameters for the slider:

Value	0
Minimum	0
Maximum	2.0
Snap	(checked)
Snap Value	0.001
Hide	(unchecked)

Use the **X Position** and **Y Position** values to set the best position for the slider in the viewport. Many users prefer to keep the slider at the upper right corner.

Set the **Width** wide enough so you can easily see and work with the slider.

Using Track View to Set Up a Controller

Now that the slider has been created, we need to set it up so that it will be driven by the vertical movement of your mouse in a motion capture scenario. This time we're going to use the Track View window to assign a controller.

The Track View window's primary purpose is to provide a visual representation of your animation. Track View can also be used to assign and set up controllers.

1. With the slider selected, access the Quad menu and choose **Track View Selected**.

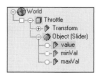
This opens the Track View window with the **Throttle** track already displayed and all other tracks hidden. Track view displays items in an indented hierarchy at the left side of the window. To expand an item, click the **[+]** next to it.

2. Expand the **Throttle** listing, then expand the **Object (Slider)** listing.

3. Highlight the **Value** track.

4. Click the **Assign Controller** button on the Track View toolbar.

Clicking this button here has exactly the same effect as clicking it on the **Motion** panel. But on the **Motion** panel you have access only to **Position**, **Rotation** and **Scale**, while in Track View you have access to all the object's parameters.

5. Select the **Float Motion Capture** controller and click **OK**.
6. On the Motion Capture properties window, click the **Value** button. Choose **Mouse Input Device** and click **OK**.
7. In the Mouse Input Device rollout, set the **Axis** to **Vertical** and **Scale** to 0.001.

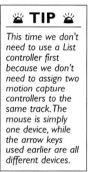

☼ TIP ☼
This time we don't need to use a List controller first because we don't need to assign two motion capture controllers to the same track. The mouse is simply one device, while the arrow keys used earlier are all different devices.

Figure 6.24 *Motion Capture properties window*

8. Close the Motion Capture properties window, and close Track View.
9. Save the scene with the filename **RoverControl04.max**.

Preparing to Test the Motion Capture Setup

Now that we've gotten everything in place, it is time to test the setup and verify that it works.

1. Go to the **Utilities** panel.
2. On the Utilities rollout, click the **Motion Capture** button.

A Motion Capture rollout appears beneath the Utilities rollout. This is where you will do your testing and recording.

3. Locate the Tracks group.

Here you will find a listing of all the Motion Capture controllers that have been setup. Next to each controller is a small box. Currently all the boxes are deselected. These boxes act at a sort of checkbox that lets the system know what will and will not be involved in a test or recording.

4. Activate all the controllers by checking all the checkboxes.
5. Activate the Top viewport and zoom/pan the view so the dummy object is in the center and is large enough for you to see any positional changes. Also, verify that the **Throttle** slider is visible in the viewport.

Testing the Motion Capture Setup

To test our setup, we will use the Test feature, which starts the motion capture session but does not record keys.

1. On the Motion Capture rollout, click the **Test** button in the Record Controls group.
2. Press the **<Up>** and **<Down>** arrow keys on the keyboard.

   In the Top viewport, watch the dummy object more forward and backward.

   If the object is not moving, it's time to start troubleshooting. Go back to the earlier steps in the setup and look for what went wrong.

   When you let go of the key, do you see the dummy snap back to its origin? This is why we were not able to directly connect the remote control car to this particular motion capture rig.

3. Press the **<Left>** and **<Right>** arrow keys on the keyboard. The dummy should rotate slightly to the left and right.
4. Roll your mouse forward and back while watching the Throttle slider.

   The value indicator should move to the left and right. Note how the actual value is changing.

   The slider will move very slowly. This will help us maintain control of the remote control car after we have written all the scripts and are ready to start recording. If you would like the slider to move quicker, simply go back to the value track on the Throttle slider and adjust the motion capture Scale setting from 0.001 to .01. This will give the throttle more kick.

5. To cancel the motion capture test, simply right-click.
6. Save the scene with the filename **RoverControl05.max**.

Creating Custom Attributes

One of the issues to deal with in this type of setup is that the rover doesn't know where it was at frame 0, so it can't go back there when it needs to.

To correct this problem we will create custom *attributes*, or parameters, for the mainControlNode object. These attributes will be used to record its position and rotation of the object at frame 0.

We will need four attributes, one for the initial position on each of the XYZ axes, and one for the initial Z rotation. We also need an attribute for the rover scale so it will be easy to scale later on when it is merged into the final scene.

1. Select the **mainControlNode** object.
2. Choose *Animation/Add Custom Attribute* from the menu.

   The Add Parameter dialog appears.

Figure 6.25 *Add Parameter dialog*

The Add Parameter dialog allows you to create custom parameters. These parameters themselves don't do anything, but you can connect them to other real values in the scene and manipulate the real values using the custom parameters.

3. On the Add Parameter dialog, enter the following settings to create the custom attribute called **initPosX**, which will hold the initial position along the X axis.

   On the Attribute Style rollout:
   Parameter Type   Float
   UI Type   Spinner
   Name   initPosX

The Add Parameter dialog doesn't close when you click Add, so you can enter many attributes without having to reopen the window each time.

   On the Float UI Options rollout, set the following parameters.
   Range   -99999 to 99999
   Default   0.0
   Alignment   Center

4. Click **Add** to add the parameter.

5. Enter four more parameters, changing only the parameters shown below. Be sure to click **Add** after you enter each parameter.

It is very important that you set the Default value for the scaleFactor attribute to 1. If you leave it at 0, when you connect the attribute to the rover's scale later on, the rover will scale to 0 size and disappear from the scene.

   Name   initPosY
   Range   -99999 to 99999

   Name   initPosZ
   Range   -99999 to 99999

   Name   initRotZ
   Range   -180 to 180

   Name   scaleFactor
   Range   0 to 50
   Default   1

6. Close the Add Parameter dialog.

Let's take a look at where the new attributes can be found.

7. Go to the **Modify** panel and scroll to the bottom of all the available parameters. A Custom Attributes rollout appears at the bottom of the panel. Expand this rollout.

Here you will find all the attributes that were added a few moments ago. The parameters appear on the Modify panel because we added the attributes to the object's base level.

Wiring the Rover's Scale

The custom attributes don't do anything on their own. — they have to be *wired* (connected) to another value in order to have an effect on the scene.

Here we will wire the **scaleFactor** attribute to the rover's scale. The other attributes will be used by the script later in this chapter.

A parameter that holds one number, such as **scaleFactor**, can only be wired to another parameter that consists of one number. This presents a problem as the rover's scale consists of three values, one for each of the X, Y and Z axes.

We need to separate the rover's scale values into three separate tracks, one for each axis, so we can wire the **scaleFactor** value to each axis individually. The **ScaleXYZ** controller can be used for this purpose, so we'll start off by assigning this controller to the **mainControlNode**.

1. Select the **mainControlNode**.

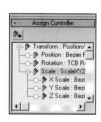

2. On the **Motion** panel, highlight the scale track, click **Assign Controller** and assign a **ScaleXYZ** controller to the track.

On the Assign Controller rollout, you can expand the **ScaleXYZ** controller to see the separate tracks, one for each axis.

Next we will use the *Wire Parameters* option to connect the **scaleFactor** attribute to each of the new scale tracks.

3. From the menu, choose *Animation/Wire Parameters/Wire Parameters*. From the pop-up menu that appears, choose *Object (Dummy)/Custom_Attributes/ scaleFactor*.

4. Click on the **mainControlNode**. On the pop-up menu that appears, select *Transform/Scale/X Scale*.

The Parameter Wiring dialog appears. On the left, the **scaleFactor** attribute is highlighted. On the right side of the dialog, the **X Scale** parameter is highlighted.

5. In the center of the dialog, click the right arrow, then click **Connect**.

This makes a one-way connection between the two highlighted parameters. Changes to **scaleFactor** will affect the **X Scale**, but changes to the **X Scale** will not affect the **scaleFactor** attribute.

Technical Setup for the Rover

☼ **TIP** ☼

You can also access this dialog by choosing Animation/Wire Parameters/Parameter Wire Dialog from the menu. If you do this, you will need to navigate through the parameters on each side to select parameters to wire.

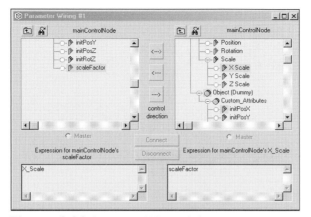

Figure 6.26 *Parameter Wiring dialog*

6. On the right side of the dialog, highlight the **Y Scale** track, click the right arrow button, then click **Connect**. Repeat this operation for the **Z Scale** track.

7. Close the dialog.

    We've got it all wired up now, so let's test it out.

8. On the **Modify** panel, increase or decrease the **scaleFactor** attribute. The rover should change in size as the attribute is changed. Change the **scaleFactor** back to 1 when you have finished testing the connection.

9. Save the scene with the filename **RoverControl06.max**.

MAXSCRIPTS TO CONTROL THE ROVER

Now that we have the motion capture controllers set up and the custom attributes added to the mainControlNode, it is time to complete the process of making it move. To do this we are going to need to write a simple script.

If you aren't familiar with scripting, it is recommended that you do the tutorials in *Chapter 5 Introduction to MAXScript* before starting the scripts in this chapter.

IMPROVING PERFORMANCE

To test the scripts, we will need to move the rover around. To see what's going on, you need to have your system performing as fast as it possibly can.

To help increase performance, there are a number of things that can be done:

- Turn off **Use NURMS Subdivision** for all objects.

- If you have applied any materials with maps, make sure the maps are not being displayed in any viewports.

- Hide parts of the rover. The lights, battery, door, struts, springs, and fuel tank, and even the body and engine are not necessary to verify that this rover is moving appropriately. Selecting and hiding each of these items will give you the best performance increase.

Before continuing, complete all the steps above to help your system operate at its best during the scripting phase.

WHAT ARE WE SCRIPTING?

Before we begin writing the script, we need to an understanding of what we want to accomplish.

- We need to force a script to execute every time there is a change to the current **3ds max** animation time. Basically, if the time slider moves, the script needs to run.
- We need to read the positional value of the dummy (RC_Controller) and then multiply it by the current value of the Throttle slider.
- To make the rover move, we will add the result calculated in the previous step to the current position of the rover. The key thing is to make sure we do this in the mainControlNode's "local space."
- Finally, we will need to rotate the mainControlNode to match the rotation of the dummy. This will give us our steering power.

When approaching a script like this, it is best to write down the functionality on paper first. Keep it simple. There is no reason to try to code it at this point, you simply want to capture the overall view of what is to happen. Once you are happy with the description that you have come up with, start breaking it down into code. Remember, MAXScript executes code linearly, so try to write it that way; one step at a time.

TUTORIAL 6.3 ROVER CONTROL SCRIPTS

We'll work on the script in small bite size chunks. This will help keep you focused on each task at hand.

Starting the Script

To start the script, we'll write some opening comments so other users will know who wrote the script and what it does.

1. On the **Utilities** panel, click **MAXScript**, then click **New Script**.

   The MAXScript editor appears.

Figure 6.27 *MAXScript editor*

TIP

The first and last lines are commented lines with equals signs, designating the beginning and end of the opening comments. These characters could just as easily have been ampersands (&) or other characters.

2. Type the following into the editor.

   ```
   --***************************************************
   --Script: Rover Callback. ms
   --Author: [Your Name here]
   --Created: [Today's Date]
   --Modified Last: [Today's Date]
   --Script Description: MasterController will be a basic function setup to
   --be run every time the time slider is changed. It will read information
   --from an object called RC_Controller (a dummy) and a Throttle slider
   --and apply it to the rover to make it move. The end result will be the
   --capability to drive the rover around the screen in real time.
   --***************************************************
   ```

3. Press **<Enter>** three times to put some blank lines after the opening comments.

Detecting Time Slider Changes

Our first task is to write code that will detect when the time slider changes.

1. Type the following into the MAXScript editor:

   fn masterController =
   (
)

2. Press **<Enter>** twice to create some more white space.

   This function, **masterController**, needs to be *registered* with **3ds max** so it will run every time the time slider changes. A special command called **registerTimeCallback** tells **3ds max** to run a specific function every time there is a change in the time slider.

3. Type the following:

   registerTimeCallback masterController

4. From the top of the script editor, choose **File/Save** from the menu and save the file as **Rover Callback.ms**.

   If we execute the code, what will happen each time the time slider is changed? If you said, "Nothing," then you are partially correct. The **masterController** function will get called but since there is no code inside its function, nothing will happen. So before we evaluate this script, lets add some functionality to the **masterController** function.

5. Between the opening and closing parentheses type the following:

   print "There has been a time change!"

   Every time **masterController** gets called it will print the message in the Listener. Let's try it out.

6. On the MAXScript rollout, click the **Open Listener** button. Keep this window in view so that you can watch and see if anything gets printed to it when you move the time slider.

7. From the script editor window, select **File/Evaluate All** from the menu.

    This causes the script to be executed, which in turn will register the **masterController** function with **3ds max**.

8. Drag the time slider and watch the Listener window. You should see the appropriate message displayed over and over. If not, check your code and correct any errors, then try again.

    In causing the function to print to the Listener window, we are using yet another method of testing our script as we go along. This particular command will not, of course, be part of the final code, but it was important in helping us make sure the code was all working properly.

Adding Functionality to the Master Controller

Now that you have seen how we can register a function so that it will run every time the time slider is changed, let's add some real functionality!

1. Delete the **print** statement that you added a minute ago.

2. Enter the following code in between the **masterController's** opening and closing parentheses:

    ```
    moveValue=$RC_Controller.pos*$Throttle.value
    -- multiplies dummy position by throttle value
    -- to figure the new amount to be added

    set coordsys local
    move $mainControlNode moveValue
    set coordsys world
    rotate $mainControlNode $RC_Controller.rotation
    ```

    This code will do the following:

    moveValue = $RC_Controller.pos * $Throttle.value
    Multiplies the position of the dummy RC_Controller by the current value of the throttle and saves the result in the variable **moveValue**.

    set coordsys local
    Sets the reference coordinate system to Local. We do this so when we move the mainControlNode it moves according to its local axes and position, not the world axes and position.

    move $mainControlNode moveValue
    With the Reference Coordinate System set to Local, this line will move the mainControlNode from its current position by the additional amount found in the variable **moveValue**.

set coordsys world
Sets the reference coordinate system back to world, which is necessary for proper rotation.

rotate $mainControlNode $RC_Controller.rotation
Sets the **mainControlNode's** rotation to match the rotation of the dummy object. Keep in mind that this will also rotate the **mainControlNode's** local coordinate system, which is what we want. This will make it so the **mainControlNode** will now travel forward in the new direction in which it has been pointed.

We have now completed all of the steps we set out to perform with scripting.

3. Evaluate the updated script.

Testing the New Script

The script is monitoring the positions of the RC_Controller and the Throttle values. These two objects have been set up to respond to motion capture.

So now comes the fun part. Let's test the scripts with the motion capture system.

1. On the **Utilities** panel, click **Motion Capture**.
2. Verify that all of the tracks are highlighted.
3. Make the Top viewport active and zoom out a little so that you can see a larger area.

   If you press the **Test** button now, you will see the Throttle respond to your mouse and RC_Controller respond to your Up/Down/Left/Right arrows, but the mainControlNode will not move. Why?

   Don't forget that your script will only respond when the time slider moves. To have the time slider move while you are running a motion capture test, you will need to check the **Play During Test** check box

4. Check the **Play During Test** checkbox.
5. Press **Test**.
6. Move your mouse forward to give the mainControlNode a little gas.
7. Use the **<Up>** arrow key to drive the mainControlNode forward and the **<Left>** and **<Right>** arrow keys to turn it left and right. Try to keep the mainControlNode in view.

   We could probably stop the book right here and be happy with our new toy, but we still have work to do before the job is done.

   When you are done playing with the motion capture system, right-click to cancel the motion capture test.

If the rover goes crazy when you start the test, just stop the test and try again. Scripts do that sometimes so don't take it personally.

Resetting the Rover's Default Position

We still need to make the rover go back to its default position after the motion capture test is over.

1. Select the **mainControlNode** if it is not already selected.
2. On the Move Transform Type-In dialog, set the object's absolute world positions on the **X**, **Y** and **Z** axes back to 0.
3. On the Rotate Transform Type-In dialog, set its **Z** absolute world rotation back to 0.
4. Save the scene with the filename **RoverControl07.max**.

Did Your Rover Go the Wrong Direction?

If the rover responded appropriately to the keystrokes, then you can skip this section. On the other hand, if the rover responded opposite to the keys you pressed, don't panic. You can simply change the numbers in the motion capture controllerdialogs.

In other words, when you press the **<Up>** arrow you had originally told the motion capture controller that the Range to be applied was 20 units. To make the rover move in the correct direction, you will need to change the value to -20. Below are the steps required to flip the position motion capture controllers:

1. Select the **RC_Controller.**

2. On the **Motion** Panel, expand the **Position** track.
3. Highlight the first **Position Motion Capture** track, then right click and select **Properties** from the pop-up menu.
4. In the dialog that is presented, scroll down to the bottom and change **Range** to -20.
5. Repeat steps 3 and 4 for the second **Position Motion Capture** sub-track, except change its **Range** to 20.
6. Retest the motion capture and verify that everything is now moving in the correct direction.
7. Save the scene.

Unregistering the Callback Function

Now that you've seen the rover actually move, it's time to make the tracks respond to the movement. We want it to seem that the rotation of the tracks is driving the rover forward, left, or right. In order to do this, we are going to add several functions to our script.

Before we start writing the code required to control the tracks, we will need to enhance the script so updates can be made without any problems.

When a callback function is created, it is assigned an internal number and its name is tossed out. From then on, it must be referred to by this number. This means that you will not be able to redefine the **masterController** function or its content without reregistering the script.

There are two approaches that we can take to fix this. First, we can simply unregister the callback function each time the script is evaluated by the developer. When the script is then reregistered, all of the new content in the callback function will work without a problem. Or, we could have the callback function call some other function. This way, every time we need to make changes, we simply change the called function.

In our script, we are going to un-register **masterController** each time the script is evaluated. We're doing this so that later on we will be able to add additional code to the masterController without fear of it not working properly.

1. Change the code below the opening comments so it is the same as below:

   ```
   unregisterTimeCallback masterController
   fn setMainControlNode =
   (
       moveValue=$RC_Controller.pos*$Throttle.value
       -- multiplies dummy position by throttle value
       -- to figure the new amount to be added
       set coordsys local
       move $mainControlNode moveValue
       set coordsys world
       rotate $mainControlNode $RC_Controller.rotation
   )
   fn masterController =
   (
       setMainControlNode( ) -- position and rotate the rover
   )
   registerTimeCallback masterController
   ```

2. Save the script.

   There's one other thing that we've done in the above script. We put all of the code responsible for handling the movement of the rover into its own function called **setMainControlNode**.

   It's best to keep related segments of code in their own functions. This will help make troubleshooting easier when problems arise.

Making the Tracks Rotate

Now let's make the tracks rotate. To do this, we will create a function named **matchTracks** and have it determine how far the tracks need to rotate, and then rotate them by that amount.

This new function needs to be placed before the **masterController** function in the script. We will be adding the actual call to this function inside the **masterController** function, and if the **matchTracks** function is not already defined, the script will fail.

1. Add the following code to your script just before the **setMainControlNode** function.

   ```
   fn matchTracks =
   (
       distInc=$RC_Controller.pos[2]*$Throttle.value
       --this is later in time and may not always be the same as
       --moveValue!
       C=497
       --This value will change depending on the length of your box.
       --If you adjusted the Stretch parameter on the Path Deform,
       --you will need to multiply that number by the length of your
       --box and use that number instead.
       percentValue=(100/C as float)*distInc
       -- this figures how much we will need
       -- to add to the current percentage
       $tracks01[2][1][1].value+=percentValue -- now add it.
       $tracks02[2][1][1].value+=percentValue
   )
   ```

   We will need to call this function from somewhere in order for it to do anything.

2. Inside the **masterController** function, add the following line directly beneath the **setMainControlNode( )** function call:

   `matchTracks( ) -- handle the track rotation`

   Your **masterController** function should now look like this:

   ```
   fn masterController =
   (
           setMainControlNode( ) -- position and rotate the rover
           matchTracks( ) -- handle the track rotation
   )
   ```

3. Save the script.

The **matchTracks** function is designed to read the current value of the Throttle and multiply it by the current Y position of RC_Controller, store the result in the variable distInc and later multiply that value by 100 divided by the circumference of our tracks. This will give us a percentage value that we can use to rotate the tracks with.

The first thing that you might find confusing is the reference to **$RC_Controller.pos[2]**. The designation **pos** is a shortened way of referring to **position**. The **[2]** refers to a subtrack. In our case, the position track has only three subtracks: X position, Y position, and Z position. Based on this information, you should be able to see how the number 2 would relate to the Y position.

Earlier, we set up a motion capture controller that would control the Y position of the RC_Controller object. By reading the Y position of the RC_Controller, we can determine if the rover should be moving in a positive or negative direction. But this is not enough information if we want to be able to control the speed as well. This is why we are multiplying the RC_Controller's Y position by the Throttles current value. As you slide the mouse forward, the Throttle's slider value increases. The larger the slider value, the faster the tracks will rotate.

Finally, we use the circumference of the tracks in an equation that will determine how far the tracks should be moved. The circumference is really just the length of the box that we used for the tracks earlier. You should have written this number down just prior to converting the box to an Editable Poly.

There is one additional consideration. If you used the Stretch parameter, you will need to multiply your length by the Stretch value you assigned. We used 0.5 for our Stretch value, so in the script we are using a setting of 497.

The result of the equation is then added to the current Percent parameter for each of the Path Deform modifiers. This will cause the tracks to rotate around their paths in the proper direction and amount.

Handling the Tracks When the Rover Turns

At this point our script will handle the rotation of the tracks in both a forward and reverse direction, but only when the rover is moving in a straight line. What about when it is turning? Well, the tracks should rotate in opposite directions — the track on the outside of the turn should rotate forward and the track on the inside should rotate in reverse.

To handle these calculations, we are going to add a new function called rotationControl. This new function will constantly monitor RC_Controller's Z rotation track and take care of the opposite rotations if the Z rotation track is anything other than zero. This will be easy to setup because if you recall, when we created RC_Controller, we made sure that its position and rotation X, Y, and Z were all set to zero.

1. Add the following immediately after the matchTracks function:

```
fn rotationControl =
(
    rotationQue=$RC_Controller.rotation.z
    C=497
    distInc=(rotationQue*1000) as float
    percentValue=(100/C as float)*distInc
    --right:
    if rotationQue<0 then
    (
        $tracks01[2][1][1].value-=percentValue
        $tracks02[2][1][1].value+=percentValue
    )
    --left:
    if rotationQue>0 then
    (
        $tracks01[2][1][1].value-=percentValue
        $tracks02[2][1][1].value+=percentValue
    )
)
```

We will call this function from inside the **masterController** function.

2. Add the following statement immediately after the **matchTrack()** function call:

   rotationControl( ) -- handle the tracks when the rover turns

3. This step is for verification only. Your **masterController** function should now look like this:

```
fn masterController =
(
    setMainControlNode( ) -- position and rotate the rover
    matchTracks( ) -- handle the track rotation
    rotationControl( ) -- handle the tracks when the rover turns
)
```

4. Save and evaluate the script, and test the script with another motion capture test.

5. After you have completed the test, don't forget to reset the absolute world values for both position and rotation.

   In the above function, the first line stores the current Z rotation of RC_Controller in a variable called **rotationQue**.

   The next three lines work exactly the same way they did in the **matchTracks** function. They calculate the distance that the tracks will need to be moved so that the motion looks correct.

   It might seem odd that we are multiplying the **rotationQue** variable by 1000. This is because the angle of rotation is returned as a very small number.

Next, we simply check to see if **rotationQue** is less than or greater than 0. Based on the results, we will add the **percentValue** to the proper PathDeform modifier's percent parameter and subtract it from the other.

Measuring the Wheels

In order to work with the wheel rotation when the rover is traveling in a straight line, we must first take a closer look at the wheels.

The rover wheels are divided into two categories: the four wheels attached to the rover itself, and the four wheels inside the track supports. This gives us eight different wheels to deal with.

Keeping up with the wheels are very important. You will need to know which ones are on the front, the back, which ones were mirrored, and which ones were just cloned. This is important because we are going to be dealing with their individual circumferences as well as their local coordinate systems. The wheels that have been mirrored will have to be rotated in a negative direction for them to turn the same way as their counterparts on the other side of the rover.

The first step is to measure each wheel's radius. We will later use the radius to calculate the circumference of each wheel, which will in turn tell us how many times the wheel should turn when the rover moves.

> ### ☼ TIP ☼
> For those who slept through this class, we remind you that the circumference is the length of the circle around each wheel.

1. Make sure the names of your wheels match the illustration below. Otherwise, the script will not work correctly.

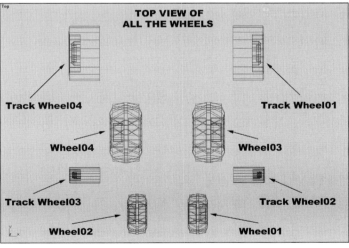

Figure 6.28 *Wheel names*

2. On the **Create** panel, click **Shapes**. Click **Circle**.

3. Create a circle that matches the circumference (outside) of one of the wheels.

4. Write down the name of the wheel you are measuring along with the circle's **Radius** value, then delete the circle.

Repeat this step for each wheel.

These are our numbers. We include them so you can see how we fit them into the script later on, so you can do the same with your own numbers.

Name	Radius
Wheel01	28
Wheel02	28
Wheel03	41.5
Wheel04	41.5
TrackWheel01	37.5
TrackWheel02	11
TrackWheel03	11
TrackWheel04	37.5

Making the Wheels Rotate When the Rover Travels Straight

To make the wheels rotate, there is no need to create a new function. We will be adding the appropriate code into the matchTracks function.

 TIP

The additional parenthesis groupings around the bulk of the new section help to keep the code together and easy to follow.

1. Add the following code to the **matchTracks** function. The code should be added just before the closing parenthesis of the function.

    set coordsys local
    (
        C=2*pi*28
        angleInc=(360/C)*distInc
        rotate $Wheel01 -angleInc z_axis
        rotate $Wheel02 angleInc z_axis
        C=2*pi*41.5
        angleInc=(360/C)*distInc
        rotate $Wheel03 -angleInc z_axis
        rotate $Wheel04 angleInc z_axis
        C=2*pi*11
        angleInc=(360/C)*distInc
        rotate $TrackWheel01 -angleInc z_axis
        rotate $TrackWheel04 angleInc z_axis
        C=2*pi*37.5
        angleInc=(360/C)*distInc
        rotate $TrackWheel02 -angleInc z_axis
        rotate $TrackWheel03 angleInc z_axis
)
    set coordsys world

    Your **matchTracks** function should now look like this:

```
fn matchTracks =
(
    distInc=$RC_Controller.pos[2]*$Throttle.value
    -- this is later in time and may not always be the same as
    -- moveValue!
    C=497
    -- the length of the box for the treads if there is stretching then
    -- multiply by that value
    percentValue=(100/C as float)*distInc
    -- this figures how much we will need
    -- to add to the current percentage
    $tracks01[2][1][1].value+=percentValue -- now add it.
    $tracks02[2][1][1].value+=percentValue
    set coordsys local
    (
        C=2*pi*28
        angleInc=(360/C)*distInc
        rotate $Wheel01 -angleInc z_axis
        rotate $Wheel02 angleInc z_axis
        C=2*pi*41.5
        angleInc=(360/C)*distInc
        rotate $Wheel03 -angleInc z_axis
        rotate $Wheel04 angleInc z_axis
        C=2*pi*11
        angleInc=(360/C)*distInc
        rotate $TrackWheel01 -angleInc z_axis
        rotate $TrackWheel04 angleInc z_axis
        C=2*pi*37.5
        angleInc=(360/C)*distInc
        rotate $TrackWheel02 -angleInc z_axis
        rotate $TrackWheel03 angleInc z_axis
    )
    set coordsys world
)
```

Each of these commands is explained a little later in this section.

2. Save the script.

3. Evaluate the script and run another motion capture test. Verify that the wheels are turning in the proper direction when you make the rover move forward and backward.

4. If the wheels seem to be rotating in the opposite direction, then change the direction of the wheels by reversing the commands for the clockwise and counterclockwise directions. Change your code so that the it matches the following section:

TIP

Watch the test in a shaded view with Edged Faces enabled to really see what's going on.

```
C=2*pi*28
angleInc=(360/C)*distInc
rotate $Wheel01 angleInc z_axis
rotate $Wheel02 -angleInc z_axis
C=2*pi*41.5
angleInc=(360/C)*distInc
rotate $Wheel03 angleInc z_axis
rotate $Wheel04 -angleInc z_axis
C=2*pi*11
angleInc=(360/C)*distInc
rotate $'Track Wheel01' angleInc z_axis
rotate $'Track Wheel04' -angleInc z_axis
C=2*pi*37.5
angleInc=(360/C)*distInc
rotate $'Track Wheel02' angleInc z_axis
rotate $'Track Wheel03' -angleInc z_axis
```

5. After the test, don't forget to reset the position and rotation of the **mainControlNode**.

The new script lines that you added to the **matchTracks** function are fairly straightforward but do include a little fancy math. Let's take a closer look at what we did.

For the wheels to rotate properly we must rotate them in their local coordinate systems. To do this, we first set the coordinate system to Local with the command:

set coordsys local

Now, every rotation that we do will happen in the objects local coordinate system. All transformations will continue this way until we set the coordinate system back to world (which we do after rotating all of the wheels):

set coordsys world

Next, we divide the code up into four similar blocks. Each block will handle the rotation of two tires that are of the same radius.

If you think about how much a wheel rotates when a car moves, the circumference and number of rotations are directly related to the distance traveled. For example, if a wheel has a circumference of 24 inches, then it will rotate once (360 degrees) each time the car moves 24 inches forward. With this information, we can use the circumference to calculate the number of degrees the wheel rotates.

The first line in each block calculates the circumference of the wheel. The circumference can be calculated with a math formula that you should know from your days in school:

Circumference = 2 * PI * Radius

The value PI is a magic number equal to about 3.1416 that is often used in geometric calculations dealing with circles, cylinders and spheres. MAXScript has a PI value built into it, so you can simply state it as **pi**.

The first line calculates the circumference of the wheel and places it in the variable **C**.

The circumference is then used to determine the angle increment for the wheel's rotation as the rover drives. The equation for calculating the number of degrees of rotation is:

Degrees to rotate = (360 / circumference) * distance

If this doesn't make sense to you, try plugging in a few values that you know for sure will work. For example, if a car with a 24-inch wheel moves 12 inches, the wheel should make half a rotation, or 180 degrees. Work with this equation until you are convinced it works.

Calculating the angle increment in the script was done with this equation, using the following code:

angleInc=(360/C)*distInc

With the angle increment calculated and stored in the variable **angleInc**, all we need to do is simply rotate the wheel. The rotate command works with the following syntax:

rotate [object] [degrees] [axis]

This is used in the code with these commands:

rotate $Wheel01 -angleInc z_axis
rotate $Wheel02 angleInc z_axis

Note that the first command shown above makes **angleInc** negative. While modeling the rover earlier, we used **Mirror Selected Objects** to mirror the wheels onto the other side of the rover. The mirrored wheels must be rotated on a negative angle in order to rotate in the same direction.

Making the Wheels Rotate When the Rover Turns

Now that we have the wheels rotating properly when the rover moves forward and backward, we need to address what the wheels do when the rover turns.

Earlier, we wrote a special function to handle the tracks rotating in opposite directions when the rover turned. If the tracks were rotating opposite one another, it would only make since that the wheels also rotate opposite one another.

1. To have the wheels rotate in opposite direction, you will need to add the same code you did above twice to the **rotationControl** function. We have to add this code twice to accommodate for a left or a right turn. Below is the entire **rotationControl** function after the code has been added:

```
fn rotationControl =
(
    C=497
    rotationQue=$RC_Controller.rotation.z
    distInc=(rotationQue*1000) as float
    percentValue=(100/C as float)*distInc
    --right:
    if rotationQue<0 then
    (
        $tracks01[2][1][1].value-=percentValue
        $tracks02[2][1][1].value+=percentValue
        set coordsys local
        (
            C=2*pi*28
            angleInc=(360/C)*distInc
            rotate $Wheel01 -angleInc z_axis
            rotate $Wheel02 -angleInc z_axis

            C=2*pi*41.5
            angleInc=(360/C)*distInc
            rotate $Wheel03 -angleInc z_axis
            rotate $Wheel04 -angleInc z_axis

            C=2*pi*11
            angleInc=(360/C)*distInc
            rotate $'Track Wheel01' -angleInc z_axis
            rotate $'Track Wheel04' -angleInc z_axis

            C=2*pi*37.5
            angleInc=(360/C)*distInc
            rotate $'Track Wheel02' -angleInc z_axis
            rotate $'Track Wheel03' -angleInc z_axis
        )
        set coordsys world
    )
    --left:
    if rotationQue>0 then
    (
        $tracks01[2][1][1].value-=percentValue
        $tracks02[2][1][1].value+=percentValue
        set coordsys local
        (
            C=2*pi*28
            angleInc=(360/C)*distInc
            rotate $Wheel01 -angleInc z_axis
            rotate $Wheel02 -angleInc z_axis
```

```
                C=2*pi*41.5
                angleInc=(360/C)*distInc
                rotate $Wheel03 -angleInc z_axis
                rotate $Wheel04 -angleInc z_axis

                C=2*pi*11
                angleInc=(360/C)*distInc
                rotate $'Track Wheel01' -angleInc z_axis
                rotate $'Track Wheel04' -angleInc z_axis

                C=2*pi*37.5
                angleInc=(360/C)*distInc
                rotate $'Track Wheel02' -angleInc z_axis
                rotate $'Track Wheel03' -angleInc z_axis
            )
          set coordsys world
       )
    )
```

2. Save the script.
3. Evaluate the script.
4. Run another motion capture test and verify that the wheels are rotating properly.

   This testing phase can be really tricky. It does not take long for your eyes to start playing tricks on you. If this happens and you are really not sure which way things are rotating, create some sort of polygon object and link it to the wheel in question. Make sure you model something that will help determine the direction of rotation. Something pointy like a cone works well.

5. Once you have completed testing, don't forget to reset the position and rotation of the **mainControlNode**.

CREATING A USER INTERFACE

At this point you have created a script that allows you to drive the rover around your screen. This is pretty cool but does not offer much in terms of productivity for two reasons.

POSITION RESET

Currently the script does not handle resetting the rover's position after the motion capture test has been completed. This means that the next time the test is run, the rover will start from its current location. This problem will be solved via a script.

MOTION CAPTURE PLAYBACK

Each time you have run a motion capture session, you have done it through a test. So what happens when we actually record a motion captures session? We generate lots of keys for the following tracks:

- The Throttle's value track
- The RC_Controller's Y position track
- The RC_Controller's Z rotation track

After the motion capture session is over, the keys are dumped to the above-mentioned tracks. Dragging the time slider back and forth will cause the rover to do things that you may not expect. Remember, the animation was recorded linearly from the moment the motion capture session started until the moment it ended. Dragging the time slider slowly forward will cause the rover to respond fairly closely to the animation you recorded. But the moment you drag backward, things start to go bad.

You must keep in mind how our functions worked in the previous script. We monitored RC_Controller's Y axis position track for any positive or negative movement. We then added this movement to the rover. If the number was positive, the rover moved forward. Likewise, if the number was negative, the rover moved in reverse.

Now, let's say the animation was simply the rover moving forward from point A to B. As you drag the time slider forward, the callback function is called and a positive value is added to the rover, and all is good. But as soon as you start to drag the time slider backward, the rover will continue to go forward. This is because the RC_Controller never went negative so there is still a positive value on it, even if the animation is playing backward.

Another side of the same problem occurs when you try to render the scene. If you were to do a quick render of one frame, everything would appear to be in the same place as it is in viewports. But the moment you tried to render an animation, the final animation would show the rover just sitting there. This is because the script is not executed for each frame. In other words, the time slider is not moving, and animation is rendered according to the object positions at frames on the time slider.

SCRIPTS TO THE RESCUE

What we are going to do to overcome these problems is build a user interface that will allow you to determine if the callback script is going to run or not. This way, you could stop the callback function from running when the time slider is dragged. This will prevent the rover's position and rotation from being altered.

We will also create a routine that will save the motion capture animation to keyframes. This will keep us from having to rely on RC_Controller and the callback script.

Finally, we will throw in a few additional features to reset the position and delete the saved animation so the script will provide everything we need.

TUTORIAL 6.4 ROVER CONTROL USER INTERFACE

We will start the second script with just enough code to get a bare-bones interface up and running. We will then start adding all of the needed UI controls.

Creating a Utility

1. On the **Utilities** panel, click **MAXScript,** then click **New Script**.
2. Type the following code in the Script Editor:

   utility treadsUtil "Rover MoCap"
   (
)

3. Save the script with the name *Rover UI.ms*.
4. Select **Rover MoCap** from the pulldown list and check that the Rover MoCap rollout appears.

   The only UI control contained in this rollout is the default Close button.

5. Click **Close**.

Adding the Main Console Rollout

We are now going to include a new rollout called Main Console. This is where you will embed all the UI controls so you can easily use them.

1. Update the script so that it looks like the following:

   utility treadsUtil "Rover MoCap"
   (
       rollout rll_main "Main Console"
       (
           checkbox chkbxSwitch "Enable Script"
)
       on treadsUtil open do addrollout rll_main treadsUtil
       on treadsUtil close do removerollout rll_main treadsUtil
)

2. Save and evaluate the script.

   You should now see a rollout beneath the Rover MoCap rollout called Main Console. This new rollout should contain a single checkbox labeled Enable Script

3. Click **Close**.

As we mentioned earlier, we do not always want the callback function **masterController** to be executed when the time slider is moved. We have added a checkbox called **Enable Script** to the Main Console rollout that can be used to determine whether the **masterController** function will execute all the other functions from the other script. This control will be handled differently than other controls. Instead of writing an event block, we will rely on the callback script to check its value. We will tie the two scripts together in the next section.

The **on treadsUtil open** and **on treadsUtil close** lines make sure the rollout is available and open whenever the utility is called and opened, and erased from memory when the utility is closed.

Bridging the Scripts

1. Return to the *Rover Callback.ms* script. It should still be open in another editor window.

    The **masterController** function currently contains three function calls:

    setMainControlNode( )
    matchTracks( )
    rotationControl( )

    These function calls should execute only if the **Enable Script** checkbox has been checked in the Main Console rollout. To set this up we will use an **if** statement.

2. Edit your **masterController** function so that it looks like the following:

    fn masterController =
    (
        if treadsUtil.rll_main.chkbxSwitch.state!=false then
        (
            setMainControlNode( ) -- position and rotate the rover
            matchTracks( ) -- handle the track rotation
            rotationControl( ) -- handle the tracks when the rover turns
)
)

3. Save and evaluate the script.

    Now, every time the callback occurs, **masterController** will first check to see if the **Enable Script** checkbox is checked. If a value of True is returned (indicating that the box is checked), the three functions will get executed. If the value is false, the **masterController** will terminate without calling any of the functions.

    The above **if** statement is very similar to the one we used in *Chapter 5 Introduction to MAXScript*. The main difference is the use of the **!=** operator to compare the two statements. This operator means *not equal to*. The code under the if statement is executed only if the results are *not equal to false*, which is a roundabout way of saying *true*.

Implementing a Reset Button

After each motion capture session, you will need to send the rover back to its original position and rotation before you can start another session. It would be tedious to have to keep moving the rover back manually or typing in its position and rotation in a Transform Type-In dialog each time. We will set up a Reset Position button to do this for us automatically.

Earlier, we created a series of custom attributes for the **mainControlNode** called **initPosX**, **initPosY**, **initPosZ**, and **initRotZ**. Once you have the rover positioned for starting the motion capture session, you can type its position and rotation parameters into these custom attributes on the **Modify** panel. The following script will then move the rover back to these values when the Reset Position button is clicked.

1. In the **Rover UI.ms** script, update the rollout **rll_main** function to match the one below:

   ```
   rollout rll_main "Main Console"
   (
       checkbox chkbxSwitch "Enable Script"
       button btnResetPos "Reset Position"

       on btnResetPos pressed do
       (
           -- Note: the next 3 lines should be on the same line
           -- in your script. We don't have room to put them all on
           -- the same line here. You can leave out this comment.
           $mainControlNode.pos=[$mainControlNode.initPosX,
           $mainControlNode.initPosY,
           $mainControlNode.initPosZ]

           $mainControlNode.rotation.z=$mainControlNode.initRotZ
       )
   )
   ```

   If you have the time and feel like a challenge, write an additional button into the UI that will copy the current positions and rotation into these attributes anytime you click it. This will allow you to change the starting position and rotation quickly without having to type them in on the **Modify** panel.

2. Save and evaluate the script.

   Let's test the script.

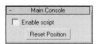

3. Select **Rover MoCap** from the **Utilities** pulldown menu, and verify that the interface looks right. You should have a Main Console rollout with a Reset Position button.

4. Select the **mainControlNode** and move it to another location in the scene.

5. Click the **Reset Position** button.

   Currently, **initPosX**, **initPosY**, **initPosZ**, and **initRotZ** are set at their default values of 0, so the rover should jump back to the 0,0,0 point in the scene.

   There is a possibility that the rover will flip over or rotate strangely when the **Reset Position** button is clicked. This is due to the type of rotation used by **3ds max**, which sometimes behaves unpredictably. If the script makes the rover rotate or flip when you didn't ask it to, sometimes re-evaluating the script a few times can make it start working properly.

   If the script worked correctly, go ahead and test the functionality of the two scripts together.

4. Check the **Enable Script** checkbox and then run a motion capture test session. After driving the rover around for a bit, stop the test, return to the **Utilities** panel and click the **Reset Position** button.

   Currently, **initPosX**, **initPosY**, **initPosZ**, and **initRotZ** are set at their default values of 0. If you've done everything correctly, the rover should immediately jump back to its starting position of 0, 0, 0.

5. If that tests correctly, uncheck the **Enable Script** checkbox and run another motion capture test. If you've typed everything correctly, the rover should just sit there. The only thing that should be moving is the RC_Controller dummy.

Creating the Bake Arrays

This entire setup relies on the motion capture of data applied to a dummy and a slider. A callback function executes each time there is a change in time and will read the position of the dummy and slider and make the appropriate calculations so that our rover appears to be moving on its own.

Once the rover is moving the way we want it to, we'll need to generate keyframes for the data. The process of setting keyframes based on motion capture data is called *baking* the animation. Once the animation is set as keyframes, you can't get it back as motion capture data any more. What does this have to do with baking? Well, have you ever tried to unbake a cake? You get the idea.

In this section we will create a series of arrays that will store all pertinent information regarding the rover, then later write this information out to keyframes on the appropriate tracks.

We also need a checkbox to determine whether data should be written to these arrays, and when the array data is emptied.

1. In the **Rover UI.ms** script, update the rollout **rll_main function** so the script looks like the following:

```
rollout rll_main "Main Console"
(
    checkbox chkbxToggle "Enable MoCap"
    checkbox chkbxSwitch "Enable Script"
    button btnResetPos "Reset Position"
    on chkbxToggle changed state do
    (
        if chkbxToggle.checked then
        (
            tickArray=#( )
            posArray=#( )
            rotArray=#( )
            percentArray01=#( )
            percentArray02=#( )

            bakeWheel01Array=#( )
            bakeWheel02Array=#( )
            bakeWheel03Array=#( )
            bakeWheel04Array=#( )
            bakeWheel05Array=#( )
            bakeWheel06Array=#( )
            bakeWheel07Array=#( )
            bakeWheel08Array=#( )
        )
    )
    on btnResetPos pressed do
    (
        -- Note: the next 3 lines should be on the same line
        -- in your script. We don't have room to put them all on
        -- the same line here. You can leave out this comment.
        $mainControlNode.pos=[$mainControlNode.initPosX,
        $mainControlNode.initPosY,
        $mainControlNode.initPosZ]

        $mainControlNode.rotation.z=$mainControlNode.initRotZ
    )
)
```

2. Save and evaluate the script.
3. Select **Rover MoCap** from the **Utilities** pulldown menu.
4. Verify that the rollout now contains the new checkbox.
5. Click **Close**.

TIP

Don't rush to test the script. At this point we have not added the code that will actually put information into the arrays. This will come in the next section.

In the above code, we have added a new checkbox labeled **Enable MoCap** and its related event. The event is handled by the line **on chkbxToggle changed state do** which executes its embedded code anytime the checkbox is checked or unchecked. Once it has been clicked, the event checks the status of the checkbox to see if it was checked or unchecked. If the checkbox was checked, a series of arrays are emptied of any data that they contain.

The arrays are being used as a holding place for the current time, position, rotation, and percentage values. Later, the information that has been stored in these arrays can be extracted and turned into keyframes.

Putting Information into the Arrays

Now that we have the arrays in place and a UI options that will allow you to tell the script to record information into the arrays, we need to write the appropriate code to put information into the arrays. We will be adding this code to the **Rover Callback.ms** script because it is the one that is executed every time there is a change to time.

As of right now, the **Rover Callback.ms** script will not be able to see the array variables that have been added to the UI script. For these arrays to be seen inside this script, we will need to make them global.

1. In the **Rover UI.ms** script, add the following code to the very beginning of the script. It is very important that these lines of code come before all others.

   ```
   global tickArray=#( )
   global posArray=#( )
   global rotArray=#( )
   global percentArray01=#( )
   global percentArray02=#( )
   global bakeWheel01Array=#( )
   global bakeWheel02Array=#( )
   global bakeWheel03Array=#( )
   global bakeWheel04Array=#( )
   global bakeWheel05Array=#( )
   global bakeWheel06Array=#( )
   global bakeWheel07Array=#( )
   global bakeWheel08Array=#( )
   ```

2. Save and evaluate the script.

   By adding the above code, any script can now see the arrays, including the **Rover Callback.ms** script. Now let's add the code required to bridge the two scripts together even further than before:

   In the **Rover Callback.ms** script, we need to add the code that will check the state of the Enable MoCap checkbox and respond appropriately if it is checked. If the checkbox is checked, the arrays need to be populated with the right data.

3. Update the **masterController** function so that it looks like the one below.

```
fn masterController =
(
    if treadsUtil.rll_main.chkbxSwitch.state!=false then
    (
        if treadsUtil.rll_main.chkbxToggle.state==true then
        (
            t=currentTime --get current time
            bakePos=$mainControlNode.position
            -- get the mainControlNode position
            bakeRot=$mainControlNode.rotation
            -- get the mainControlNode rotation
            bakePercent01=$tracks01[2][1][1].value
            --the % for the left track
            bakePercent02=$tracks02[2][1][1].value
            --the % for the right track
            bakeWheel01=$wheel01.transform
            bakeWheel02=$wheel02.transform
            bakeWheel03=$wheel03.transform
            bakeWheel04=$wheel04.transform
            bakeWheel05=$'Track Wheel01'.transform
            bakeWheel06=$'Track Wheel02'.transform
            bakeWheel07=$'Track Wheel03'.transform
            bakeWheel08=$'Track Wheel04'.transform
            --- drop all of them into arrays

            append tickArray t
            append posArray bakePos
            append rotArray bakeRot
            append percentArray01 bakePercent01
            append percentArray02 bakePercent02
            append bakeWheel01Array bakeWheel01
            append bakeWheel02Array bakeWheel02
            append bakeWheel03Array bakeWheel03
            append bakeWheel04Array bakeWheel04
            append bakeWheel05Array bakeWheel05
            append bakeWheel06Array bakeWheel06
            append bakeWheel07Array bakeWheel07
            append bakeWheel08Array bakeWheel08
        )
        setMainControlNode() -- position and rotate the rover
        matchTracks() -- handle the track rotation
        rotationControl() -- handle the tracks when the rover turns
    )
)
```

4. Save and evaluate the script. Do not test the code at this time, just make sure the evaluation returns no errors.

The above code will store the percentage values for the tracks, the position and orientation of the mainControlNode (the rover's parent), and the orientation of the wheels.

We also record the current time into an array called **tickArray**. This way, we have a reference as to what time all of the other data was recorded. Later, we will match each position and orientation values with the time array to bake the correct animation so that we will have proper viewport playback, preview, and final render.

TIP
A tick is a special time unit used by 3ds max, equivalent to a 4800th of a second. It is a more accurate way of keeping time than using frames numbers.

After all of the values have been placed in a temporary variable, the variable is then appended to the appropriate array. We do this with the **Append** command. The syntax of the **Append** command is seen below:

Append [array] [element]

By appending the data to the array, we do not erase anything that is already being stored there.

The **Rover Callback.ms** script is now complete. The remaining setup will occur in the **Rover UI.ms** script.

Baking the Animation

We only have two more things to write for the rover's technical setup to be complete.

First we are going to write the code that will handle baking the array data into keyframes. Then we are going to add a button that will provide you with a quick way to delete all the key that are generated by baking, a quick reset button that clears everything if you (or the client) doesn't like the animation.

1. To create the button that will bake the animation, add the following code to the UI script, inside the rollout **rll_main function**. Place the button between the **Enable Script** checkbox and the **Reset Position** button.

    button btnBake "Bake animation"

2. Next, add its event block inside the same function. Add it right after the **Reset Position** button. This will make it the first event block defined within the function.

```
on btnBake pressed do
(
    animate on
    for b=1 to tickArray.count-1 do
    (
        at time tickArray[b]
        (
            $mainControlNode.rotation=rotArray[b]
            $mainControlNode.position=posArray[b]
            $wheel01.transform=bakeWheel01Array[b]
            $wheel02.transform=bakeWheel02Array[b]
            $wheel03.transform=bakeWheel03Array[b]
            $wheel04.transform=bakeWheel04Array[b]
            $'Track Wheel01'.transform=bakeWheel05Array[b]
            $'Track Wheel02'.transform=bakeWheel06Array[b]
            $'Track Wheel03'.transform=bakeWheel07Array[b]
            $'Track Wheel04'.transform=bakeWheel08Array[b]

            $tracks01[2][1][1].value=percentArray01[b]
            $tracks02[2][1][1].value=percentArray02[b]
        )
    )
)
```

3. Save and evaluate the code. If there are any errors while evaluating the code, look for any misspellings.

   The keyframes are created by first turning on the Animate button and making some sort of change to a parameter. Activating the animate button in a script is accomplished with the command **animate on**.

   Next, we extract the data back out of the array variables through the use of a loop. The length of the loop is determined by the number of elements in **tickArray - 1**. The minus one keeps us from using the last element in the array.

   At the very end of a motion capture session, the time slider is reset back to 0. This causes the callback script to be invoked and the last bit of information is recorded with a noted time of frame 0, but the rover's position does not move. This means that our array now contains two entries with a time of 0: The very first element and the last element. All we need to do is simply ignore the last element, so we subtract one from **tickArray**.

   The **for** statement uses the index variable **b** as the element number for the array. Each time we go through the loop, we save the time in **tickArray** with the **at time tickArray[b]** command.

Next, we simply set the mainControlNode, wheels, and percent values of the track to the information that was stored in the arrays at the current element, determined by **b**. Since the animate button is on, keys are recorded.

Pretty slick, if we do say so ourselves.

Deleting Keyframes

The final code that we will add to the UI script will provide you with a way to delete all the keyframes that are generated during the motion captures session and the baking operation. This is handy if you are unhappy with the final animation and you need to redo it.

1. To create the button to delete the animation, add the following code to the UI script, inside the rollout **rll_main function**. Place the code just after the **'Bake Animation'** button:

    button btnDelAnim "Delete Animation"

2. Add the following event code into the **rll_main function**. Add this code after the bake animation event:

    ```
    on btndelanim pressed do
    (
        deletekeys $Throttle.value.controller
        deletekeys $RC_Controller.position.controller
        deletekeys $RC_Controller.rotation.controller
        deletekeys $mainControlNode.position.controller
        deletekeys $mainControlNode.rotation.controller
        deletekeys $Tracks01[2][1][1].controller
        deletekeys $Tracks02[2][1][1].controller
        deletekeys $wheel01.transform.controller
        deletekeys $wheel02.transform.controller
        deletekeys $wheel03.transform.controller
        deletekeys $wheel04.transform.controller
        deletekeys $wheel05.transform.controller
        deletekeys $wheel06.transform.controller
        deletekeys $wheel07.transform.controller
        deletekeys $wheel08.transform.controller
    )
    ```

3. Save and evaluate your script. If there are any errors presented in the Listener, find and correct them.

4. Choose the **Rover MoCap** option in the **Utilities** pulldown menu and make sure the new button appears on the Main Console rollout.

Using the Script

The workflow for driving the rover is fairly straightforward, but involves several steps. It is very easy to forget a step and this can lead to problems. Below is an overview of the procedures involved:

You must always have the **Rover Callback.ms** script running. This means that after opening the max scene, you will need to reevaluate this script, otherwise the rover will just sit there.

By default, a new scene only has 100 frames. This is not enough time to really see what the rover is doing.

1. Set your animation length to at least 500 frames by clicking **Time Configuration** and changing the **Length** parameter.
2. On the Main Console rollout on the **Utilities** panel, make sure the **Enable MoCap** and **Enable Script** checkboxes are both checked. Otherwise, the rover won't move when you run the motion capture session.
3. On the Motion Capture rollout, make sure that the **Out** parameter matches the new end frame of your animation. This is the point at which the motion capture session will end:

Prepare yourself to record by making sure you have an idea of what you are going to do. Will you be driving forward, reversing, turning? How fast will you go?

4. Press the **Start** button in the Record Controls section, and use your mouse and keyboard to perform the motion capture motions.
5. Once the motion capture is complete, return to the Rover MoCap utility. Uncheck **Enable MoCap** and **Enable Script**.

Every so often the script will get confused, and the rover will do all sorts of crazy things. If this happens, relax! Delete the animation and try again.

6. Click **Bake Animation**.

7. Play the animation with the **Play Animation** button in the view controls area of the **3ds max** user interface.
8. If you are unhappy with the animation, go to frame 0 and click **Delete Animation** on the **Utilities** panel.

When using the **Delete Animation** button, the time slider must be at frame 0. Otherwise, the RC_Controller could be at some position other than 0,0,0.

If you find that your rover takes off the moment you hit **Test** or **Start** in the motion capture session, check the RC_Controller's absolute position on the Move Transform Type-In. All of positions (X, Y, and Z) should be 0.

9. When you have finished playing with the setup, delete the animation and save the entire scene as **Rover - Setup.max**.

SUMMARY

In this chapter, you learned how to:

- Use the *PathDeform modifier to make an animatable track of unusual shape
- Control the rover with a dummy object
- Set up and perform motion capture with the mouse and keyboard
- Create and wire custom parameters
- Write scripts to turn the rover wheels and tracks appropriately when the rover is moved
- Write a custom user interface to control the motion capture process
- Write a script to save motion capture data as keyframes

Later on, you'll get to use these scripts to create the motion for the rover. But for now, your next task is to set up the alien for animation.

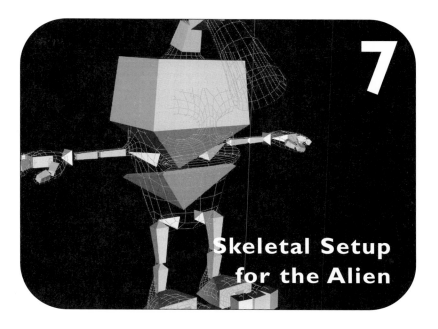

7
Skeletal Setup for the Alien

In the commercial we are currently developing, we are responsible for making the alien interact with the remote control and rover as a living creature. In this chapter, you will learn how to setup specialized tools that will make animation of the alien as easy and fast as possible.

TECHNIQUES IN THIS CHAPTER
In this chapter, you will learn how to:

- Create a character's skeletal structure with Bones objects
- Set up and use inverse kinematics (IK) chains
- Create, place and link helpers for using IK chains
- Create new parameters for controlling the alien
- Assign controllers for rotating joints
- Mirror a character setup

TUTORIALS IN THIS CHAPTER

This chapter focuses on creating the alien skeleton and creating basic controllers for each joint.

- *Tutorial 5.1 Alien Skeleton* shows you how to create the skeleton using Bones objects.

- *Tutorial 5.2 IK Solvers and Point Helpers* walks you through the process of setting up point helpers and IK solvers to control each limb.

- *Tutorial 5.3 Hand Control System* and *Tutorial 5.4 Foot Control System* cover the assigning of controllers that will later be used to animate the skeleton.

- *Tutorial 5.5 Mirroring and Completing the Skeleton* mirrors the one side of the skeleton you have created to the other side, and sets up controllers for the new side.

A FEW THINGS YOU SHOULD KNOW BY NOW

At this stage of our project, we're working with some advanced features. To save us from repeating instructions on things you already know (or should already know), let's review a few basic techniques in **3ds max**. This way, we won't have to bore you with the details in the tutorials, and you can just do your thing.

- When you want to copy an object, click **Select and Move** on the Main Toolbar, hold down the **<Shift>** key and move the object. When the Clone Options dialog appears, choose the **Copy** option and enter a name for the new object, then click **OK**. In the tutorials in this chapter, when you should perform this procedure, we'll just say "Copy and move."

- To access the Move Transform Type-In dialog, click then right-click the **Select and Move** button. In the tutorials in this chapter, we'll just say "On the Move Transform Type-In dialog,..." and expect you to know how to get there.

- To access the Rotate Transform Type-In dialog, click then right-click the **Select and Rotate** button. The same goes for this dialog as for the Move Transform Type-In dialog.

- To access the Quad menu, right-click in a viewport. We'll just say "Access the Quad menu."

- To switch a viewport to shaded or wireframe mode, right-click the viewport label and choose *Smooth + Highlights* or *Wireframe* from the pop-up menu. We'll just tell you to "Switch the Perspective viewport to *Wireframe* mode," for example.

- To set up and use snaps, right-click the **3D Snap** button. On the Grid and Snap Settings dialog, check the types of snaps you want to use and uncheck all others. Close the dialog and click **3D Snap** to turn it on. To turn off snapping, click **3D Snap** to turn it off. As an example, we'll just say "Set up snaps for **Vertex** only, and turn on snaps."

CHARACTER ANIMATION

Modeling the character was only the first step in character animation. The process of breathing life into a character requires proper pre-planning and execution of that plan.

Before you begin, you will need to figure out the answers to a few questions:

- How much animation will be required of the character?
- How complex is the animation? Do you need only simple arm and head movements, or will the character run down a long corridor, get shot, fall and roll to a stop?
- Who will be animating the character? Will it be you, or will you be developing the character setup for another animator?

The answers to these questions are very important in determining the best approach for the character's setup.

CHARACTER RIGS

Before the character is animated, a control scheme called a *character rig* is set up to make animation easier and more predictable. This is especially important when more than one person is going to work on the animation.

A character rig can take many forms. Here are a few examples:

- **Skeletal Structure.** Unless the model will be animated directly, a character model requires a skeletal structure. The skeleton is then used to deform the model. A series of linked objects, such as a skeleton, is called a *hierarchy*.

- **Skinning.** Skinning associates the mesh with the skeleton bones so the mesh will deform when a bone is moved or rotated. If a skeleton is used, skinning is always required.

- **Animation with Forward Kinematics (FK).** This is the simplest form of character control for animation. The animator simply moves and rotates bones manually, going from key pose to key pose. This is one of the easiest rigs to use but can be time-consuming if you want the animation to look good.

- **Animation with Inverse Kinematics (IK) Solvers.** This form of control takes a little more time to set up, but less time to get decent results once the animation process begins. With IK, you simply animate the last object on the linked chain (the child) and the rest of the chain follows along accordingly. For example, you can move a foot and have the thigh and calf move and rotate accordingly. Compare with FK, where you would have to rotate and move the thigh, calf and foot individually.

  You may well wonder, if IK works so spectacularly, why doesn't everyone use it all the time? With IK, you lose some control over exact rotation of elbows and knees. If the final location of the child is the most important thing, and you are not that concerned with exact positioning of bones as long as it looks good, then IK is the ticket. But if you need precise control over the placement of each bone, then FK is your best bet.

- **Control Systems.** A *control system* is a setup that automates an animation process to some degree. The controls are in the form of custom user interface elements such as parameters and sliders. An example would be a slider created to move a character's lower jaw up and down. This saves the animator a great deal of time when animating repetitive motion.

Choosing the correct character rig depends on the way you answered some of the earlier questions. If the character isn't required to perform much animation then simple FK works just fine, and you need not spend a lot of time developing complex control systems. But if the character is required to do a fair amount of complex animation, it's worth it to set up an IK structure and design a control system of some kind.

Generally, characters will use different combinations of the above techniques. It really depends on who will be doing the final animation. If that person is you, then only you know what it would take to make the animation process easy. If you were doing the setup for someone else, it is very important to discuss what the character will be doing and which methods the animator would like to use.

In setting up the alien, we will be using inverse kinematics combined with various control systems. Before you have completed the entire production, you will create a sophisticated control system that will simplify the process of animation.

TUTORIAL 5.1 ALIEN SKELETON

To control the character, you typically need some sort of skeletal structure inside the character model. Rather than animate the character model directly, the skeleton is animated to deform the character model. Through skinning, the model is associated with the skeleton.

We are going to start out by placing bones on the alien's left side, then later we'll mirror the bone structure.

Preparing to Create Bones

Before you begin adding bones to a character, there are a few things that you can do to help simplify the overall process. In this case, we are going to add a material that will allow us to view the alien as a wireframe (even in a shaded view) and then we will freeze it so that we don't accidentally select the alien. Next we will freeze the eyes but we will continue to see them in shaded view.

1. Open the file **Alien Model.max** that you created in **Chapter 3**, or load this file from the **Scenes/Alien** folder on the CD.

2. Open the **Material Editor**, and click an empty sample slot.
3. In the Shader Basic Parameters rollout, check the **Wire** checkbox, and change the **Diffuse** color swatch to a very dark shade of gray.
4. Assign the material to the alien, and close the Material Editor.
5. Select the alien.

6. On the **Display** panel, on the Display Properties rollout, uncheck the **Show Frozen in Gray** checkbox if it is checked.
7. On the Freeze rollout, click **Freeze Selected**.

The alien's body is no longer selectable and remains a wireframe. This allows us to easily see through the alien so that we are able to see where the skeleton bones are being placed.

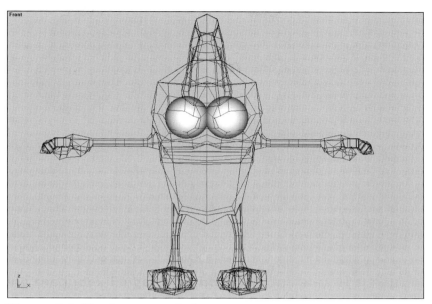

Figure 5.1 *Alien frozen as a wireframe*

8. Select both of the eyeballs and click **Freeze Selected**. In the case of the eyeballs, we don't want to view them as wireframe.

9. Finally, make both the Front and Left viewports shaded. This will help us keep track of the bones we are creating.

Creating the Pelvis Bone

The standard tools of **3ds max** include special objects called Bones. These are long, thin objects designed to be used as bones in a skeletal structure. As you create the bones, they are automatically linked together.

Using **Bones** to create a skeletal structure saves a lot of time.

1. Activate the Front viewport and click the **Min/Max Toggle** to maximize it.
2. On the **Create** panel, click the **Systems** button.
3. Click the **Bones** button.
4. Click slightly to the right of the alien's centerline to mark the starting location of the left pelvis bone. As you move your mouse (after the click) you will see the bone. Move a little higher than the starting point of the bone and click. Right-click twice to complete creation of the bone and turn off the **Bones** tool.

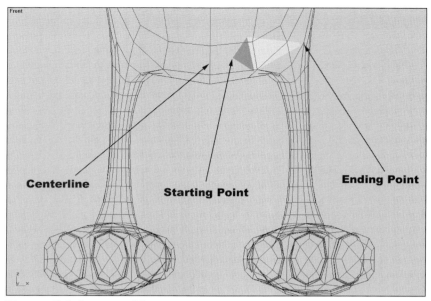

Figure 5.2 *Location of first bone*

Note that another bone was automatically created at the end, and this bone is still selected.

5. Press **<Delete>** to delete the selected bone.

6. Press the **<L>** key to change the view to the Left viewport.

☀ **TIP** ☀

This is quicker than minimizing the viewport, activating the Left viewport and maximizing again.

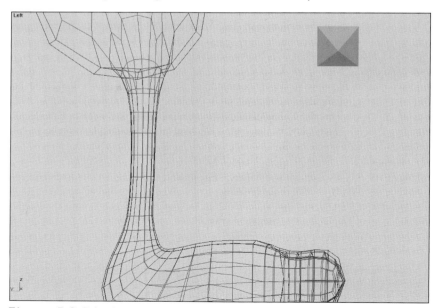

Figure 5.3 *Pelvis bone positioned in front of alien*

The bone has been placed in front of the alien.

7. Select the newly created bone.

8. Use **Select and Move** to move the bone on the X axis until it is lined up inside the pelvis.

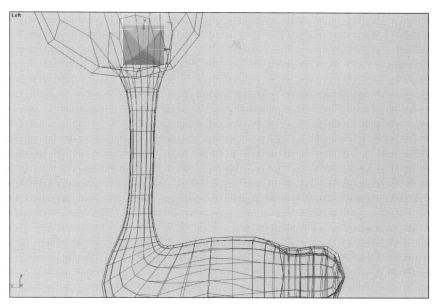

Figure 5.4 *Pelvis bone properly positioned*

Creating the Leg Bones

Next you will create a series of bones connected to the first bone. By clicking on the hip bone to start the bone creation process, the new bones will automatically be linked to the hip bone.

1. Click **Bones** again.
2. Move your cursor over the pelvis bone until the cursor changes, then click to start the new bone. Move the cursor to the knee area, slightly closer to the front of the leg, and click to set the bone. Move the cursor and click at the ankle and again near the base of the toes, then right-click to complete the creation of the hierarchy. Use the following image as a reference for bone placement.

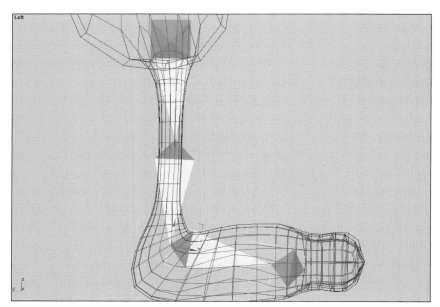

Figure 5.5 *Leg bones*

> **TIP**
> Don't delete the tail bone at the end this time.

 3. Verify that the last bone in the hierarchy is selected. Use **Select and Rotate** to rotate it so it points at the tips of the toes.

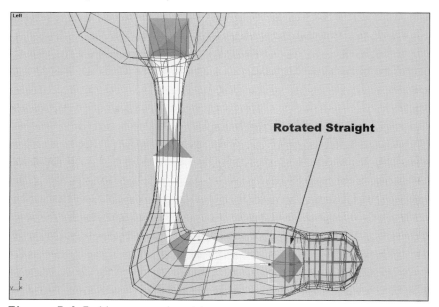

Figure 5.6 *End bone rotated straight*

Creating the Toe Bones

The toe bones will not be connected to the foot bones in our skeleton. When making the toe bones, take care not to click on the foot when starting any of the bones.

1. Press **<T>** to change the viewport to the Top viewport.
2. In the Top viewport, create three individual bones, one for each toe. Use the image below as a reference. After you right-click to complete each toe bone, delete the tail bone at the end of the toe.

   Take care not to overlap the middle toe bone with the last bone on the foot. It is not a problem if the middle toe and the last foot bone are not evenly aligned.

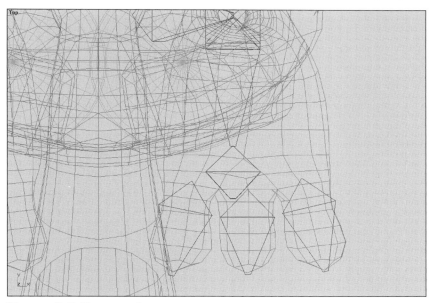

Figure 5.7 *Three new bones*

3. Click **Min/Max Toggle** to return to the four-viewport display.
4. In the Perspective viewport, move the toe bones down inside the alien model foot.
5. Save the scene with the filename *AlienBones01.max*.

Shaping the Bones to Fit the Model

Now we are going to adjust the thickness of the bones to better fit the model. This will give us two advantages:

- If all of the bones' thicknesses are about the width of the actual geometry, we can animate using just the bones (and not other objects added to the scene) since the bones will give us a good idea of the model's volume.

- When the Skin modifier is applied later on, the envelopes (regions of effect) for each bone will be more accurately sized and placed, meaning less manual adjustment for us.

The easiest way to adjust a bone's thickness is to adjust its Width, Height, and Taper parameters on the Modify panel. However, there may be times when these parameters just don't give you the shape and size you want. In this case, you can apply an Edit Mesh modifier to the bone and modify its sub-objects (vertices, polygons) as you would with any mesh object.

We will use the latter approach with the foot bone.

1. Select the foot bone.
2. On the **Modify** panel, apply an **Edit Mesh** modifier to the foot bone.

   This modifier gives you access to most of the parameters available for an Editable Mesh object, without changing the object's base type. This means the object is still a bone, but you can use sub-objects to edit it.

3. Access the **Vertex** sub-object level.
4. Select and move vertices around until the bone fills most of the foot. Use the image below as a reference.

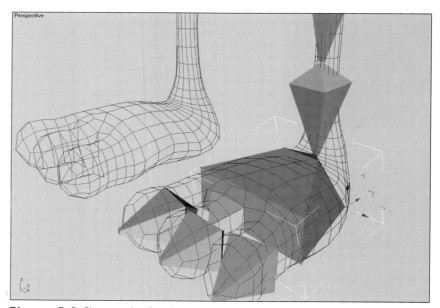

Figure 5.8 *Shaping the foot bone*

5. For each toe bone, select it and and adjust its parameters so that it fills up most of the geometry area that it will be controlling. Here are the parameters we used:

   Width 15.0
   Height 12.0
   Taper 8.0%

> **TIP**
> The bones are referred to by the joint at which they originate.

6. Adjust the parameters for the hip and knee bones (the second and third bones created) so that they better fit the geometry. The following settings were used for the images that follow. You may require different settings. The goal is to fill out the geometry with the bones.

   Hip Bone

   Width 10.0
   Height 10.0
   Taper 8.0%

   Knee Bone

   Width 8.0
   Height 8.0
   Taper 48.0%

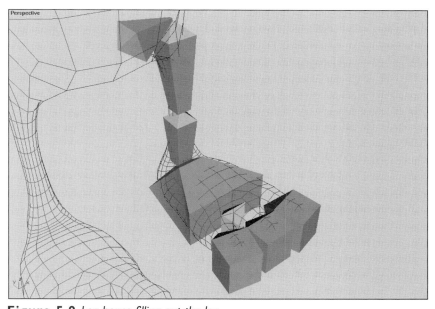

Figure 5.9 Leg bones filling out the leg

7. Using the next image as a reference, set the bone names as follows:

   Bone - Pelvis
   Bone - Hip
   Bone - Knee
   Bone - Ankle
   Bone - Take me out (Ball)
   Bone - Toe (Inner)
   Bone - Toe (Middle)
   Bone - Toe (Outer)

☼ **TIP** ☼

We have included the words **Take me out** in the ball bone so that later we can quickly identify and remove this bone from the skinning process.

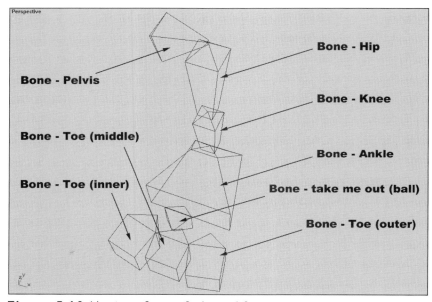

Figure 5.10 Naming reference for leg and foot

8. Save the scene with the filename **AlienBones02.max**..

Creating the Clavicle

In the next set of steps you will create the bone for the neck-to-shoulder area. This bone is called the *clavicle*.

 1. Activate the Front viewport and click **Min/Max Toggle** to bring it to full screen size.

2. On the **Create** panel, click **Bones** again.

3. Before creating the bone, change the bone parameters to the following if necessary:

   Width 10
   Height 10
   Taper 100%

4. Create a bone at the alien's clavicle as shown in the image that follows. Right-click twice to complete the bone and exit the **Bone** tool.
5. Delete the tail bone at the end.

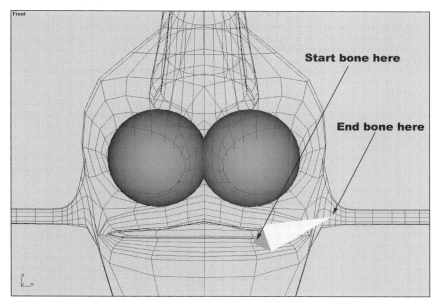

Figure 5.11 *Placement of clavicle bone*

6. Press **<T>** to switch to the Top viewport.

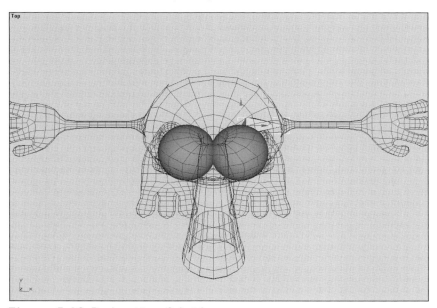

Figure 5.12 *Final position of clavicle*

You will find the bone out in front of the alien.

 7. Use **Select and Move** to move the bone back along the Y axis until it is lined up with the arm.

Creating the Arm Bones

1. Click **Bones** again.
2. Click on the end of the clavicle bone and move the cursor until you reach the elbow area. The elbow area should be slightly back from the clavicle/shoulder area. Click at the elbow to create the bone, then move the cursor a short distance to the wrist and click to create the forearm bone. Right-click to complete the bones creation process.

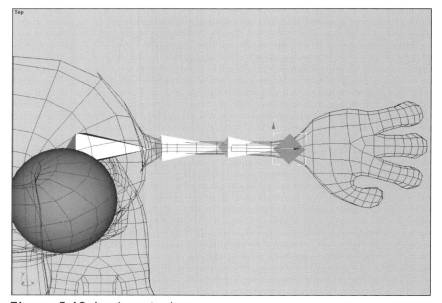

Figure 5.13 *Arm bones in place*

Adjusting the Arm Bone Sizes

 1. Click **Min/Max Toggle** to return to the four-viewport display.

2. Activate the Perspective viewport.

 This will allow you to use **Arc Rotate** to rotate around the bones and see how the fit is coming along.

3. Select the upper arm bone.
 4. Go to the **Modify** panel.

5. Adjust the **Width**, **Height**, and **Taper** parameters until the bone matches the geometry area that it occupies. The following settings were used in this example:

   Width   5.0
   Height   6.0
   Taper   21.0%

6. Repeat this process for the forearm bone. Approximate settings are:

   Width   5.0
   Height   5.0
   Taper   22.0%

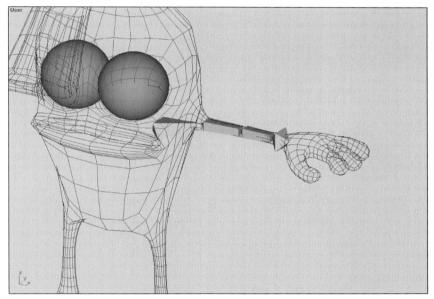

Figure 5.14 *Upper and lower arm bones resized to fit model*

Creating the Hand and Finger Bones

1. Activate the Top viewport and click **Min/Max Toggle**.
2. Click **Bones**.
3. Place your cursor over the ending joint of the hand until the cursor changes shape, indicating that you are continuing the hand hierarchy. Click to start the bone, then move the cursor to the middle of the hand and click. Right-click twice to complete the bone and turn off the Bone tool.
4. Delete the tail bone.

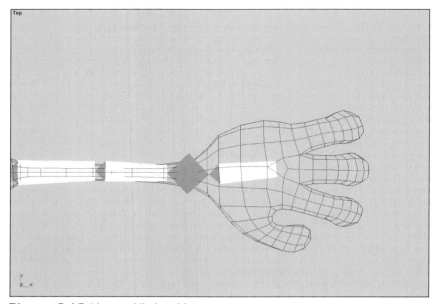

Figure 5.15 *New middle hand bone*

> ☃ **TIP** ☃
>
> *Make sure the knuckle joint is after the separation for the fingers to ensure the best deformation around the knuckle area later on.*

5. Create three additional bones for the pinky finger, using the following image as a guide. Delete the tail bone at the end.

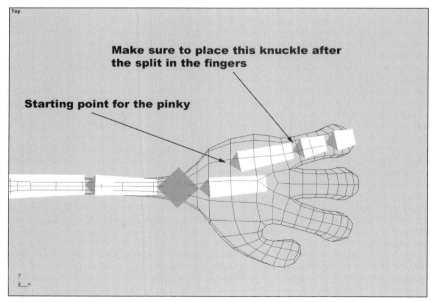

Figure 5.16 *Pinky bones placement*

6. Create the rest of the fingers with two bones for each finger, including the thumb. The last bone should extrude slightly past the end of the finger (or thumb) tip. Delete the extra bone at the end of each finger and the end of the thumb..

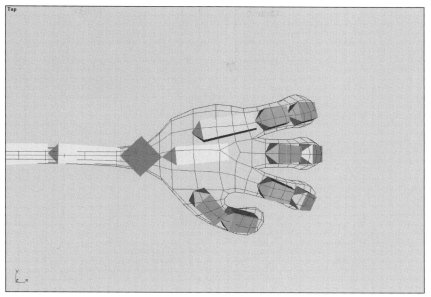

Figure 5.17 *Finger bones*

Adjusting the Hand/Finger Bone Sizes

1. Click **Min/Max Toggle** to return to the four-viewport display.

2. Activate the Perspective viewport.

3. Click **Select and Rotate** and verify that the **Reference Coordinate System** is in **Local** mode.

4. Rotate and move all of the finger bones until they are inside and properly matching the geometry.

   Make sure you perform the following rotations:

   - The first index finger bone needs to be rotated slightly around its X axis (counter clockwise).

   - The first pinky bone needs to be rotated slightly around its X axis (clockwise).

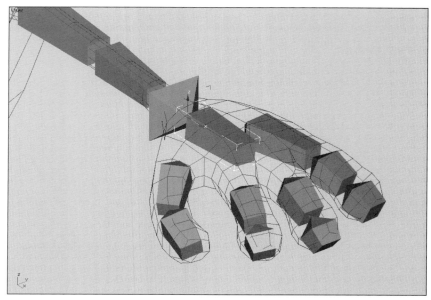

Figure 5.18 *Finger bones positioned*

5. Select the bone at the center of the hand.

> ☼ **TIP** ☼
>
> Don't worry if you see the center hand bone overlapping part of the thumb and pinky bones. We have made this bone large to make the skinning process easier later on.

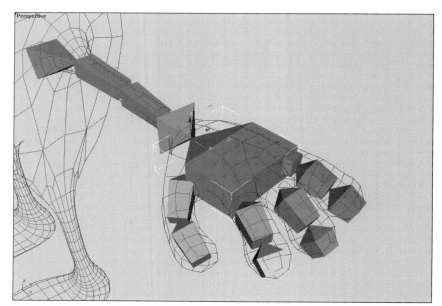

Figure 5.19 *All hand bones*

6. On the **Modify** panel, use the **Width**, **Height**, and **Taper** parameters to scale the bone up to the size of the hand. Approximate values used are:

   Width   7.0
   Height 20.0
   Taper   2.5%

7. On the Main Toolbar, click **Select and Non-uniform Scale** and set the **Reference Coordinate System** to **Local**.

8. Scale the bone along its X axis until it fills the hand.

Naming the Arm Bones

In this section, use **Arc Rotate** in the Perspective viewport to view and select the arm/hand bones.

1. Select each of the following bones and name them as follows. Use the figure that follows as a reference for renaming the bones.

   Bone - Clavicle
   Bone - Upper Arm
   Bone - Lower Arm
   Bone - Wrist
   Bone - Hand
   Bone - Thumb A, Bone - Thumb B
   Bone - Index A, Bone - Index B
   Bone - Middle A, Bone - Middle B
   Bone - Pinky A, Bone - Pinky B, Bone - Pinky C

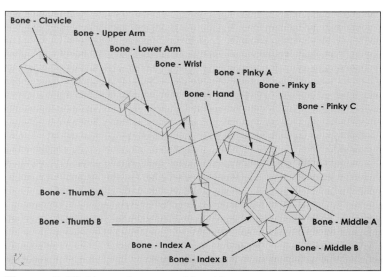

Figure 5.20 *Names for arm and hand bones*

2. Save the scene with the filename *AlienBones03.max*.

Completing the Body Skeleton

1. On the **Create** panel, click the **Systems** button. Click **Bones**.
2. In the Left viewport, place four bones in the alien, two in his body and two in his nose. Use the following illustration to see where to place them. Be sure to delete the tailbone when you have finished creating the chain of bones.

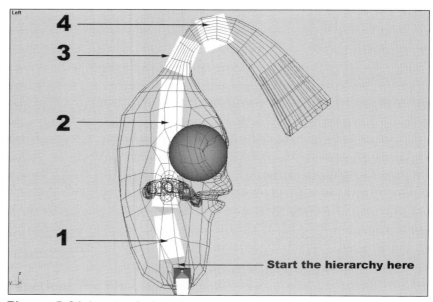

Figure 5.21 *Location for body bones*

3. In the Perspective view, name and adjust the size of the four new bones until they fit the geometry. Use the figure that follows for reference.

   Bone - Back A
   Width   65.0
   Height 51.5
   Taper   8.0%

   Bone - Back B
   Width   84.0
   Height 55.0
   Taper   12.0%

   Bone - Nose A
   Width   16.0
   Height 15.0
   Taper   30.0%

Bone - Nose B
Width   20.0
Height   19.0
Taper   17.0%

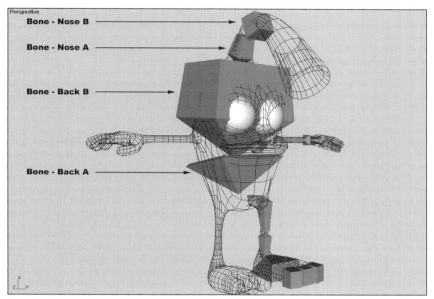

Figure 5.22 *Names and approximate sizes for body and nose bones*

4. Save the scene with the filename **AlienBones04.max**.

   Now that we have half the skeleton complete, we will take an efficient approach and completely set up the arm and leg before we duplicate and mirror them to the other side of the alien.

ALIEN CONTROL SYSTEM

One of the most efficient ways to animate a skeleton is with inverse kinematics (IK). IK uses a controller called a *goal* to maneuver the last object in the chain, which in turn controls the entire chain. This causes the skeleton's limbs to react to movements of hands and feet, much in the way a puppet does.

For example, a goal at the hand can be moved to have the entire arm bend accordingly. If you wanted the character to wave hello, you would simply pick up the hand and wave it around to have the arm follow along. Using this approach, we will be able to quickly animate the alien simply by positioning IK goals at keyframes.

IK SOLVERS

3ds max assigns a goal to a linked chain when an *IK solver* is applied to the chain. The IK solver is a utility that solves (figures out) how the bones should move and rotate when the goal is moved. IK chains are very easy to set up and control. An IK solver is set up from the top of the chain to the bottom, such as from the thigh to the foot.

3ds max has three different IK solvers available. We are going to use the most practical one, the HI Solver. This type of IK solver works with any type of IK chain regardless of its number of bones.

Setting up an IK solver creates an *IK chain* from the first to the last bone in the chain. The goal is positioned at the pivot point of the last bone in the chain.

POINT HELPERS

Although you can pick up a goal and move it around, working with IK chains is easier if you use helpers of some kind. You then link each goal to a helper, then move the helper around instead of the goal.

In our control system, each goal will be linked to a point helper. Later on, we will set up a user interface to make it easy and quick to select the appropriate point helper.

TUTORIAL 5.2 IK SOLVERS AND POINT HELPERS

In this tutorial, we will set up the alien skeleton so it can be controlled with point helpers and IK solvers. This will make the alien very easy to animate.

Setting up the Leg IK

1. Select the hip bone, named **Bone - Hip**.
2. Activate the Perspective view.
3. From the **Animation** menu, choose *IK Solvers/HI Solver*.

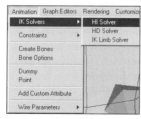

Figure 5.23 *Applying an HI Solver*

As you move your cursor around your Perspective view, you will see a dotted line leading back to the hip bone.

4. Click on **Bone - Ankle**, as shown in the next figure.

This action creates an IK chain running from Bone - Hip to Bone - Ankle. The goal appears as crosshairs at the Bone - Ankle pivot point.

The new IK chain is selected by default right after you create the chain.

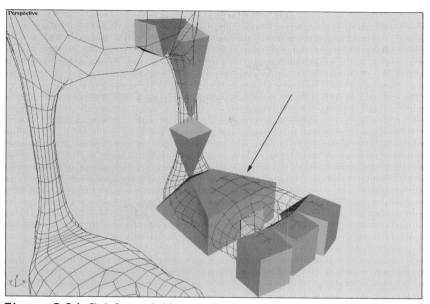

Figure 5.24 *Click Bone - Ankle to make the IK chain*

5. On the **Motion** panel, change the name of the IK chain to **IK - Ankle**.
6. Select **Bone - Ankle**. Using the same technique, create an IK chain with an HI Solver running to **Bone - Ball**, as shown in the next figure.

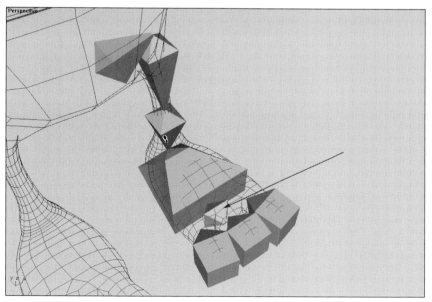

Figure 5.25 *IK chain from Bone - Ankle to Bone - Ball*

7. Name the new IK solver **IK - Ball**.

Creating a Point Helper

1. Set up snaps for **Pivot** only, and turn on snaps.

   This will enable you to create and place point helpers right at the bone pivot points, which is where the goals are located.

2. On the **Create** panel, click **Helpers**. Click **Point**.
3. Create a point helper at the location of the **IK - Ankle** goal. This will be at the pivot point of **Bone - Ankle**.
4. Name the helper **Point - Ankle Rotate**.
5. Create another point helper at the location of **IK - Ball**.
6. Name the helper **Point - Ball Rotate**.
7. Copy and move the helper over to the very tip of the middle toe. Name the new helper **Point - Toe Rotate**.

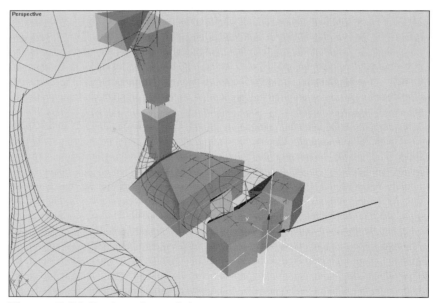

Figure 5.26 *Duplicated point helper*

8. Copy and move **Point - Toe Rotate** to the heel of the foot. Name the new helper **Point - Heel Rotate**.

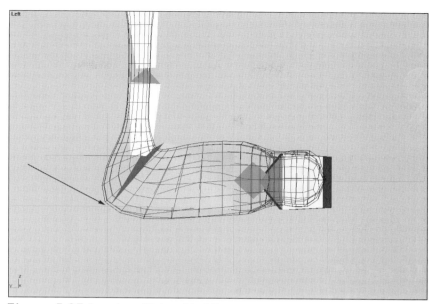

Figure 5.27 *Location of Point - Heel Rotate*

Pointing the Toes in the Right Direction

Next, we want to use a Look-At constraint to make the toe ball point at the toes.

1. Select **Bone - Take me out (ball)**.

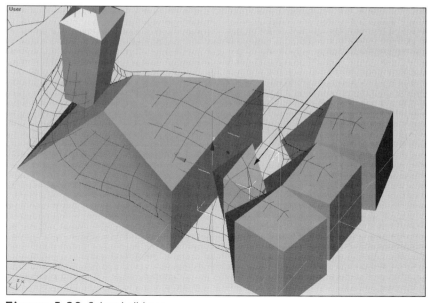

Figure 5.28 *Select ball bone*

2. From the menu, choose **Animation/Constraints/Look-At Constraint**.
3. In the Perspective viewport, click **Point - Toe Rotate**.

   This will cause the ball bone to always point toward the toe rotation point helper.

4. Turn off snaps.

Linking the Toes to the Ball

Now that we have created a series of point helpers that will serve as our main control points, we need to link everything together. By linking one object to another we essentially give the child object a new pivot point.

1. Select the three toe bones.

2. Click the **Select and Link** button on the Main Toolbar.
3. Click and drag from one of the selected toe bones to **Bone - Take me out (ball)**. The icon will change indicating that you are ready to link. Release the mouse and the ball bone will become the parent of the three toe bones.

4. Click the **Select** button on the Main Toolbar.

Finishing the Foot Links

Using the linking technique applied above, complete the following links:

1. Link **IK - Ankle** to **Point - Ball Rotate**.
2. Link **IK - Ball** to **Point - Toe Rotate**.
3. Link **Point - Ball Rotate** to **Point - Toe Rotate**.
4. Link **Point - Toe Rotate** to **Point - Heel Rotate**.
5. Link **Point - Heel Rotate** to **Point - Ankle Rotate**.

   If you like, you can open up a Schematic View to see if all the links are correctly made.

6. From the menu, choose **Graph Editors/Schematic View/Open Schematic View**.

Your Schematic View should look similar to the following figure. If it does not, you can use Unlink Selection to unlink some of the points, and try again.

> **TIP**
> *If you find it difficult to select a particular parent object with the mouse, you can easily select a parent object by clicking Select and Link then pressing <H> to select the parent from a list.*

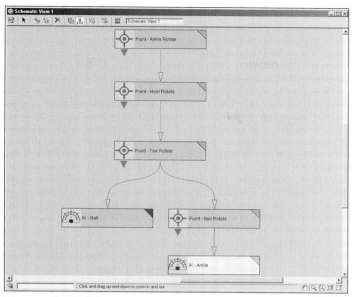

Figure 5.29 *Schematic View showing point helper and IK hierarchy*

7. Save the scene with the filename ***AlienBones05.max***.

Creating the Knee Target

1. Select **Point - Toe Rotate**. Copy and move it up so the new point is level with the knee.

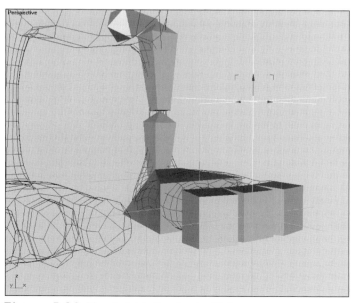

Figure 5.30 *New point helper level with knee*

2. Name the new point helper **Point - Knee Target**.
3. Select **IK - Ankle**.
4. Go to the **Motion** panel.

5. On the IK Solver Properties rollout, click **Pick Target**, then pick **Point - Knee Target**.

   This will cause the ankle to swivel (rotate) with the knee when the knee is animated.

6. Select **Point - Knee Target** and link it to **Point - Toe Rotate**.

Freezing the IK Solvers

1. If the **Select and Link** button is still active, click **Select object** on the Main Toolbar.
2. Press the **<H>** key to open the Select by Name dialog. Select **IK - Ankle** and **IK - Ball**.

3. Go to the **Display** panel.
4. Click **Freeze Selected** to freeze the selection.

   This will prevent you from accidentally selecting these points in the future.

5. Save the scene with the filename *AlienBones06.max*.

Testing the Leg IK Setup

Now that our IK setup is complete for one leg, let's test it.

1. Select **Point - Ankle Rotate**. In the Perspective view, move the point object on the YZ plane to see how it affects the rest of the leg. The knee should bend when you move the point upward. Undo the move when you have finished testing.

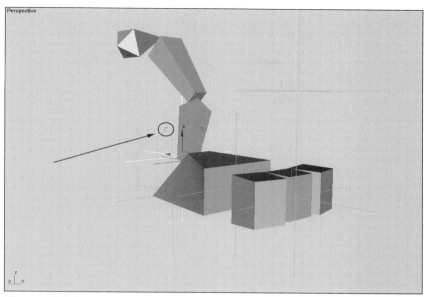

Figure 5.31 *Moving leg by moving Point - Ankle Rotate*

2. To test the rotation of the heel, select **Point - Heel Rotate**. In the Perspective view, rotate it around its local X axis to make the foot rotate. Undo the rotation when you have finished.

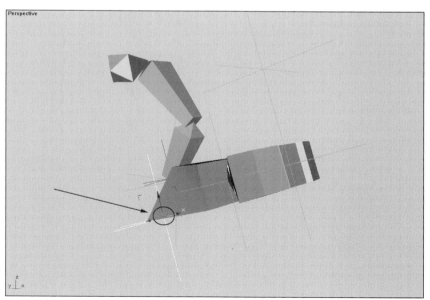

Figure 5.32 *Rotating around Point - Heel Rotate*

3. To test the rotation around the ball of the foot, select **Point - Ball Rotate** and rotate it around its local X axis. Undo the rotation when finished.

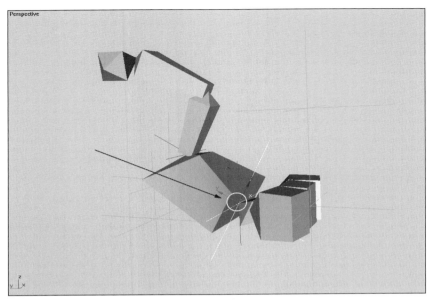

Figure 5.33 *Rotating around Point - Ball Rotate*

4. To test the rotation on all the toes, select **Point - Toe Rotate**. In the Perspective view, rotate it around its local X axis. Undo to reset it to its original rotation.

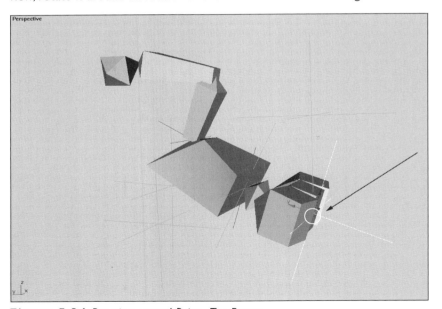

Figure 5.34 *Rotating around Point - Toe Rotate*

5. To test the toes individually, select any of the three toes. With the **Reference Coordinate System** set to **Local**, rotate its around its Y axis to control its wiggle. Undo the rotation. Repeat for the other two toes.

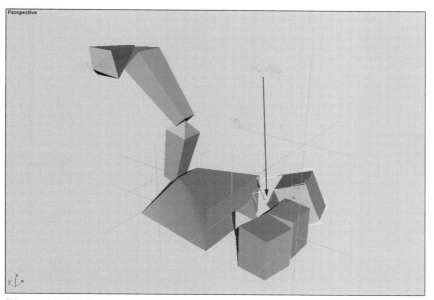

Figure 5.35 *Rotating toes one on one axis*

6. To test the ankle rotation, select **Point - Ankle Rotate**. With the **Reference Coordinate System** set to **Local**, rotate it around its Y axis. Undo the rotation.

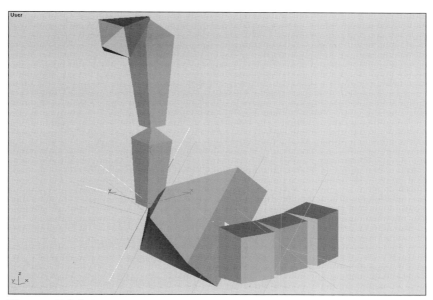

Figure 5.36 *Rotating ankle point helper around Y axis*

Bit of a problem here -- the toes aren't rotating with the rest of the foot! Let's fix this up to work correctly.

Making the Toes Rotate with the Foot

1. Select **Bone - Take me out (ball)**.
2. Go to the **Motion** panel.
3. Click the **Rotation** button located on the PRS Parameters rollout.
4. On the Look-At Constraint rollout, locate the Select Upnode section. In this section, uncheck the **World** checkbox. The button next to it will become active.
5. Click the button (currently labeled **None**) next to the World checkbox and select **Point - Ankle Rotate** in the Perspective viewport.
6. Select **Point - Ankle Rotate** and test its rotation around the Y axis again.

   The toes now rotate along with the foot.

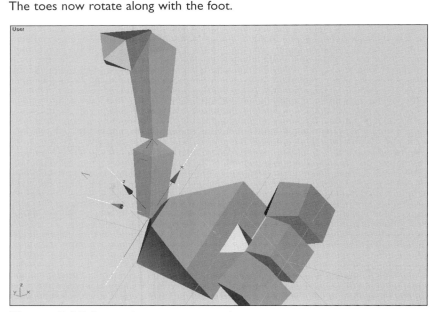

Figure 5.37 *Foot and toes rotate properly*

Adding IK to the Arm

1. In the Perspective viewport, select **Bone - Upper Arm**.
2. On the *Animation* menu, choose *IK Solvers/HI Solver*. Click on **Bone - Wrist**. Name the new chain **IK Chain Wrist**.

3. Verify that snaps are still set to **Pivot** only, and turn on snaps.
4. On the **Create** panel, click **Helpers**. Click **Point**.
5. Place a point helper at the pivot point for **Bone - Lower Arm**.

6. Turn off snaps.

7. Move the **Point** back on the Y axis just behind the elbow.

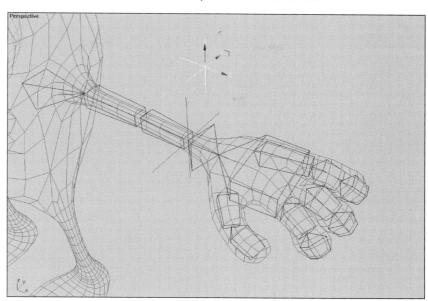

Figure 5.38 *New point helper moved directly behind elbow region*

8. Name the new point helper **Point Elbow Target**.

Making the Wrist Swivel with the Elbow

1. Select **IK Chain Wrist**.

2. Go to the **Motion** panel.
3. Click the **Pick Target** button.
4. Click on **Point Elbow Target** in the Perspective viewport.

   Now the wrist will always swivel (rotate) when the elbow swivels.

Linking the Fingers to the Hand

1. Click **Select and Link**.
2. Link **Bone - Pinky A** to **Bone - Hand**.
3. Link **Bone - Middle A** to **Bone - Hand**.
4. Link **Bone - Index A** to **Bone - Hand**.
5. Link **Bone - Thumb A** to **Bone - Hand**.

6. Click **Select and Move** to turn off the **Select and Link** button.
7. Save the scene with the filename *AlienBones07.max*.

Setting up the Hand Control Point

1. Activate the Perspective view.

2. Make sure snaps are set to **Pivot** only, and turn on snaps.

3. On the **Create** panel, click **Helpers**, then **Point**. Place a point helper at the **Bone - Wrist** pivot point, and name the new point helper **Point Hand Controls**.

4. Turn off snaps.

5. Select **IK Chain Wrist**.

6. Click the **Select and Link** button. Press **<H>** on the keyboard, select **Point Hand Controls** and click **Link**.

    Now we will be able to move the arm around by moving the **Point Hand Controls helper**.

7. Select and freeze the **IK Chain Wrist**.

Testing the Arm IK Setup

1. Select **Point Hand Controls**.

2. In the Perspective view, move the selected point helper on the XY plane.

Figure 5.39 *Arm/hand moved by moving Point Hand Controls helper*

3. Select the **Point Elbow Target**.

4. Move the helper up and down and watch how the arm responds.

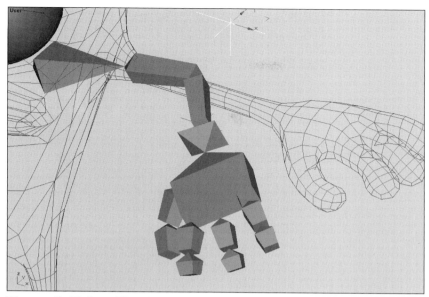

Figure 5.40 *Point Elbow Target moved up on Y axis*

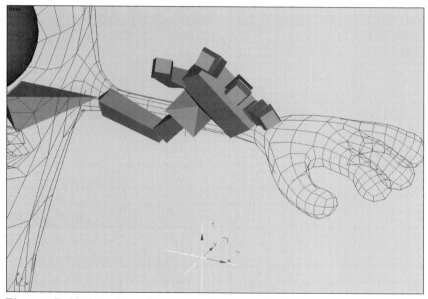

Figure 5.41 *Point Elbow Target moved down on Y axis*

5. Undo all of the operations until the arm is back to its original position.
6. Save the scene with the filename **AlienBones08.max**.

   The point helpers and IK solvers are all set up for the character. Go ahead and play with the setup a little bit to get a feel for how it works. Just be sure to revert to your saved version before continuing.

TUTORIAL 5.3 HAND CONTROL SYSTEM

Now that the IK chains and point helpers have been set up, we can create a custom control system for the hands. This control system will be used later on to animate the alien's arms, hands and individual fingers.

Perhaps you are accustomed to animating objects in **3ds max** by turning on the **Animate** button and transforming (moving, rotating, scaling) the objects. There is another way to set up animation, with *animation controllers*.

In **3ds max**, each object has three *tracks* for transformation, aptly named Position, Rotation and Scale. By default, each track is assigned a controller that allows you to animate it by turning on the **Animate** button and transforming the object. However, you can assign different controllers to one, two or all three tracks to control that aspect of the object's transformation with the controller.

In our case, we will assign a controller that can then be controlled by other parameters, making it easy for you to animate the alien.

We will start by adding *custom attributes* to the main control point for the hand.

Adding Custom Attributes to the Hand Control

1. With **Point Hand Controls** selected, choose *Animation/Add Custom Attribute* from the menu. The Add Parameter dialog appears.

Figure 5.42 *Add Parameter dialog*

The Add Parameter dialog allows you to create custom parameters or attributes. These parameters themselves don't do anything, but you can connect them to other real values in the scene and manipulate the real values using the custom parameters. These parameters will be used in conjunction with animation controllers later on.

2. On the Add Parameter dialog, add the attributes listed as follows.

Each attribute will have the following parameters set the same way:

Parameter Type	Float
UI Type	Spinner
Width	160
Default (Range)	0
Alignment	Center
Add Attribute to Type	Object's Base Level

Add the following attributes using the parameter settings above for each one. Click the **Add** button after setting each Name/Range pair to create the attribute.

Name Index Knuckle
Range 0 - 10

Name Index Middle
Range 0 - 10

Name Middle Knuckle
Range 0 - 10

Name Middle Middle
Range 0 - 10

Name Pinky Knuckle
Range 0 - 10

Name Pinky Middle
Range 0 - 10

Name Pinky Cup
Range 0 - 10

Name Thumb Base
Range 0 - 10

Name Thumb Spread
Range -5 to 5

Name Thumb Middle
Range 0 - 10

Name Wrist Up Down
Range -10 to 10

Name Wrist Side to Side
Range -10 to 10

Name Wrist Rotation
Range -10 to 10

Name Finger Spread
Range -5 to 10

3. Close the Add Parameter dialog.

Setting up the Thumb Controller

Now that all our custom attributes are created, we'll set up a controller for the thumb rotation.

1. In the Perspective viewport, select **Bone - Thumb A**.

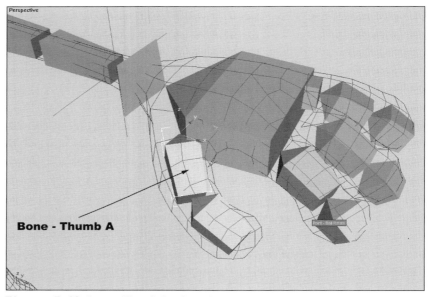

Figure 5.43 *Bone - Thumb A selected*

2. Go to the **Motion** panel.
3. Expand the Assign Controller rollout.

    Here is where you assign animation controllers to specific tracks.

4. Highlight the **Rotation** track.

    The default controller for the Rotation track is TCB Rotation. We will change this controller to a different one that can be more easily controlled with our system.

5. Click the **Assign Controller** button at the top of the Assign Controller rollout.

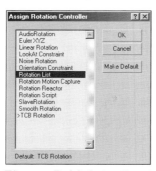

Figure 5.44 *Assign Rotation Controller dialog*

6. On the Assign Rotation Controller dialog, select **Rotation List** and click **OK**.

   The Rotation List controller isn't really a controller on its own. Its main purpose is to set up two or more slots so you can assign two or more controllers to the same track.

   Note that the default controller **TCB Rotation** has been retained, but there is now an additional subtrack called **Available**.

7. Expand the **Rotation List** controller and highlight the **Available** subtrack.

8. Click the **Assign Controller** button.

9. Select **Euler XYZ** as the controller and click **OK**.

   The Euler XYZ controller has three subtracks, one for each of the X, Y and Z axes. This type of controller allows you to control one, two or all three axes individually. Compare with the default TCB Rotation controller, which does not give you control over individual axes. This is why we are using the Euler XYZ controller instead.

Wiring the Thumb with a Reactor

Now that we have our chosen controller in place, we can *wire* the thumb's Y rotation to another value, which we will use to control it later on. Wiring is simply the process of connecting two parameters together so changing one will change the other. One way to accomplish this is to use a Reactor controller, which causes a parameter to react when another parameter is changed.

1. Click **Select and Rotate** on the Main Toolbar.

2. Verify that the **Reference Coordinate System** is set to **Local**.

   We want to wire the **Thumb Base** custom attribute to the rotation of **Bone - Thumb A** around its local Y axis.

3. In the Assign Controller rollout, expand the **Euler XYZ** track and highlight the **Y Rotation** subtrack.

4. Click the **Assign Controller** button.

5. Select a **Float Reactor** controller from the Assign Rotation Controller dialog and click **OK**.

   The Reactor Parameters dialog appears.

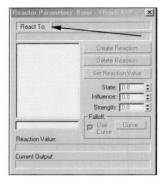

Figure 5.45 *Reactor Parameters dialog*

The Reactor controller causes a parameter to change as a reaction to another parameter or value within the scene. In this case, we will make the thumb's Y rotation react to one of the custom attributes we created earlier, Thumb Base.

6. In the Reactor Parameters dialog, click the **React To** button. Click on the **Point Hand Controls** object. On the pop-up menu that appears, select *Object (Point Helper)/ Custom_Attributes/ Thumb Base*.

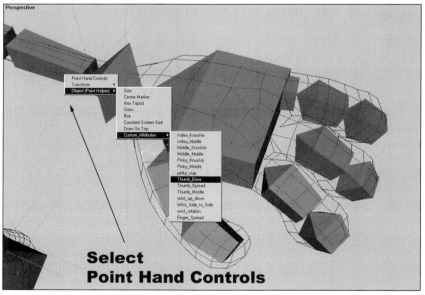

Figure 5.46 *Assigning Thumb Base custom attribute as parameter for reaction*

Establishing the Reaction Relationship

So far, we have set up a reaction relationship between the Y rotation of the thumb base and the custom attribute Thumb Base. We still need to tell the Y rotation track what to do when Thumb Base changes, and when to do it.

Setting up the following relationship parameters is critical to making our custom setup work.

1. Select **Point Hand Controls**.
2. On the **Modify** panel, locate the custom attribute **Thumb Base**. Set the **Thumb Base** attribute to 10.
3. Click the **Create Reaction** button on the Reactor Parameters dialog.
4. Adjust the **State** in a positive direction until the thumb is at an extreme position that you are happy with, such as 1.0.

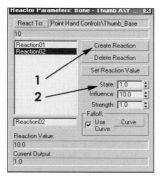

Figure 5.47 *Create Reaction, then adjust State value to establish best relationship*

5. Click the **Set Reaction Value** button to record the relationship.

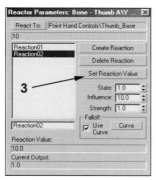

Figure 5.48 *After finding State that works best, click Set Reaction button*

6. Close the **Reactor Parameters dialog**.

You have just created a reactive scenario where **Bone - Thumb A** rotates around its local Y axis when the **Thumb Base** custom attribute is changed.

When you first assigned the **Point Hand Controls/Thumb Base** as the controlling parameter, the initial values of **Thumb Base** and the current rotation value of **Bone - Thumb A** were recorded into **Reaction01**. You then added a second reaction so you would have a minimum - maximum relationship.

7. To test the reaction, use the spinners to adjust the value of **Thumb Base**. You should see the thumb bone react appropriately. Set the value back to 0.

Wiring the Thumb Spread

You will now use another Reactor controller to control the thumb's Z rotation. This axis of rotation controls how far the thumb rotates away from the hand. We call this the *thumb spread*. You will use the same technique you used earlier to assign the Reactor controller to the Y axis rotation.

1. Select **Bone - Thumb A** and go to the **Motion** panel.
2. On the Assign Controller rollout, highlight the **Z Rotation** subtrack under the **Euler XYZ** track.
3. Click **Assign Controller** and assign a **Float Reactor** controller to this track.
4. Click the **React To** button and select **Point Hand Controls**. From the pop-up menu, select *Object (Point Helper)/Custom Attributes/Thumb Spread*.
5. Select **Point Hand Controls** and go to the **Modify** panel. Set the **Thumb Spread** parameter to -5.
6. On the Reactor Parameters dialog, click **Create Reaction**. Adjust the **State** to approximately -0.75, and click **Set Reaction Value**.
7. Set **Thumb Spread** to 5.0. Click **Create Reaction**, adjust **State** to approximately 0.5, and click **Set Reaction Value**.
8. Close the Reactor Parameters dialog.
9. Test the reaction by adjusting **Thumb Spread** to different values. When you have finished, set **Thumb Spread** back to 0.

Wiring the Remaining Finger Controllers

The custom attribute Finger Spread will cause the fingers to spread in toward one another or out from one another. We will connect each finger one at a time.

Now that you know how to set up tracks and assign a Reactor controller, you will be given basic instructions and parameters, but not every instruction in detail. If necessary, refer back to the last four sections for details on these procedures.

For each of these bones, you will set up a **Rotation List** controller for the **Rotation** track, then click **Available** and choose **Euler XYZ**, then expand the **Euler XYZ** controller and highlight one of the subtracks for a specific axis. Assign the **Float Reactor** controller to the specified track.

Then you will click React To and choose **Point Hand Controls,** then select *Object (Point Helper)/Custom Attributes/[custom attribute]* from the pop-up menu.

Next, you will set the custom attribute value on the **Modify** panel, click **Create Reaction** on the Reactor Parameters dialog, set the **State** spinner and click **Set Reaction Value**.

After you have set up the reaction, close the Reactor Parameters dialog and change the custom attribute parameter on the **Modify** panel to test the reaction relationship. When testing is complete, change the custom attribute back to 0 before continuing.

1. Assign the **Rotation List** and **Euler XYZ** controllers for each bone listed below, then assign a **Float Reactor** controller to the **Z Rotation** track using these values and the **Finger Spread** custom attribute.

    Where two reactions are created for the same bone, you can simply set the **Modify** panel value and click **Create Reaction** again for each one.

Bone Name	Modify panel value	State value
Bone - Index A	-5.0	0.5
Bone - Index A	10	-0.55
Bone - Middle A	-5.0	0.18
Bone - Middle A	10	-0.25
Bone - Pinky B	-5.0	-0.22
Bone - Pinky B	10	0.44

    Next you will set up the controllers for the finger curls. To give you better control over the finger curling action, we have broken the curl into two places, one at the knuckle and one at the middle of the finger.

    In the following step we will set up controllers for the custom attributes Index Knuckle, Middle Knuckle, and Pinky Knuckle so they control the right bones.

    There is no need to set up the **Rotation List** and **Euler XYZ** controllers for these bones as they have already been set up. We will use the controllers that we put in place in the previous step.

2. Set up the **Float Reactor** controller for the **Y Rotation** track using these values and the specified custom attribute.

Bone Name	Custom Attribute	Modify panel value	State value
Bone - Index A	Index_Knuckle	10	1.0
Bone - Middle A	Middle_Knuckle	10	1.0
Bone - Pinky B	Pinky_Knuckle	10	1.0

    The middle finger bones have never had controllers assigned to them, so we will need to assign the **Rotation List** and **Euler XYZ** controllers before moving into the Float Reactor controllers.

3. Make sure the following bones have been assigned the **Rotation List** and **Euler XYZ** controllers, then assign a **Float Reactor** controller to the specified track using these values and the specified custom attribute.

Bone Name	Rotation Track	Custom Attribute	Modify panel value	State value
Bone - Index B	Y	Index_Middle	10	1.0
Bone - Middle B	Y	Middle_Middle	10	1.0
Bone - Pinky C	Y	Pinky_Middle	10	1.0
Bone - Pinky A	X	Pinky_Cup	10	-0.6
Bone - Thumb B	Z	Thumb_Middle	10	1.0

☝ **TIP** ☝
Like us, the alien has the ability to rotate his pinky a little bit on the X axis to cup it into his palm. The Pinky_Cup custom attribute controls this rotation.

Wiring the Wrist Controllers

Here you will use the same techniques to set up the wrist controllers.

1. Assign the **Rotation List** and **Euler XYZ** controllers for each bone listed below, then assign a **Float Reactor** controller to the specified track using these values and the specified custom attribute. This will set the wiring for rotating the wrist up and down.

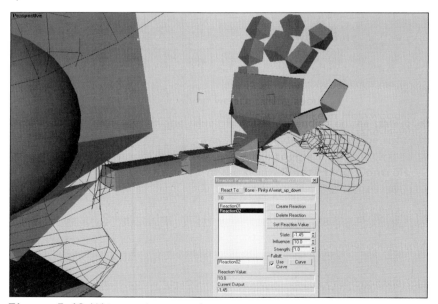

Figure 5.49 *Wrist rotation upward*

Bone Name	Rotation Track	Custom Attribute	Modify panel value	State value
Bone - Hand	Y	Wrist_up_down	10	-1.45
(Set **Modify** value, click **Create Reaction** again)			-10	1.45

2. Assign the **Rotation List** and **Euler XYZ** controllers for each bone listed below, then assign a **Float Reactor** controller to the specified track using these values and the specified custom attribute. This will set the wiring for the remaining wrist controls.

Bone Name	Rotation Track	Custom Attribute	Modify panel value	State value
Bone - Hand	Z	Wrist_Side_to_Side	10	0.7
(Set **Modify** value, click **Create Reaction** again)			-10	-0.7
Bone - Wrist	X	Wrist_rotation	10	1.25
(Set **Modify** value, click **Create Reaction** again)			-10	-2.25

   The hand wiring setup is now complete.

3. Save the scene with the filename *AlienBones09.max*.

4. Feel free to play with the new wiring system to rotate the hand, thumb and fingers. When you have finished experimenting, reload the scene.

TUTORIAL 5.4 FOOT CONTROL SYSTEM

As you've seen with the arm, it's easy to set up control systems so we can simply adjust a custom attribute and in turn affect numerous other objects.

With the arm, we placed all of the control on Point Hand Controls object. Seeing as we will do most of the leg animation through the Point - Ankle Rotate helper, it would only make sense to add our custom attributes to this object.

As with the hand control system, the first order of business is to set up some custom attributes.

Adding Custom Attributes to Point-Ankle Rotate

1. Select **Point - Ankle Rotate**.

2. Select *Animation/Add Custom Attribute* from the menu.

3. On the Add Parameter dialog, add attributes listed below. Each attribute will have the following parameters set the same way:

Parameter Type	Float
UI Type	Spinner
Width	160
Default (Range)	0
Alignment	Center
Add Attribute to Type	Object's Base Level

   Add the following attributes using the parameter settings above for each one. Click the **Add** button after setting each Name/Range pair to create the attribute.

   Name   footRoll
   Range   -5 to 10

Name bigToe
Range -5 to 5

Name middleToe
Range -5 to 5

Name littleToe
Range -5 to 5

4. Close the Add Parameter dialog.

Assigning the Heel Rotation Controllers

By now you are an old hand at setting up controllers.

The heel rotation will be set up slightly differently from the hand and finger controls. Here you will assign two Euler XYZ controllers to the heel so you will still be able to use the Select and Rotate transform to rotate the heel with one of the controllers while using the Reactor controller on the other.

1. Select **Point - Heel Rotate**.
2. On the **Motion** panel, assign a **Rotation List** controller to its **Rotation** track.
3. Expand the **Rotation List** track and select the **TCB Rotation** track.
4. Replace this controller with an **Euler XYZ** controller.
5. Assign another **Euler XYZ** controller to the track labeled **Available**.

Setting up the Heel Reactor Controller

1. Expand the second **Euler XYZ** controller.
2. Assign a **Float Reactor** controller to the **X Rotation** subtrack.
3. On the Reactor Parameters dialog, click the **React To** button and select **Point - Ankle Rotate**.
4. From the pop-up menu select *Object (point helper)/Custom Attributes/footRoll*.
5. Select **Point - Ankle Rotate**. On the **Modify** panel, set **footRoll** to -5.
6. In the *Reactor Parameters* dialog click the **Create Reaction** button.
7. Set the **State** spinner to -0.4 and click the **Set Reaction** button.
8. Close the Reactor Parameters dialog.

9. Test the setting by changing the **footRoll** parameter on the **Modify** panel. Reset the parameter to 0 when you have finished.

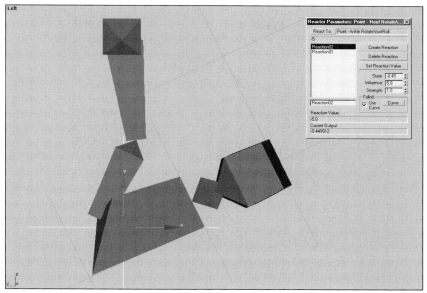

Figure 5.50 *Rotation of foot with footRoll set to -5*

Setting up the Remaining Foot Controllers

The remaining foot controllers will be set up in the same way, with two Euler XYZ controllers on the Rotation track. You will assign the Float Reactor controller to the second Euler XYZ controller in each case.

1. Set up the second **Euler XYZ** controller for the **Point - Ball Rotate** object as follows:

Bone Name	Rotation Track	Custom Attribute	Modify panel value	State value
Point - Ball Rotate	X	footRoll	5	0.45

2. When you have finished, leave the **footRoll** value at 5.

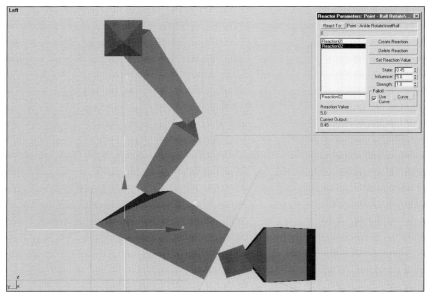

Figure 5.51 *Rotation of foot with footRoll set to 5*

3. Set up the second **Euler XYZ** controller for the **Point - Toe Rotate** object:

Bone Name	Rotation Track	Custom Attribute	Modify panel value	State value
Point - Toe Rotate	X	footRoll	10	0.45

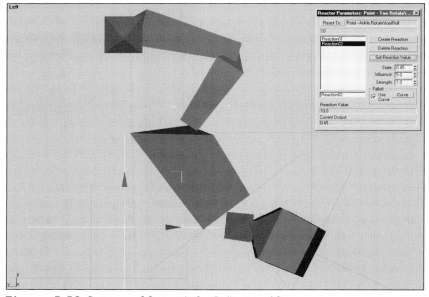

Figure 5.52 *Rotation of foot with footRoll set to 10*

4. Set up the remaining controllers as follows:

Bone Name	Rotation Track	Custom Attribute	Modify panel value	State value
Bone - Toe (Inner)	Y	bigToe	-5	0.4
			5	-0.4
Bone - Toe (Middle)	Y	middleToe	-5	0.4
			5	-0.4
Bone - Toe (Outer)	Y	littleToe	-5	0.4
			5	-0.4

The leg setup is now complete.

5. Save the scene with the filename *AlienBones10.max*.

TUTORIAL 5.5 MIRRORING AND COMPLETING THE SKELETON

Now that the limbs are completely set up, we want to mirror our setup to the other side of the alien body. The controller setups will need to be adjusted after mirroring, and the new object names will need to be changed.

Mirroring the Leg

1. Unfreeze **IK-Ankle** and **IK-Ball**.
2. In the Front viewport, use a marquee to select all of the leg and foot assembly.

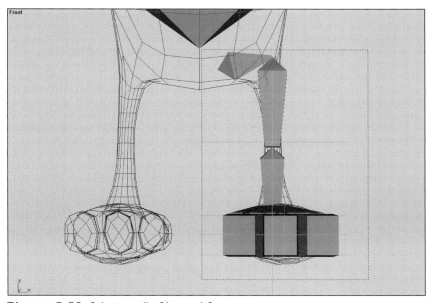

Figure 5.53 *Selecting all of leg and foot*

3. With the Front viewport active, click **Mirror Selected Objects** on the Main Toolbar.

4. On the Mirror: Screen Coordinates dialog, set the following settings:

   Mirror Axis   X
   Clone Selection   Copy
   Mirror IK Limits   Checked

   Adjust the **Offset** until the new duplicated leg lines up with the leg geometry (approximately -52.5).

   Click **OK**.

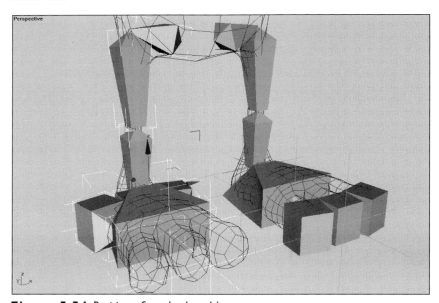

Figure 5.54 *Position of newly cloned leg*

Correcting the Mirrored Toes

After you have positioned the new leg you may have noticed a problem with the toes. This is caused by the Look-At Constraint that was applied to the ball bone.

1. Select **Bone - Take me out (ball)01**.
2. Select the **Motion** panel.
3. Click the **Rotation** button located under the PRS Parameters rollout.

4. On the Look-At Constraint rollout, locate **Select Look-At Axis**. Check the **Flip** checkbox.

   This flipped the toes, but they are still backward.

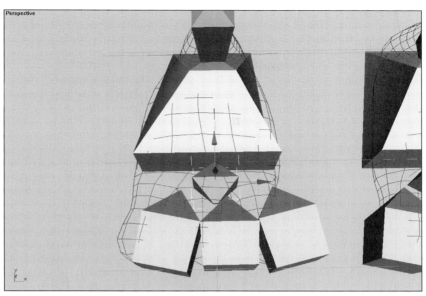

Figure 5.55 *Toes flipped in wrong direction*

5. Under the Source Axis section, check the **Flip** checkbox.

The toes are now positioned correctly.

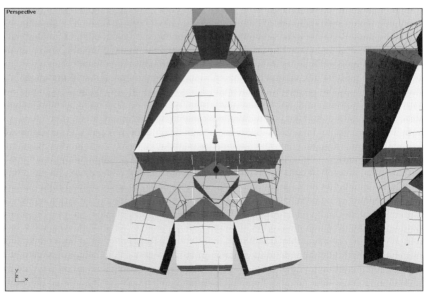

Figure 5.56 *Toes are now correct*

At this point the custom attributes on the leg do not work properly. All of the Float Reactor controllers need to be set up again. This will be done shortly.

Mirroring the Arm

1. On the **Display** panel, unfreeze **IK Chain Wrist**.
2. In the Front viewport, use a marquee to select all the arm and hand bones.

3. Use **Mirror Selected Objects** to mirror the arm in the same way you mirrored the leg.

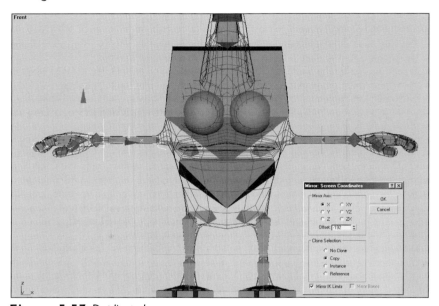

Figure 5.57 *Duplicated arm*

4. While all of the new objects for the arm are still selected, right-click in the Perspective viewport to activate it without losing your selection, and verify that the arm and fingers are positioned properly. If the fingers are not aligned properly, simply move the entire selection along the X axis until it fits properly in the alien model.
5. Save the scene with the filename *AlienBones11.max*.

Setting up the New Arm and Leg

When an object is copied in **3ds max**, all custom attributes and controllers are also copied. Unfortunately, all of the reaction relationships that we've set up were not duplicated. We are going to have to reassign all of the reaction relationships.

At times, **3ds max** will get confused when you first duplicate an object with custom attributes that are involved in a reaction scenario. A work around is to restart **3ds max** and reopen your scene.

1. Be sure to save the scene, then close and reopen **3ds max**. Reload the scene.

Now we will need to reestablish the Reactor connections. Look back at the previous tutorial if you have trouble with these steps.

> **TIP**
> When the arm and leg were mirrored, all objects were named with the original names plus the suffix 01. You will rename these objects later on to specify the appropriate side of the body.

2. For each of the objects listed below, go to the **Modify** panel and make sure all custom attributes are set to 0. Then go to the **Motion** panel and set up the Float Reactor connections again.

   Use the Reactor settings from the previous tutorial, starting with the section titled ***Wiring the Thumb with a Reactor*** on page 367. For each object, the Euler XYZ controllers will be in place, but the Float Reactors will need to be set up again for all the new objects.

   Bone - Thumb A01
   Bone - Thumb B01
   Bone - Index A01
   Bone - Middle A01
   Bone - Pinky B01
   Bone - Index A01
   Bone - Middle A01
   Bone - Pinky B01
   Bone - Index B01
   Bone - Middle B01
   Bone - Pinky C01
   Bone - Pinky A01
   Bone - Hand01
   Bone - Wrist01
   Point - Heel Rotate01
   Point - Ball Rotate01
   Point - Toe Rotate01
   Bone - Toe (Inner)01
   Bone - Toe (Middle)01
   Bone - Toe (Outer)01

3. Save the scene with the filename *AlienBones12.max*.

Creating the Root Point Helper

With the arms and legs competed, it's time to link all of the various pieces together. For the root of our hierarchy, we are going to use another point helper.

1. Verify snaps are set to **Pivot** only. Turn snaps on.

2. On the **Create** panel, click **Helpers**. Click **Point**.

3. In the Front viewport, click on **Bone - Back A**.

   This will place a point helper at the rotation point of this bone.

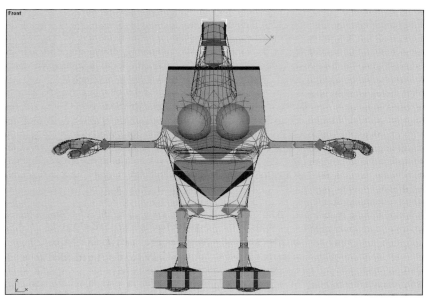

Figure 5.58 *Placing a point helper at Bone - Back A pivot point*

 4. Turn off snaps.

5. On the **Modify** panel, change the point helper's name to: **Point Alien Root**.

6. On the Parameters rollout, set the **Size** to: 75.

We have made the point helper large enough so that it protrudes through the alien's body. This will make selecting the point object much easier, which will result in a faster animation process.

 7. Click **Select and Move**.

8. In the Front viewport, move **Point Alien Root** down on the Y axis so that is even with the pelvis bone.

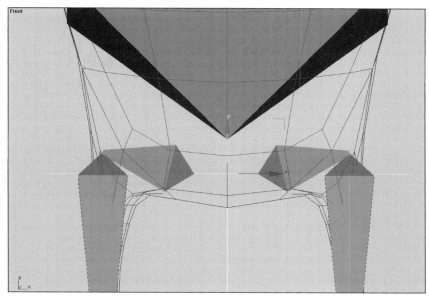

Figure 5.59 *Location of root point helper*

Linking the Pieces

 1. On the **Display** panel, click **Unfreeze All**.

 2. Use **Select and Link** to link the following objects:

   Bone - Clavicle to **Bone - Back B**
   Bone - Clavicle01 to **Bone - Back B**
   Left - Eye to **Bone - Back B**
   Right - Eye to **Bone - Back B**

   In Schematic View, your new hierarchy should look similar to the following figure.

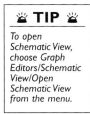

> ☀ **TIP** ☀
> To open
> Schematic View,
> choose Graph
> Editors/Schematic
> View/Open
> Schematic View
> from the menu.

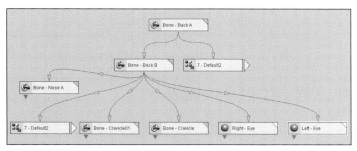

Figure 5.60 *Schematic View of current hierarchy*

3. Link the following objects:

   Bone - Back A to **Point Alien Root**
   Bone - Pelvis01 to **Point Alien Root**
   Bone - Pelvis to **Point Alien Root**

Testing the Setup

1. Select **Point Alien Root**.
2. In the Perspective view, move the helper up and down on the Z axis.

   This should cause the alien body to move up and down while the hands and feet stay still. If this happens the alien skeleton is set up properly.

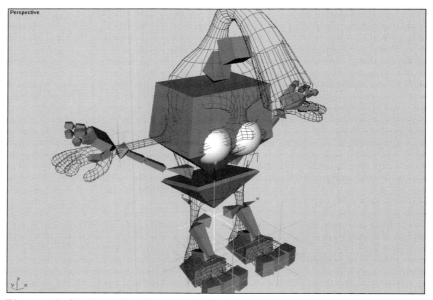

Figure 5.61 *Point Alien Root moved down on Z axis*

3. Undo any changes so that **Point Alien Root** is back to its default position.
4. Save the scene with the filename *AlienBones13.max*.

Giving the Objects Proper Names

Now that we have duplicated all of the bones and control systems and linked everything together, we need to rename our objects. The purpose of renaming the objects is so you can easily distinguish between body parts from the left and right sides.

So far you have been naming your objects with the appropriate names except for a left or right indicator, so you don't have a lot of work ahead of you. This process will take you no more than ten minutes and should not be skipped!

You would be surprised how many people fall into bad habits of not properly naming their objects. If you skip this step, you will run into trouble later on when you can't find objects or get confused and animate the wrong side of the alien.

We have left this task for last as it is often easier to rename all the objects after everything has been created and duplicated. Any name changes will ripple through controllers and custom attributes so the correct names will show up in all parts of **3ds max**.

When renaming a number of objects, the easiest method is to use Schematic View. In Schematic view, you will need to expand some of the objects to see their linked child objects. Next, click each object icon located to the left of the object's name to select that object, and then rename it on the **Modify** panel.

Go through all of your objects and identify the ones that are left side objects (those without the 01 suffix) and add the prefix **L -** to each one. For right side objects (those with the suffix 01), add the prefix **R -** and remove the 01 from the end of the object name. See the list below to clarify the object naming conventions.

When you are finished, you should have the following named objects in your scene:

Alien Body
Bone - Back A
Bone - Back B
Bone - Nose A
Bone - Nose B
L - Bone - Ankle and **R - Bone - Ankle**
L - Bone Clavicle and **R - Bone Clavicle**
L - Bone - Hand and **R - Bone - Hand**
L - Bone - Hip and **R - Bone - Hip**
L - Bone - Index A and **R - Bone - Index A**
L - Bone - Index B and **R - Bone - Index B**
L - Bone - Knee and **R - Bone - Knee**
L - Bone - Lower Arm and **R - Bone - Lower Arm**
L - Bone - Upper Arm and **R - Bone - Upper Arm**
L - Bone - Middle A and **R - Bone - Middle A**
L - Bone - Middle B and **R - Bone - Middle B**
L - Bone - Pelvis and **R - Bone - Pelvis**
L - Bone - Pinky A and **R - Bone - Pinky A**
L - Bone - Pinky B and **R - Bone - Pinky B**
L - Bone - Pinky C and **R - Bone - Pinky C**
L - Bone - Take me out (ball) and **R - Bone - Take me out**
L - Bone - Thumb A and **R - Bone - Thumb A**
L - Bone - Thumb B and **R - Bone - Thumb B**

L - Bone - Toe (Inner) and **R - Bone - Toe (Inner)**
L - Bone - Toe (Middle) and **R - Bone - Toe (Middle)**
L - Bone - Toe (Outer) and **R - Bone - Toe (Outer)**
L - Bone - Wrist and **R - Bone - Wrist**
L - IK - Ankle and **R - IK - Ankle**
L - IK - Ball and **R - IK - Ball**
L - IK Chain Wrist and **R - IK Chain Wrist**
L - Point - Ankle Rotate and **R - Point - Ankle Rotate**
L - Point - Ball Rotate and **R - Point - Ball Rotate**
L - Point - Heel Rotate and **R - Point - Heel Rotate**
L - Point - Knee Target and **R - Point - Knee Target**
L - Point - Toe Rotate and **R - Point - Toe Rotate**
L - Point Elbow Target and **R - Point Elbow Target**
L - Point Hand Controls and **R - Point Hand Controls**
Left - Eye and **Right - Eye**
Point Alien Root

Hiding What Is Not Used

You are almost ready to start skinning the alien, but first, there is just one more thing that we need to do: hide the objects we will not be using. We have used several point helpers in the scene along with a few IK chains to indirectly control various bones. We need to hide everything that will not be directly used for animation.

1. Press **<H>** on the keyboard to open the Select by Name dialog.

2. Select the following objects:

   L - Point - Ball Rotate
   R - Point - Ball Rotate
   L - Point - Heel Rotate
   R - Point - Heel Rotate
   L - Point - Toe Rotate
   R - Point - Toe Rotate
   L - IK - Ankle
   R - IK - Ankle
   L - IK - Ball
   R - IK - Ball
   L - IK Chain Wrist
   R - IK Chain Wrist

3. On the **Display** panel, click the **Hide Selected** button.

Placing the Bones in a Selection Set

Creating a named selection set allows you to quickly select a set of objects later on by picking the name from a pulldown list. The Named Selection Sets entry area is located to the left of the **Open Track View** button on the Main Toolbar.

1. On the **Display** panel, check the **Geometry** checkbox and highlight **IK Chain Object** and **Point**.

   This will hide everything except the bones.

2. In the Perspective view, use a marquee to select all of the bones.

3. On the Main Toolbar, enter **skeleton** in the **Named Selection Set** entry area, and press **<Enter>**.

Shading the Alien with a Solid Material

The very last step is to display the alien as shaded in preparation for the next chapter.

1. On the **Display** panel, uncheck the **Geometry** checkbox so you are able to see the alien's body again.

2. Open the **Material Editor**.

   The material that is currently assigned to the alien shows as wireframe in one of the sample slots.

3. Select the wireframe material sample slot.

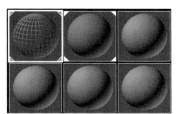

Figure 5.62 *Wireframe material in sample slot*

4. On the Shader Basic Parameters rollout, uncheck the **Wire** checkbox.

   This will cause the alien to become fully shaded.

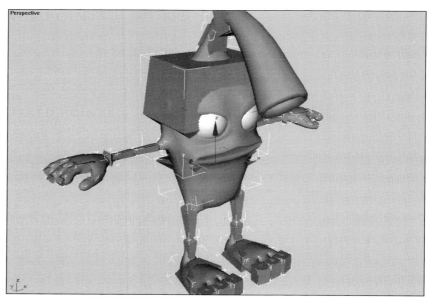

Figure 5.63 *Alien shaded*

The alien is now ready for the skinning and deformation setup, which comes in the next chapter.

5. Save the scene with the filename **Alien - Setup.max**.

SUMMARY

In this chapter, you have learned how to:

- Create a skeleton from prelinked Bones objects
- Use IK solvers to create IK chains between the skeleton's main joints
- Create, place and link point helpers, and link each IK chain to a helper
- Create custom parameters
- Assign controllers for rotating individual axes
- Wire the rotation of point helpers (and thus IK chains) to custom attributes using the Float Reactor controller
- Mirror a character skeleton

The alien skeleton is now ready to use in deforming the mesh. This is accomplished with the Skin, Morpher and Flex modifiers, as you'll see in the next chapter.

Alien Skinning and Deformation Setup

In the commercial we are currently developing, we are responsible for making the alien interact with the remote control and rover as a living creature. In this chapter, you will learn how to setup specialized tools that will make animation of the alien as easy and fast as possible.

TECHNIQUES IN THIS CHAPTER

In this chapter, you will learn how to:

- Use the Skin modifier to associate bones with the mesh
- Adjust the Skin modifier settings for proper deformation
- Create facial expressions
- Organize facial expressions with the Morpher modifier
- Use the Flex modifier to add bounce to an object
- Direct the alien's eyes to always look in a specified direction

TUTORIALS IN THIS CHAPTER

The tutorials in this chapter deal primarily with setting up deformation tools that will be used later on when animating the alien.

- **Tutorial 8.1 A Skinning Primer** will show you the basics of skinning. If you've never used the Skin modifier before, it is highly recommended that you do this tutorial before attempting to skin the alien.

- **Tutorial 8.2 Skinning the Alien** walks you through the steps of applying the Skin modifier and adjusting the effects of the bones on the mesh for proper deformation.

- **Tutorial 8.3 Morph Target Setup** sets up the alien's facial expressions for the animation process later on.

- **Tutorial 8.4 Eye Control System** shows you a quick setup for quick and easy control of the alien's eye movements.

SKINNING

Skinning is the process of associating bones with a mesh. You can then animate the bones to deform the mesh. Skinning is the primary method used for animating characters.

THE BASICS OF SKINNING

In **3ds max**, skinning is accomplished by applying the Skin modifier to the mesh. You can then specify the objects to be used as bones for deforming the mesh. These bones can be **3ds max** Bones, primitive objects, or even splines.

When the Skin modifier is applied, vertices in the mesh are assigned to one or more bones. When the bone is moved or rotated, its assigned vertices move along with it accordingly.

By and large, the default setup doesn't give you a perfect skinning solution. Some vertices won't be assigned to any bone and will get left behind when you move the bones around. Some vertices will need to be assigned to different bones or to two or more bones to get the right result.

Envelopes

When the bones are selected, the Skin modifier generates *envelopes* for each one. An envelope is a capsule-shaped manipulator around a bone object.

Each bone generates two envelopes, one inside the other. The vertices that fall within a bone's envelopes are affected by that bone's movement or rotation. If a vertex falls inside two or more overlapping envelopes from adjacent bones, its deformation is blended accordingly.

The default size and orientation of an envelope depends on the bone itself. If you are using regular **3ds max** Bones, the envelope extends along the longest axis of the bone.

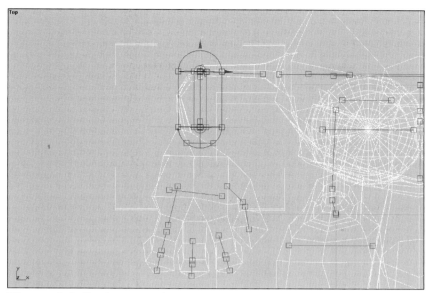

Figure 8.1 *Envelope around alien's lower arm*

To change how vertices are affected by a bone, you can adjust the size and orientation of an envelope. This is an important part of setting up the Skin modifier for proper deformation.

Bone Influence and Vertex Weights

A bone can influence a vertex with 100% of its motion, or with some portion of 100%. For example, a vertex at the elbow might be influenced 50% by the upper arm rotation and 50% by the lower arm.

The term *vertex weight* is used to describe the amount of influence a bone has over a vertex. This term is actually backward -- the bone is the one with the weight (influence). Nonetheless, **3ds max** and other 3D programs use the term *vertex weight* to mean "the amount of influence a bone has over the vertex".

You can set vertex weights manually with a numeric value, or you can set them with envelopes. With the envelope method, all vertices inside a bone's inner envelope are affected 100% by the bone's movement. For vertices inside the outer boundary, weight is assigned based on the vertex's position between the inner and outer envelope boundaries. Vertices closer to the inner envelope will receive more weight than those close to the outer envelope boundary.

Deformers for Special Occasions

Envelopes and weighting will serve to set up most parts of the skin to deform properly. However, there might be times when the mesh pinches or bulges at certain points no matter what you do. Special deformer tools built into the Skin modifier are the answer to this problem. With these deformers, you can custom-shape problem areas to deform a certain way at specific times.

For example, when an elbow or knee is bent more than 90 degrees, noticeable pinching or squashing can occur at the inside part of the joint. If you adjust the envelopes to cure this problem, other parts of the mesh then don't deform properly.

With a custom deformer, you can rotate the bones to the offending angle, then push and pull the mesh to remove the pinch. During the animation process, the custom fix will come into play automatically whenever the two adjacent bones at that joint reach or exceed the specified angle.

Deformers can also be used for special deformation needs. For example, a deformer can make a muscle bulge in a character's bicep when he bends his arm past a certain angle.

We won't be using this tool for our character as he is fairly simple and doesn't have any skinning problems that vertex weights can't solve. However, you should be aware of deformers in case you need them for your own projects. By the time you've skinned the alien with the tutorials in this chapter, you'll be in good shape to check out the procedure for using deformers in the **3ds max** manual and online help.

SKINNING THE ALIEN

The alien's low resolution makes him fairly easy to work with -- there are very few vertices to worry about. On the other hand, you will probably end up with a slightly different default envelope setup than we did after applying the Skin modifier. Your model may be slightly larger or smaller than ours, or the bones may be longer, wider or shorter. These variations all play part in how the skin is calculated, resulting in different skin solutions.

For this reason, we are not going to be able to tell you the exact values for your Skin parameters. Instead, we are going to cover each of the major problem areas and explain the various solutions. After you spend some time practicing different techniques, you'll find the art of skinning is not so hard after all.

TUTORIAL 8.1 A SKINNING PRIMER

To demonstrate the basics of skinning, we are going to walk through a simple skinning tutorial.

Building a Simple Skinning Scene

1. Reset **3ds max**.

> **TIP**
> To change to wireframe mode, you can also activate the viewport and press <F3>.

2. Right-click the Perspective viewport label and choose **Wireframe** from the pop-up menu.

3. In the Left viewport, create a cylinder. Set the cylinder parameters as follows:

Radius	15
Height	-175
Height Segments	10
Cap Segments	1
Sides	18

4. On the **Create** panel, click **Systems**. Click **Bones**.

5. In the Top viewport, create two bones. Start the first one at the cylinder's left side, and the second in the middle of the cylinder.

6. After right-clicking to end bone creation, delete the tail bone.

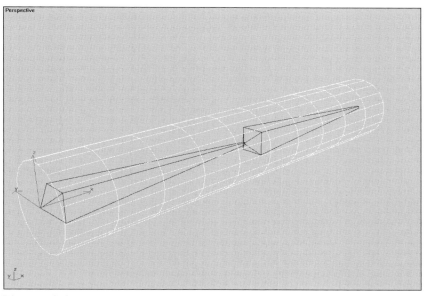

Figure 8.2 *Cylinder with two bones inside it*

7. If necessary, move the bones so they sit inside the cylinder. Check all viewports to make sure the bones are inside the cylinder from all views.

Applying the Skin Modifier

1. Select the cylinder.
2. On the **Modify** panel, apply the **Skin** modifier to the cylinder.
3. On the Parameters rollout, click the **Add Bone** button.

    The Select Bones dialog appears.

4. Select both of the bones and click **Select**.

    The moment you click the Select button, envelopes will be generated for the bones. You can't see them just yet, but you'll get a chance to work with them in a moment.

Testing the Skinning Setup

1. In the Perspective viewport, select the second bone.
2. Click **Select and Rotate**.
3. Change the **Reference Coordinate System** to **Local**.
4. Rotate the bone around its local Y axis.

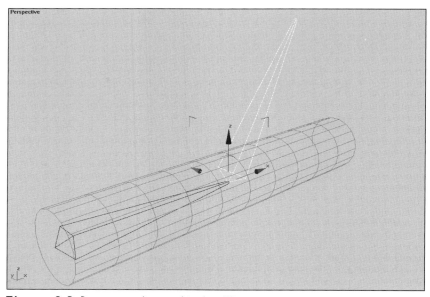

Figure 8.3 *Bone rotated around its local Y axis*

Note that the cylinder vertices did not deform when the bone was rotated. This is because the default envelopes are too small, and do not encompass the cylinder vertices.

5. Undo the rotation.

Adjusting the Envelopes

1. Select the cylinder again.

 2. On the **Modify** panel, click the **Edit Envelopes** button.

3. Highlight **Bone02** on the list below the Edit Envelopes button.

   The envelope for Bone02 appears in all viewports.

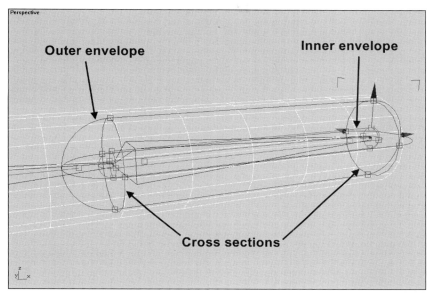

Figure 8.4 *Envelope for Bone02*

Look closely at the envelope as you will need to become familiar with its various parts. The envelope is shown as a capsule within a capsule. The inner capsule, or envelope, is a bright red color. In absolute envelope mode (the default mode), the inner envelope passes on a 100% weight to all the vertices it encompasses. The weight value drops off from the inner envelope to the outer envelope.

Both the inner and outer envelopes contain *cross sections* at either end of the envelope which define the shape of the envelope. On each cross section are four evenly spaced small squares. When **Select and Move** is turned on, you can click and drag on one of these squares to change the size of the envelope at that end.

The outer envelope is always larger than the inner envelope. If you increase the size of the inner envelope to the point where it reaches and exceeds the size of the outer envelope, the outer envelope will scale up as well.

 4. Select the outer cross section at the end of the cylinder.

 5. Use **Select and Move** to move one of the control points and increase the cross section size.

As the cross section becomes larger, the vertices affected by the envelope change color.

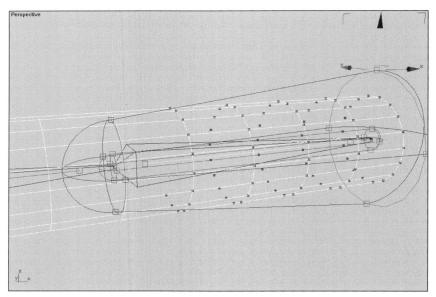

Figure 8.5 *Outer cross section enlarged*

6. Click **Edit Envelopes** to turn it off.
7. Click **Select and Rotate** and rotate the second bone again.

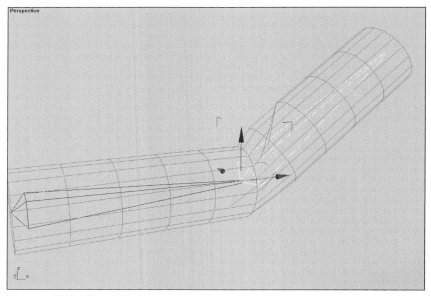

Figure 8.6 *Effect of enlarging outer cross section*

This time, vertices inside the outer envelope are affected by the bone's rotation.

8. Undo the rotation.

Viewing Vertex Weights

So far you have seen the vertices change color in a wireframe view as the envelope is enlarged and the vertex weights change. A shaded view is also helpful for viewing vertex weights.

1. Select the cylinder.
2. Click the **Edit Envelopes** button.
3. Highlight **Bone02** on the envelope list.
4. Change the Perspective viewport to **Smooth + Highlights**.

In a shaded view, the vertices affected by the selected bone are shown in different colors representing the various vertex weights. As colors become more red, the weight approaches 100%.

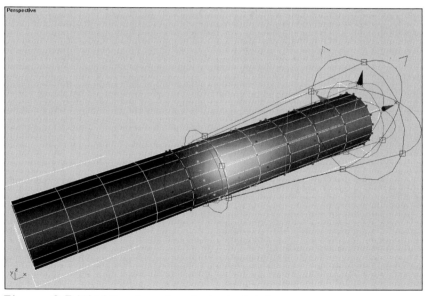

Figure 8.7 *Weighting shown by color in shaded view*

Rotating and Resizing an Envelope

The ends of the envelopes can be moved around to change the envelope influence area. At each end of an envelope is a small dark gray box. By clicking and dragging one of these boxes, you can move that side of the envelope. This will allow you make the entire envelope larger or smaller, or rotate its direction.

1. Click **Select and Move**.
2. Select the dark gray box at one end of the second bone.

   The box will turn pink to indicate that it has been selected., and a red transform manipulator appears along the envelope center.

3. Move the box around to increase and decrease the size of the envelope, and rotate its direction.
4. Undo until the envelope is positioned with the bone again.

Manually Setting a Vertex Weight

There will be times when adjusting an envelope will not produce the results you are looking for no matter how hard you try. In this case, you can assign specific weights to vertices. This means that the vertices will ignore the envelope assigned weighting and respect only the weighting you supply.

3ds max calls this type of weight an *absolute weight*. This can be confusing, as there is also a type of absolute weighting for envelopes, as discussed earlier. To differentiate between the two, we will call one absolute envelope weighting and absolute vertex weighting. Here we will use absolute vertex weighting to change the weights of individually selected vertices.

1. Change the Perspective view to *Edged Faces* mode.

2. In the Filters section on the Parameters rollout, check the **Vertices** checkbox.

   This step is necessary to allow selection of vertices on the skinned object.

3. In the Perspective view, select a vertex that you are interested in adjusting the weights for.
4. On the Parameters rollout, adjust the **Abs. Effect** parameters.

   This parameter assigns an absolute weight to the selected vertices, causing them to ignore the weighting set by the envelope.

   Now you know the basics of skinning. Use the techniques that we've just covered to see if you can blend between the two bones so when the second bone is rotated upward, the mesh deforms appropriately as shown in Figure 8.8.

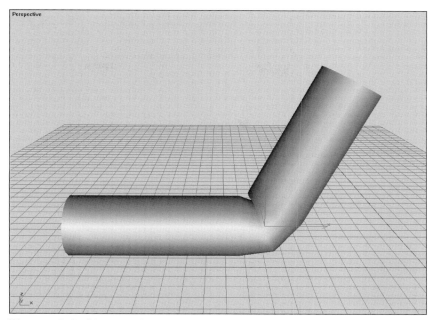

Figure 8.8 *Cylinder deforming properly when bone is rotated*

TUTORIAL 8.2 SKINNING THE ALIEN

Now that you've learned the basics of skinning, you can apply what you've learned to the process of skinning the alien.

Turning Off NURMS Subdivision

1. Either open the file **Alien - Setup.max** you created in the last chapter, or load the file **Alien - Setup.max** from the **Scenes/Alien** folder on the CD.

   When the alien was created, we used NURMS subdivision to add detail to the alien's body and make it appear very smooth. If we were to apply the Skin modifier to the alien with NURMS subdivision on, we would end up spending much more time than necessary weighting the many vertices.

   A better technique is to skin a low-resolution version of the alien and then smooth it after the Skin modifier has has calculated all of the weighting. In this way, we will not have to concern ourselves with the weights of so many vertices.

2. Select **Alien Body**.
3. On the **Modify** panel, uncheck **Use NURMS Subdivisions**.

   This will cause the alien's body to revert to low detail.

> **TIP**
>
> We will not be turning NURMS subdivision back on. After skinning and before animating, we're going to apply a MeshSmooth modifier to provide the same effect that the NURMS subdivisions did. While animating the alien, we'll be able to turn MeshSmooth on and off so we can alternate at will between fast feedback and a smoothed alien body.

Figure 8.9 *Alien after turning off Use NURMS Subdivisions*

Applying the Skin Modifier

1. On the **Modify** panel, apply the **Skin** modifier to the alien.
2. On the Parameters rollout, click the **Add Bone** button.
3. On the Select Bones dialog, type ***bone** in the top text field so all objects containing the word bone are selected.
4. Hold down the **<Ctrl>** key and click the two entries that contain **Take me out (ball)**. Click the **Select** button.

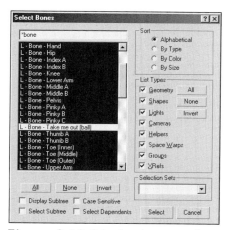

Figure 8.10 *Selecting the correct bones*

These are the two ball bones on the foot that we do not want to use in the skinning process. We named them this way so they would be easy to find.

All the selected names are added to the list box in the Parameters rollout. Envelopes are automatically generated for each bone, and the Skin modifier does its best to determine the appropriate vertices to assign each one.

Testing the Skin

The geometry has now been skinned but as you know, that does not mean it's perfect. To see what problems exist, you will need to move the alien into various positions.

1. On the **Display** panel, make sure **Points** are visible.
2. Select **Point Alien Root**.
3. In the Perspective view, move **Point Alien Root** into various positions.

   The mesh deforms when the bones are moved, but the deformation is not exactly what we had in mind. The skin is stretching in funny places.

Figure 8.11 *Moving the Point Alien Root stretches the alien skin*

4. Undo all your movements until the alien is back in his original pose.
5. Save the scene with the filename **AlienSkin01.max**.

Preparing for Skinning Poses

As you can tell from your tests, the alien's envelopes will need to be adjusted. You most likely saw stretching, pulling, and bubble gum type effects.

To fix the envelopes properly, we are going to need to move the alien into various positions so we can see how changes to the envelopes will affect the deformation. After the adjustment process is complete, we will need to return the alien back to his original pose. To achieve this style of tweaking, we are going to use animation.

We will be animating the alien in various poses. After we have completed the animation, we will begin adjusting the weights. You will find it very convenient to slide the time slider back and forth to see how the geometry is deforming as the bones are moved into various positions. This beats having to manually rotate the bones every time we want to test a parameter change.

To pose the alien, we will be using point helpers, so we don't even have to have the bones visible! Hiding the bones will eliminate clutter on the screen and make it easier to see what is going on.

1. On the **Display** panel, hide all the Bones.

2. Click the **Time Configuration** button at the lower right of the screen. On the Time Configurations dialog, set the **End Time** in the Animation section to 50 and click **OK**.

   A 50-frame animation will be sufficient for setting up our key poses.

3. Click **Select and Move**, and set the **Reference Coordinate System** to **Local** if necessary.

Setting up the First Key Pose

> **TIP**
>
> *When setting up animation poses for checking skinning, use only poses that you're sure the character will use. For example, if the character will never raise his hands over his head, don't worry about this position. Working only with poses that you know will be used saves a lot of time.*

1. Click the **Animate** button to turn it on.
2. Move the time slider to frame 10.
3. Select **L-Point Hand Controls**.
4. In the Top viewport, move the point helper so that the left elbow is bent at a 90-degree angle.

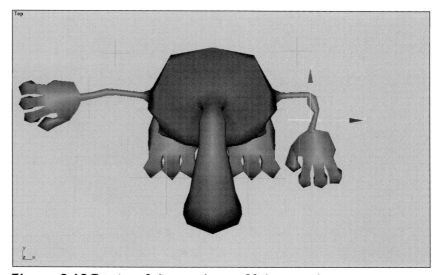

Figure 8.12 *Top view of alien arm bent at 90-degree angle*

While setting up the key poses, don't be alarmed if parts of the mesh don't follow along or if the alien begins to look very strange. We will fix all this after we have set up all the key poses.

5. Select **R - Point Hand Controls** and bend the right arm 90 degrees as well.
6. In the Left viewport, select both **L - Point - Ankle Rotate** and **R - Point - Ankle Rotate** and move them so the legs are bent at the knees by 90 degrees.

    You will most likely need to rotate the helpers as well so that the feet are not in a really strange pose.

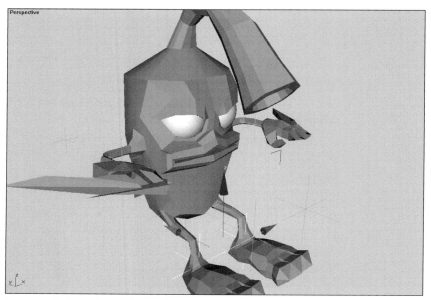

Figure 8.13 *Alien at frame 10*

Setting up the Remaining Key Poses

1. Advance the time slider to frame 20 and adjust the arms so that they are by the alien's side and the legs are straight out. You will probably have to adjust the left and right elbow targets to turn the arms properly.

Figure 8.14 *Perspective view of alien at frame 20*

2. Advance the time slider to frame 30 and position the arms so they are above the alien's head and the legs are pulled away from one another.

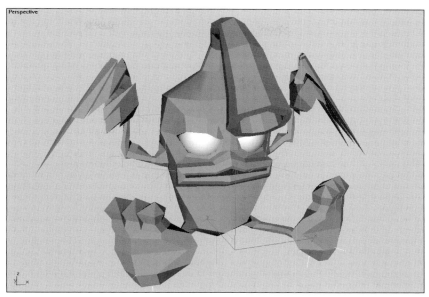

Figure 8.15 *Perspective view of alien at frame 30*

3. Advance the time slider to frame 40. Position the arms straight out with elbows bent upward at a 90-degrees angle and palms facing forward. Position the legs so the knees are bent at 90-degrees angles and the feet are rotated straight.

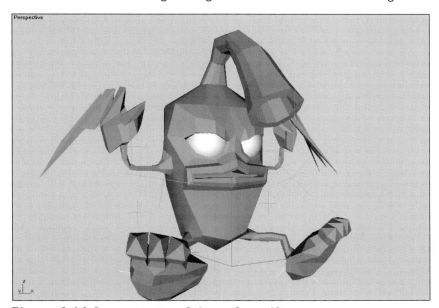

> ☼ **TIP** ☼
>
> *The position of the R - Point Elbow Target and L - Point Elbow Target will play a significant role an achieving this position.*

Figure 8.16 *Perspective view of alien at frame 40*

4. Move the time slider to frame 0.
5. Select each of the point helpers for which you have set keys.
6. On the track bar at the bottom of the viewports, hold down the **<Shift>** key and move the key dot at frame 0 to frame 50.

Figure 8.17 *Track Bar*

This will copy the alien's pose from frame 0 to frame 50 so he returns to his original pose at the end of the animation.

7. Turn the **Animate** button off.
8. Save the scene with the filename *AlienSkin02.max*.

Now you are ready to adjust vertex weights. There are a number of approaches that one can take when weighting a skinned character -- no method is the right or wrong way. After some practice, you will develop the personal technique that suits you best.

Our approach will be to start from the outside and work our way inward. In other words, we will begin with the hands, feet and nose, and will proceed from there to the limbs, and finally the torso.

As we discussed earlier, there are two ways to control vertex weights, with envelopes or with absolute weights for selected vertices. Both methods have their benefits. If you are skinning a very detailed character, you will find it much easier to manipulate envelopes than to assign absolute weights to each vertex. On the other hand, if you have a low resolution character such as our alien, setting absolute weights is quicker and yields greater accuracy.

With the alien, we will be assigning absolute vertex weights to most of the vertices. You are welcome to try working with envelopes too, but in the end you will most likely find that absolute weights much easier to work with.

To help us visualize finer detail such as finger movements, we will need to be able to see the model smoothed with the MeshSmooth modifier. After it is applied, we will turn it off for the time being, then turn it on only when we need to see how the envelopes are holding up with the finer animation.

Applying the MeshSmooth Modifier

1. Select **Alien Body**.

2. On the **Modify** panel, apply the **MeshSmooth** modifier.

   This applies the MeshSmooth modifier above the Skin modifier so it will smooth the deformed mesh.

3. On the Subdivision Amount rollout, set the **Iterations** to 1.

4. Check the **Render** checkbox next to the Iterations setting for Render Values. Set the **Render Iterations** to 2.

    This will provide an even smoother alien when we render the scene.

5. Highlight the **Skin** modifier on the modifier stack.

6. On the **Modify** panel, click the **Show end result on/off toggle** button twice or more to see what happens.

    When the button is on (the default), you can see the results of the MeshSmooth modifier. When it's off, you can only see the results up to and including the Skin modifier.

7. Turn off the **Show end result on/off toggle**.

    We will only toggle it on when we need to see how our weighting will look on the final model.

Adjusting the Hand Bone Weights

As you can see, moving the arms will result in some of the hand geometry being left behind. This needs to be the first thing corrected.

It can be quite confusing to start the delicate process of assigning vertex weights when the geometry is all messed up like this. Instead, we will start by assigning all the hand's geometry 100% to the hand bone.

Later on, we can add weights to the fingers to fine-tune the weights without the hand falling apart all the time. As you add weights to the finger bones, weights will be subtracted from the hand bone.

1. On the **Modify** panel, highlight the **Skin** modifier and click the **Edit Envelopes** button.

2. Zoom and pan the Top viewport to show a closeup of the alien's right hand.

    Even though the bones are currently hidden, we are still able to see the envelope markers for each bone.

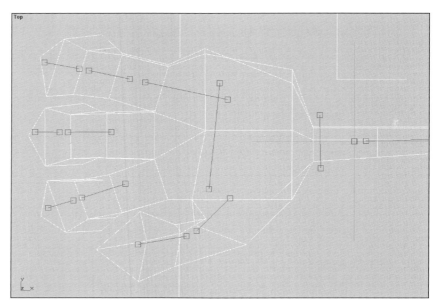

Figure 8.18 *Hand closeup in Top viewport*

3. Select the **Hand Bone** envelope marker at either end.

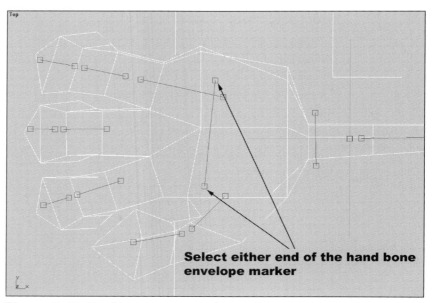

Figure 8.19 *Hand bone envelope marker*

4. On the **Modify** panel, check the **Vertices** checkbox in the Filters section.

    This will allow us to control the absolute weight of selected vertices.

Right now each of the hand/finger vertices have different colors indicating varying weights, and some vertices are not even assigned. The goal right now is for all hand/finger vertices to be weighted 100% to this bone.

5. Use a marquee to select all the vertices on the hand and fingers.

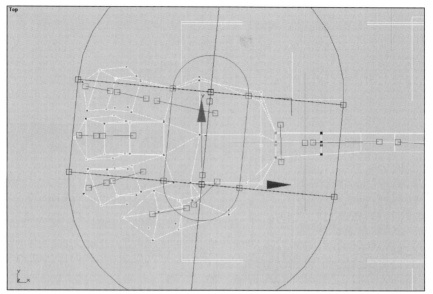

Figure 8.20 *Hand vertices selected*

6. For the **Abs. Effect** parameter in the Weight Properties section, enter a value of 100.

   All the vertex colors have just changed to red. This indicates that they are receiving 100% weight from this bone.

7. Move the time slider and see how the hand responds. This time the hand should move properly without falling apart as it did before.

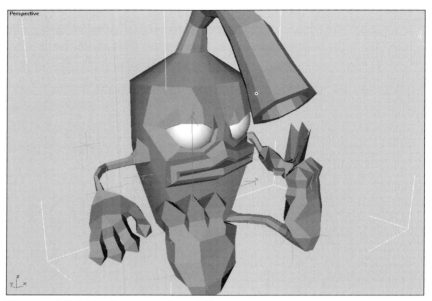

Figure 8.21 *Hand moves without falling apart*

8. Go back to frame 0. Repeat these steps on the left hand, and move the time slider to later frames to test the weight assignment. Both hands should move all together now.

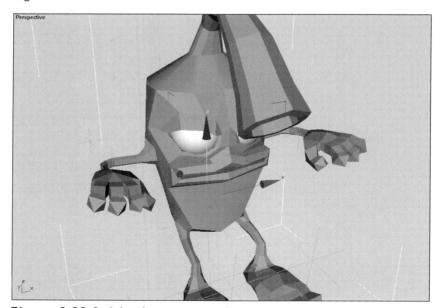

Figure 8.22 *Both hands moving properly*

9. Save the scene with the filename **AlienSkin03.max**.

Curling The Index Finger

Now that we have the hands being controlled by the hand bone, we need to start weighting each of the fingers so that the appropriate bones deform them. In this exercise, we will focus on the index finger.

When setting up our key skinning poses, we did not animate the fingers. This is because we have custom attributes that allow us to control each bone and return it back to its default position without any trouble, allowing us to check the finger curls easily and quickly without the need for key poses.

1. Turn off **Edit Envelopes**.
2. Select **R - Point Hand Controls**.
3. On the Custom Attributes rollout, set **Index Knuckle** to 10.

   This will bend the index finger. The finger geometry will not be affected since the hand bone controls all its vertices at this time, but not for long!

Assigning Index Finger Weights

1. Select the alien body again.
2. On the **Modify** panel, highlight the **Skin** modifier and click the **Edit Envelopes** button to activate envelope editing.
3. Select the tip end envelope marker that represents the end of the index finger.

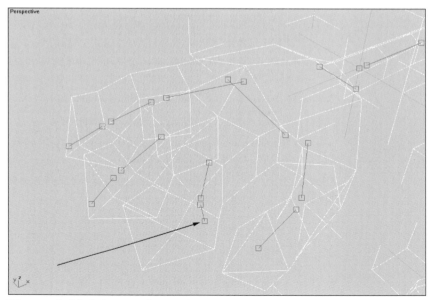

Figure 8.23 *Select tip of index finger*

4. Select the two rows of vertices closest to the end of the finger along with the tip-end vertex.

5. Enter 100 for **Abs. Effect**.

   This will cause the bone to control these vertices 100%. After entering this number, the geometry will jump to the bone.

   Figure 8.24 *Index finger jumps into position*

   Let's curl the finger a little more and see how it looks.

6. Turn off **Edit Envelopes**.
7. Select **R - Point Hand Controls**. On the **Modify** panel, set the **Index Middle** custom attribute to 10.

   The index finger should now be completely curled.

8. Select the alien again and turn on **Edit Envelopes**.
9. Turn on the **Show end result on/off toggle** so you can see how finger looks when smoothed with MeshSmooth.

   The finger has very little volume and appears flat at the bend. We'll fix this by fine-tuning the finger weights.

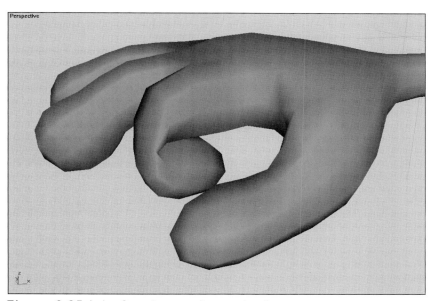

Figure 8.25 *Index finger has no volume at bend*

Fine-Tuning the Index Finger Weights

1. Change the Perspective viewport to **Wireframe**, and turn off the **Show end result on/off toggle**.

2. Select the envelope marker for the bone just prior to the end of the index finger.

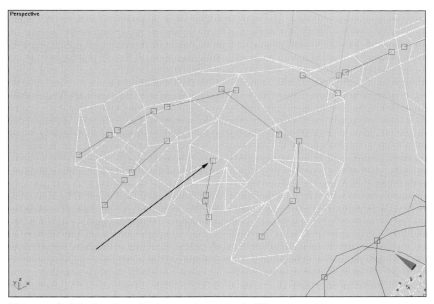

Figure 8.26 *Envelope for bone prior to end of finger*

3. Select the four vertices on top of the index finger, as shown in the next figure.

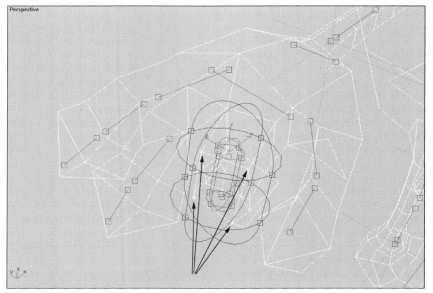

Figure 8.27 *Four vertices to select*

4. Increase the **Abs. Effect** parameter for the selected vertices until the top of the finger has some volume. A value between 35 and 50 should do the trick.

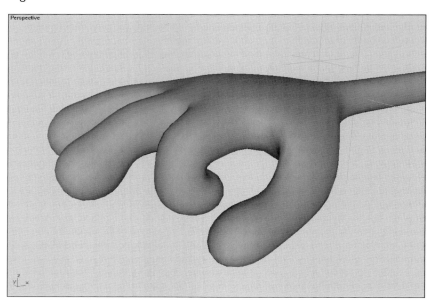

Figure 8.28 *Final index finger curled*

5. Continue adding weights to the finger vertices until the finger curls correctly. Feel free to turn the **Show end result on/off toggle** on and off to check your work frequently.

6. Use the techniques above to adjust the vertex weights on all the fingers. Be sure to save your work often.

7. When you have finished adjusting the weights, save the scene with the filename *AlienSkin04.max*.

Adjusting the Arm Vertex Weights

1. Move the time slider to frame 10.

   Here. the arms are bent at the elbows by 90 degrees.

2. If necessary, select the alien and turn on **Edit Envelopes**.

3. Turn off the **Show end result on/off toggle** if it is on.

4. In the Top viewport, select the lower arm envelope marker.

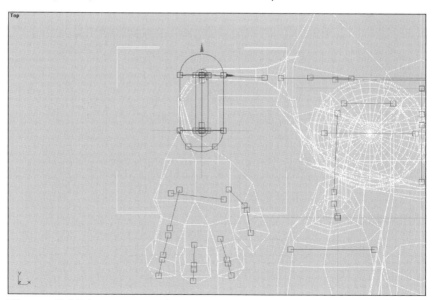

Figure 8.29 *lower arm envelope is selected*

5. Select the row of vertices around the center of the forearm.

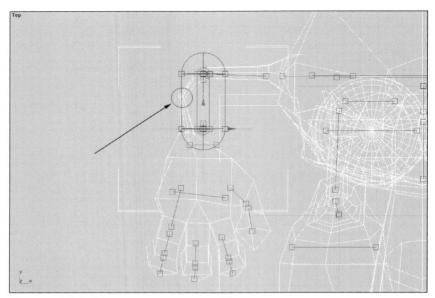

Figure 8.30 *Select vertices in center of forearm*

6. Enter a value of 100 for **Abs. Effect**.

   This will cause these vertices to jump to the forearm bone.

7. Select the wrist envelope.

8. Use a marquee to select the vertices at the wrist, and give them an **Abs. Effect** value of 100.

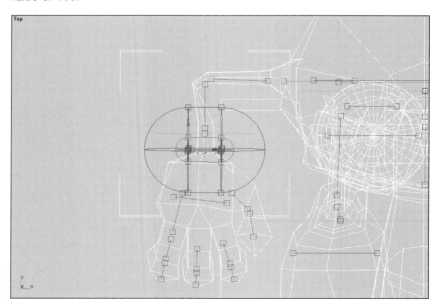

Figure 8.31 *vertices at wrist weighted 100% to wrist envelope*

9. Now that the arm is starting to take shape, you can use the same techniques to achieve a nice rounded elbow. You will need to adjust the vertex weights at the elbow for both the upper arm bone and the lower arm bone.

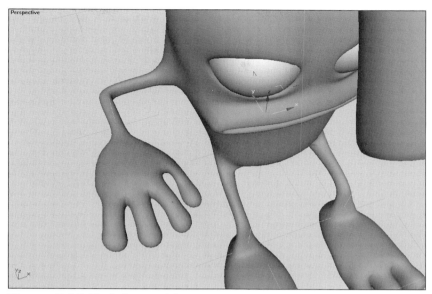

Figure 8.32 *Final elbow and forearm*

Weighting the Upper Arm and Shoulder

1. Move the time slider to frame 20.

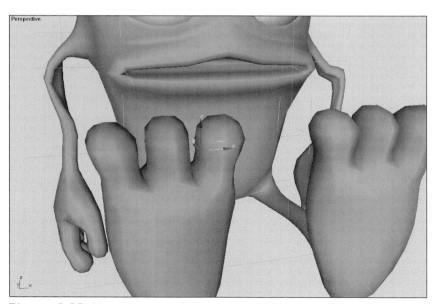

Figure 8.33 *Alien upper arm shown as smoothed*

The arms are now down by the alien's sides. Don't be surprised if the upper arm looks bad. There are most likely a few vertices that need to be weighted to the upper arm.

2. Select the upper arm envelope.
3. Weight the appropriate vertices so the arm looks more natural.

To achieve a good deformation, you will need to focus on the weighting between the upper bone and the clavicle bone. You will need to juggle back and forth between the two until you get the desired deformation.

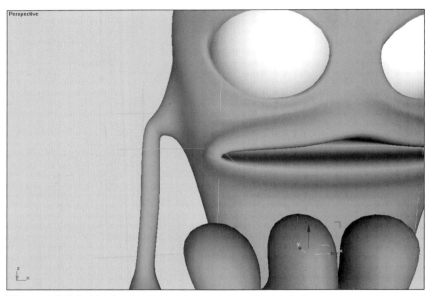

Figure 8.34 *Alien's shoulder and upper arm weighted, shown as smoothed*

4. Save the scene with the filename **AlienSkin05.max**.

Skinning the Rest of the Alien

At this point you have been introduced to enough techniques to complete the alien skinning. This is a skill that requires a lot of practice. You have the animation in place, so simply move the time slider regularly and check to see how the geometry is deforming with the alien in each of the various positions. You will find yourself often moving back and forth between key positions as you work on blending the weights just right.

The most important thing to remember is to have patience. The more you do this, the better you will get at it.

Here are some tips to help you complete the skinning process.

- For the nose, select the envelope for the last head bone and assign all the vertices past that point on the nose by 100%. The wobbliness of the nose will be achieved with the Flex modifier and not with animated bones, so you don't have to concern yourself with it.

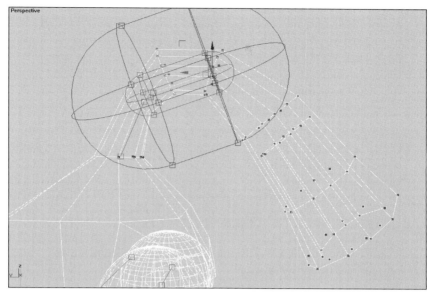

Figure 8.35 *Last envelope should own all vertices at end of alien's nose*

- The torso is divided into two main bones, an upper and a lower. Vertices from the mouth up should belong to the upper bone only to prevent the mouth and face from stretching when the head is turned. You can exclude these vertices from the lower bone by selecting the lower bone, selecting the vertices from the mouth up, and clicking **Exclude Verts** on the Parameters rollout.

When you have finished adjusting vertex weights, save the scene with the filename ***AlienSkin06.max***.

FACIAL EXPRESSIONS

For animating facial expressions in 3D, you have a number of options to choose from.

Facial motion capture. Motion capture devices can be used to capture subtleties of facial motion. This type of device works with sensors attached to a live person's face. The person talks and makes faces and the sensors capture the data, which is then passed on to the 3D software. This method can be costly and time-consuming, but is well worth it for realistic facial motion, especially for large projects such as realistic games or films. Our alien doesn't need such realistic motion and our project is comparatively short, so we won't be using this method.

Skeleton-based systems. Another method is to place bones inside the character's face that correspond to actual facial bones. The face is then skinned and the vertices weighted using techniques similar to the ones you just used to skin the alien body. This type of system is excellent for fine control over a face, particularly the mouth and eye areas, without the expense of motion capture. Again, we don't need realistic motion and our facial animation is pretty simple, so we don't have to get this fancy.

Vertex Linking. A cruder variation on the previous method is to link specific vertices to manipulators, then animate the manipulators with keyframes to move the vertices around. We could use this method for our animation, but there is an easier way to achieve the effect we want.

Morphing. Morphing is the process of gradually changing one object into another. To prepare for facial morphing, you create several copies of the face, then push and pull vertices around on each one to create various facial expressions and mouth positions. You can then morph between these expressions at specified times during the animation. You can also morph partway between expressions, and even combine expressions.

We will use the morphing method to animate the facial expressions for our project.

MORPHING BASICS

Each version of the model used in morphing is called a *morph target*. To start making morph targets, simply make a copy of the original object, and move vertices around on the copy to make a new facial expression.

Make as many copies of the original object as needed to make all your morph targets. Always keep a neutral version of the model as a base object for the morph process.

When creating morph targets, never add vertices or faces as this will prevent the morphing process from working. Remember that morphing can only take place between objects with the same number of vertices.

MORPHING WITH 3DS MAX

In **3ds max**, the Morpher modifier organizes your morph targets and makes it easy to animate from one to another. The Morpher modifier is applied to the base object, and each morph target is loaded into a *morph channel*.

Each morph channel has a percentage which can be animated. Animating between morph targets is as easy as turning on the Animate button and changing the percentages on keyframes.

Combining Targets

When creating morph targets, keep in mind that targets can be combined during the animation process. For example, suppose you want the character to be able to wink either eye or blink both eyes. You might think you should create three morph targets, one for each wink and one for both eyes blinking. In fact, you only need two morph targets, one for each eye winking. When animating the character, you can combine the two wink targets to make the character blink.

The same goes for facial expressions that use both the mouth and eyes. By combining four or five eye expressions with the same number of mouth expressions, you can in fact have over a dozen actual facial expressions.

DETERMINING FACIAL EXPRESSIONS

In working out the facial expressions you need to make for your character, you first need to figure out what emotions the character will express over the course of the animation. Work out what the eyes and mouth should do in these expressions. Then separate out the eye and mouth positions to determine your morph targets.

Of course, you always want the character to be able to blink. A few occasional eye blinks during the animation go a long way toward making a character believable.

Alien Facial Expressions

Over the course of the animation, the alien's face has to make the following expressions:

- Happy to be playing with the rover
- Puzzled, then angry when the rover stops working
- Yelling "Mom!" when he discovers the Brand X batteries

We can translate these expressions into specific eye and mouth positions.

Emotion	Eye Position	Mouth Position
Happy	Eyes open (neutral)	Mouth smiling
Puzzled	Eyes blinking	Mouth closed straight or frowning
Angry	One eye slanted	Mouth sneering
Yelling "Mom!"	Eyes closed	Mouth wide open in "O" shape
Blink	Eyes closed	

We always add the blink to our list of expressions as a blink is almost always needed in an animation, and it might not be covered by other expressions. In this case, it is covered by the "puzzled" and "yelling" expressions, but we add it to the list just to remind you for future projects.

Alien Morph Targets

To achieve these expressions, these are the morph targets you will need to create for the alien.

- Left Blink
- Right Blink
- Left eye slanted in anger
- Right eye slanted in anger
- Mouth closed
- Mouth smiling
- Mouth sneering
- Mouth frowning
- Mouth yelling "Mom!"

TUTORIAL 8.3 MORPH TARGET SETUP

In this tutorial, you will set up the morph targets to be used later on when you animate the alien. We will start by creating a copy of the alien and moving vertices to form our first morph target.

Creating the First Morph Target

1. Select the alien model.

2. On the **Modify** panel, turn off the **MeshSmooth** modifier if it is on by clicking the light bulb next to the modifier on the stack.

3. Turn off the **Show end result on/off toggle** if it is on.

4. Copy the alien to the left of itself in the Perspective viewport, and name the new object **Mouth - Smile**.

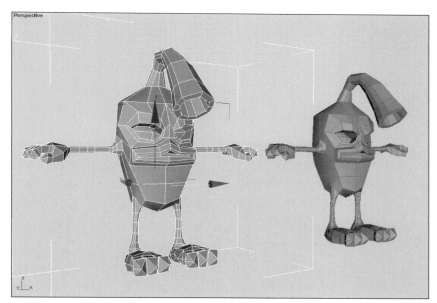

When copying the alien body, the location of the morph target in the scene is not important.

Figure 8.36 *New target is on left, original base on right*

5. With the new model selected, collapse the modifier stack.

   This converts the new alien to an Editable Poly. All the modifiers have been retained, but they cannot be edited on this object.

Shaping the Morph Target

 1. With the new model selected, access the **Vertex** sub-object level.
2. Move the vertices in the mouth area so the alien appears to be smiling.

Do not do anything that can add extra vertices or faces, such as extruding, cutting or slicing. Just move vertices.

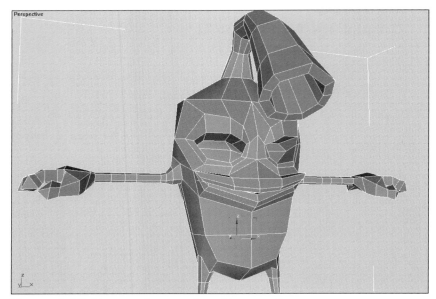

Figure 8.37 *New morph target shaped into a smile*

3. Save the scene with the filename **AlienSkin07.max**.

Creating the Remaining Targets

Copy the alien and form the remaining morph targets. Use the list below as a guide for the morph target names.

- Left Blink
- Right Blink
- Left eye mad
- Right eye mad
- Mouth - closed
- Mouth - Smile (already made)
- Mouth - Sneer
- Mouth - Frown
- Mouth - MOM!

Here are pictures of some of the morph targets we created.

TIP

When forming the morph targets for the eyes, it can be helpful to use a copy of the eye as a guide to ensure the eyelid fits properly over the eyeball.

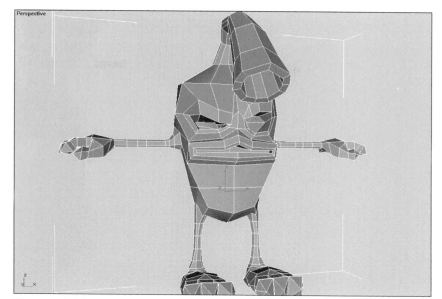

Figure 8.38 *Right eye mad*

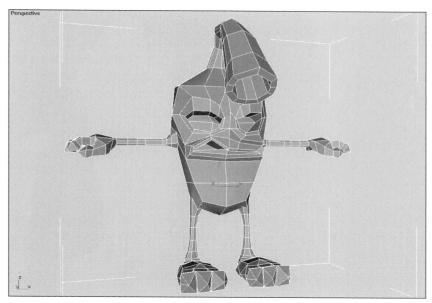

Figure 8.39 *Mouth - Sneer*

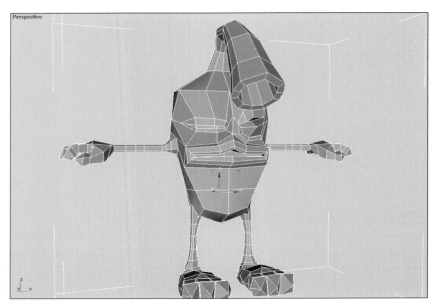

Figure 8.40 Right blink

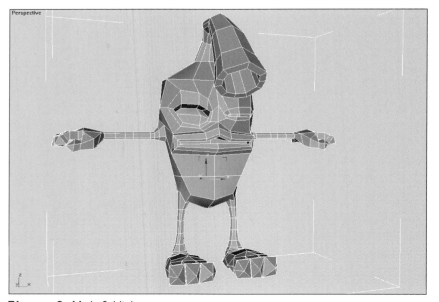

Figure 8.41 Left blink

You are welcome to create more expressions than the ones presented here. Just keep in mind that when you set up the alien control system later on, you will have to make appropriate adjustments to the script.

When you have finished creating the morph targets, save the scene with the filename **AlienSkin08.max**.

Applying the Morpher Modifier

Now that we have created all the varying morph targets, it is time to load them into the Morpher.

1. Select the base object, the original skinned alien.
2. On the **Modify** panel, apply a **Morpher** modifier to the object.

This adds the Morpher modifier on top of the MeshSmooth modifier. We want the MeshSmooth modifier to be the last one on the stack so it will smooth all the deformed geometry.

3. On the modifier stack, drag the Morpher modifier beneath the MeshSmooth modifier.

Limiting Targets to Selected Vertices

At this point we are almost ready to start loading morph targets into the Morpher modifier channels. However, loading each complete morph target object is not necessary since we only changed vertices on the torso part of the body and not the arms or legs.

We can tell the Morpher that we only want to load a selection of vertices from each morph target into the channels. This will save processing time.

To tell the Morpher modifier which vertices to load, we'll use a Mesh Select modifier.

1. Make sure the base object is selected. On the modifier stack, click on the **Skin** entry.
2. Apply a **Mesh Select** modifier.

The new modifier appears between the Skin and Morpher modifiers.

3. Expand the **Mesh Select** modifier by clicking the plus symbol **[+]** next to it.
4. Highlight the **Vertex** sub-object level beneath **Mesh Select**.

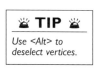

☼ **TIP** ☼

Use <Alt> to deselect vertices.

5. Select all the vertices on the torso portion of the alien's body. Do not include the arms, hands, legs, feet or nose. You can also leave out the vertices on his back.
6. Click on the **Morpher** modifier in the modifier stack.

Loading Morph Targets

1. On the Channel List rollout, locate the channel buttons which are all labeled **empty**.
2. Click the **Load Multiple Targets** button and select all the targets listed.

   This will load the targets into successive channels. Optionally, you could load each target one at a time by clicking the **Pick Object from Scene** button and then selecting the appropriate target. However, it's quicker to load them all at once with the method we used.
3. Click the **Use Vertex Selection** button in the Channel Settings section.

   This will cause the vertex selection to be used rather than the entire morph target.
4. Save the scene with the filename *AlienSkin09.max*.

   As one last setup step for the alien body, we will add a little bounce to the alien's nose. The nose needs to flop around a little when the alien moves his head.

   We could animate this motion by hand, but it can also be accomplished quickly and easily with the Flex modifier. Flex is used to simulate soft-body dynamics by applying virtual springs between each vertex on an object. These virtual springs can control how stiffly or loosely the object will bounce and flop when it is moved.

Selecting Vertices for Flex

Only the end of the nose needs to flop around, so we will apply Flex only to a selection of vertices near the end of the nose.

1. Apply another **Mesh Select** modifier to the alien.
2. On the modifier stack, drag the **Mesh Select** modifier under the **MeshSmooth** modifier.
3. Click on the new **Mesh Select** modifier, expand it by clicking the plus sign **[+]** and highlight the **Vertex** sub-object level.
4. Expand the Soft Selection rollout and check **Use Soft Selection**.

5. Select the first five rows of vertices starting from the end of the nose. Work in *Wireframe* mode to verify that you have selected them all.

   Note that the first five rows consist of three rows of vertices on the outside of the nose and two rows on the inside.
6. On the Soft Selection rollout, set the **Edge Distance** to 4 and the **Falloff** to 350.

   This will cause the vertices all the way down to the base of the nose to be affected. The vertices at the base of the nose will receive very little affect from the Flex modifier.

Applying a Flex Modifier

1. Without changing your position in the modifier stack, apply a **Flex** modifier to the alien body.

   There are numerous settings in Flex, but we are only going to use a few of them to get the look that we need.

2. Set the following settings for the Flex modifier:

   Flex 50
   Strength 20
   Sway 96
   Samples 5

   On the pulldown, change the algorithm from **Euler** to **Runge-Kutta4**

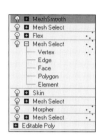

3. Finally, apply another **Mesh Select** modifier to the stack.

   Make sure that the new Mesh Select modifier is positioned above Flex and below MeshSmooth.

   Applying a Mesh Select modifier without accessing a sub-object level resets the selection to the entire object. This means the entire object will be affected by the next modifier on the stack, which is the MeshSmooth modifier.

4. Save the scene with the filename *AlienSkin10.max*.

EYE CONTROLS

The final step in setting up the alien animation system is a method for controlling the eyes. During the animation process, it would be a hassle to have to rotate the eyes every time we need them to look in another direction. To make the eye animation easier, we can force both eyes to always look at a single object out in space.

We will set up a control system where each eye is always forced to look at a specific point helper. Then the two point helpers will be linked to a dummy object. Moving the dummy will in turn rotate the eyes to always look at the point helpers.

TUTORIAL 8.4 EYE CONTROL SYSTEM

In this tutorial, you will set up a simple eye control system for the alien using point and dummy helpers.

Creating the Point Helpers

1. Create two point helpers in the scene with the names **Point - Right Eye Aim** and **Point - Left Eye Aim**.

2. Select **Point - Right Eye Aim**.

3. On the Main Toolbar, click the **Align** tool. Press **<H>** to open the Pick Object dialog, highlight **Right - Eye**, and click **Pick**.

4. On the Align Selection dialog, check **X Position**, **Y Position**, and **Z Position**. Make sure that both **Current Object** and **Target Object** are set to **Center**. Click **OK**.

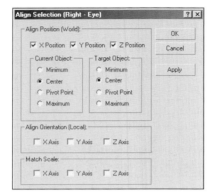

Figure 8.42 *Align Selection dialog*

This aligns the right eye point object with the center of the right eye itself.

5. In the Perspective viewport, move the point helper along the Y axis (in the Perspective viewport) so that is some distance out in front of the eyeball. See the next figure for reference.

6. In the same way, select **Point - Left Eye Aim** and align it with **Left - Eye**, then move the helper out in front of the alien. Place the two helpers the same distance from the eyes.

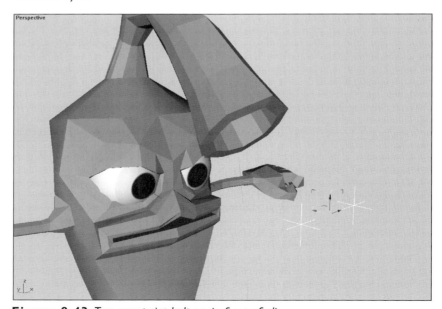

Figure 8.43 *Two new point helpers in front of alien*

Making the Eyes Look at the Point Helpers

1. Select **Right - Eye**.
2. From the menu, choose **Animation/Constraints/Look-At Constraint**.
3. Click on **Point - Right Eye Aim** to constrain the right eye to it.

   If the eye's pupil disappeared, the Look-At Constraint's axis is pointing in the wrong direction. To fix this problem, go to the **Motion** panel and try a different **Look At Axis**. With our alien, we had to adjust the axis to Z.

4. Repeat the above steps to make **Left - Eye** look at **Point - Left Eye Aim**.

Linking to the Dummy Object

1. Create a dummy helper and position it between the two point helpers. Check the position of the dummy object in all viewports to make sure it's correct. Name the new dummy object **Eye Control**.

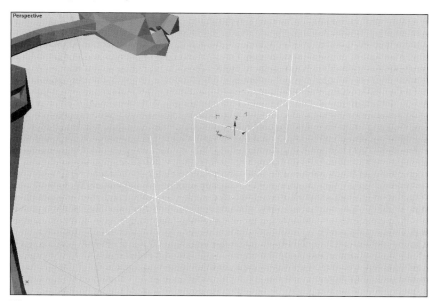

Figure 8.44 *New eye control system*

2. Use **Select and Link** from the Main Toolbar to link each of the point helpers to the dummy object.

3. Test the constraints by selecting the **Eye Control** dummy object and moving it around.

4. After you have verified that the constraints are working, undo any moves you made to the **Eye Control** dummy.

5. The alien skin and deformation setup is now complete. Save the scene with the filename **Alien - Setup and Deformation.max**.

SUMMARY

Our character is now set up and ready to be animated. As you've seen, there are many different techniques involved in setting up a character for animation. The entire goal of these setup steps is a quick and easy animation process later on.

In this chapter, you have learned:

- Skin a character with the Skin modifier
- Adjust vertex weights for skinning
- Create morph targets for facial expressions
- Set up morph channels in the Morpher modifier
- Use the Flex modifier to add bounce to the alien's nose
- Use the Look-At constraint to control the alien's eyes

There's one more important step to do to ease the animation process — you need a user interface for working with the alien. Go on to the next chapter to find out how.

This chapter focuses on the development of a special user interface that will help streamline the character animation process. This user interface will be developed using ordinary **3ds max 4** tools.

Often, character setups have numerous controls that must be animated. If you don't make your own interface to integrate all these controls, you'll find yourself going all over the **3ds max** user interface to animate the legs, the face, the feet, and so on. Having everything organized into one user interface can speed up the animation process tremendously.

TECHNIQUES IN THIS CHAPTER

In this chapter, you will learn to:

- Generate and place custom icons in a user interface
- Create a custom user interface for easily selecting from a number of objects
- Create and save poses to use over and over again
- Set up a UI for working with morph targets

This chapter will continue to build on your scripting skills, introducing some new scripting techniques that are extremely useful in character animation.

TUTORIALS IN THIS CHAPTER

This chapter has two tutorials. The emphasis is on building a custom, picture-based user interface that makes character animation quick and easy.

- ***Tutorial 9.1 Icons for User Interface*** walks you through the creation of the icons you will use in building the user interface.

- ***Tutorial 9.2 Writing the User Interface*** goes through all the code that needs to be written for the alien control system, explaining all the new commands along the way.

THE USER INTERFACE

The interface that you are going to develop is broken up into four areas. Each area and function is described below:

MAIN CONSOLE

The main console provides a graphical overview of the alien with buttons strategically placed at various locations that allow for a quick selection of major control elements. Pressing one of the buttons will perform the selection for you so that you won't have to go all over the Modify panel making selections.

At first glance you might not think that this is really needed, but after moving the alien's limbs around for a while you'll discover that all the point helpers have moved and are no longer easily identified.

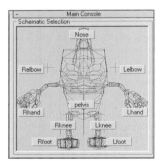

Figure 9.1 *Main Console section*

MORPH TARGETS

The Morph Targets section of the UI makes facial animation a breeze. Instead of having to open the Morpher every time you want to keyframe the weight of a morph target, you can simply do it from this section.

One special feature is a camera linked to the alien to point at his face at all times. This is simply for the purpose of always having a front-on view of the alien's face when doing facial animation.

Also, the interface provides a picture for each target showing it at 100%. This makes it easy to identify a type of deformation that you might want to introduce into the character animation. We will set up the camera a little later in this chapter.

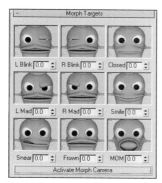

Figure 9.2 *Morph Targets section*

HAND CONTROLS

The Hand Controls section gives a graphical overview of the hand along with the corresponding spinner that directly controls the custom attributes you added earlier. This section provides a way to easily store and recall hand poses that you use frequently. Through the use of two radio buttons at the top of the rollout, you can quickly switch back and forth between the left and right hands.

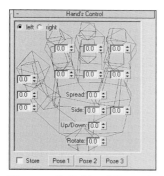

Figure 9.3 *Hand Control section*

FOOT CONTROLS

Here you are given quick access to the left and right feet so that you can quickly set the foot to a particular position or wiggle the toes, one by one.

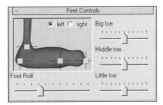

Figure 9.4 *Foot Control section*

In developing an interface like this, some preplanning is required. The first to determine is where you want the UI to be located. We saw in *Chapter 6* that there are two primary places where we can create interfaces: the Utility panel, and in a floater window.

With our interface, we'll put everything in a floater. This way, we'll be free to work in expert mode with the **3ds max** interface hidden, giving us lots of room on the screen.

TUTORIAL 9.1 ICONS FOR USER INTERFACE

Before we begin writing the code for the user interface, we will need to create several pictures. These pictures will be used as icons in the interface to help you and any other users quickly identify key areas of our character. In this way, you won't have to read through the numerous buttons and labels while hunting for a particular character control, and you'll be able to find the UI elements much more quickly.

These pictures will need to be cropped and sized to accommodate the look of the interface. Photoshop or another paint package is required to do this.

Preparing the Alien for Pictures

1. Load the *Alien - Setup and Deformation.max* file you created in **Chapter 8**, or load the file *Alien - Setup and Deformation.max* from the **Scenes/Aliens** folder on the CD.

2. Position the alien so his arms are outstretched.

3. Select both the left and right **Point Elbow Target** objects. In the Perspective view, move the targets up on the Z axis so that you can see the back of the hand in the Front viewport.

4. Select **Point Alien Root** and move the alien down slightly on the Z axis. Move the point helper down only far enough to cause a slight bend in the alien's knees.

5. Select the two **Knee Target** point helpers, one at a time, and move them so that the knees point outward and the point helpers are clear of the geometry. This will make it easier to see them.

6. Hide everything except the geometry, bones, and five of the point helpers:

   Point Alien Root
   L - Point Elbow Target
   R - Point Elbow Target
   L Point - Knee Target
   R - Point - Knee Target

   We are hiding the point helpers at the wrist and feet because we want to keep the image as simple as possible.

7. Zoom and/or pan the Front viewport to get a good view of the alien, point helpers, and bones.

> **TIP**
> You might want to change the colors of the alien and the bones so that you can easily tell the difference between the two.

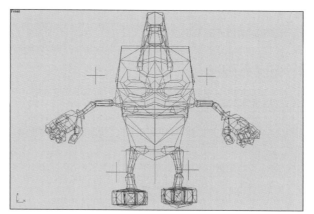

Figure 9.5 *Alien positioned in Front viewport*

8. Turn off the grid by pressing **<G>**, and change the viewport to **Wireframe** mode if necessary.

Creating the Base Picture

1. Capture the screen image to the Windows clipboard.

   If you use screen capture software, you can use your own utility. Otherwise, you can press **<Ctrl-Print Screen>** on your keyboard to capture an image of the screen and put in on the Windows clipboard.

   Next we will use a paint program to crop and size the image. The instructions here are for Photoshop, but you can use any program you like.

2. Open Photoshop and create a new image by choosing **File/New** from the menu. The width and height will already be set to match the image size that is on the clipboard. Click **OK** to open the new image.

3. Paste the captured screenshot into the new image by pressing **<Ctrl-V>** on your keyboard.

4. Use a marquee to select the area around the alien in the Front viewport.

5. Crop the image. In Photoshop, this is accomplished with *Image/Crop* or *Edit/Crop* depending on the version you are using.

   The image now needs to be the proper size.

6. Choose *Image/Image Size* from the menu. On the dialog, uncheck **Constrain Proportions** if necessary. Set the **Width** to 250 pixels and the **Height** to 240 pixels, and click **OK**.

7. Save the image with the filename *Alien_map.jpg* in the *Scripts* folder under your *3dsmax4* folder.

   The images you create must be in the same folder as your scripts, or the scripts you create later on to reference the pictures won't work.

8. Put the alien back in his default pose, either by moving the point helpers or by reloading the file without saving.

Creating the Hand Picture

Next we will create a close up shot of the hand. We are interested in capturing both geometry and bones.

1. Zoom, pan or rotate the Top viewport so that you are focused on the alien's right hand.

2. Make sure only the bones and geometry are visible. If you see any other objects, hide them.

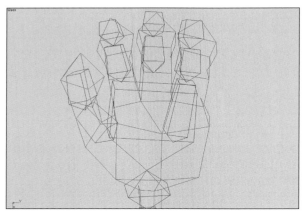

Figure 9.6 *Alien hand properly positioned*

3. Capture the image. In Photoshop, crop and/or resize the image to a **Width** of 240 pixels and a **Height** of 250 pixels.

4. Save the file in the *Scripts* folder as *Alien_hand.jpg*.

Creating the Foot Picture

The foot is the final picture that needs to be taken. This picture will need to be taken in a shaded view.

1. Zoom, pan and/or rotate the Left viewport so that you have a tight shot on the ankle and foot.
2. Hide everything except the model.
3. This step is not a requirement, but can be helpful. We have added four new point helpers and positioned them closer to the viewing angle than the foot, but in such a way that they provide an indication of the actual foot point helpers. These point helpers serve as reminders of the actual location of the point helpers you will be using.
4. You can also set the color of the alien to a bright color that will stand out, such as red. After you take the screen shot, you can change the color back.

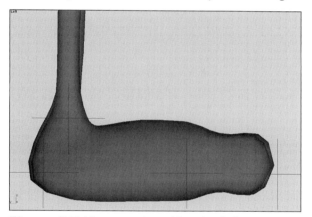

Figure 9.7 *Alien properly positioned in Left viewport*

5. Capture the image. In Photoshop, crop and/or resize the image to a **Width** of 150 pixels and a **Height** of 100 pixels.
6. Save the image in the *Scripts* folder as *Alien_foot.jpg*.

Creating the Morph Target Pictures

In this section, you will be creating nine different pictures and a camera. The pictures will be used as references for morph targets, while the camera will serve later as a non-objective view when animating the morph targets.

1. Create a free camera in the Front viewport and name it **morphCamera**.
2. Link the camera to **Bone - Back B**, the bone in the alien's head.
3. Change the Front viewport to the camera view by activating the viewport and pressing the **<C>** key on the keyboard.

4. Move the camera or work with the viewport itself until the alien's face fills most of the view, as shown in the next figure.

Figure 9.8 *Neutral facial pose*

5. Apply a temporary gray material to the alien with little or no shininess.

   This will help keep the pictures easier to understand once imbedded in the interface.

6. On the **Modify** panel, select the **Morpher** modifier on the stack.

7. Verify that all channels are set to 0. If you find one that is not, set it to 0.

   At this point, the alien should be in his neutral pose.

8. From this point forward, you will need to set each channel to 100 and all others to 0, capture the screen, go into Photoshop and crop the viewport. Then resize the images down to a **Width** of 85 and a **Height** of 60.

   Images of the nine pictures follow. Below each image is the filename to which the cropped image should be saved. Be sure to save each image in the **Scripts** folder.

Figure 9.9 *Left eye mad - Lmad.jpg*

Alien UI Control System **441**

Figure 9.10 *Right eye mad - Rmad.jpg*

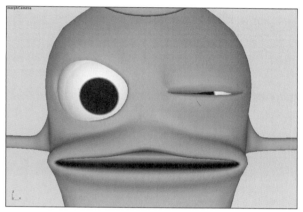

Figure 9.11 *Left eye blink - LBlink.jpg*

Figure 9.12 *Right eye blink - RBlink.jpg*

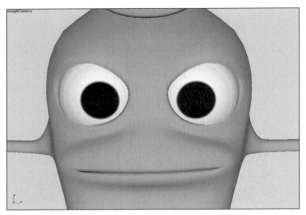

Figure 9.13 *Mouth closed - Closed.jpg*

Figure 9.14 *Sneering - Sneer.jpg*

Figure 9.15 *Frowning - Frown.jpg*

Figure 9.16 *Smiling - Smile.jpg*

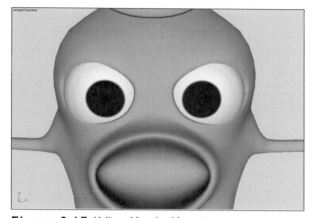

Figure 9.17 *Yelling Mom! - Mom.jpg*

9. Verify that all images are saved with the correct names.

   At this point you can reload the scene without saving, and get ready to start scripting!

TUTORIAL 9.2 WRITING THE USER INTERFACE

You will be writing this script in the MAXScript editor. After each section of code, we will explain the functionality of what you have just added. You will not always be reminded to save, so make sure that you occasionally do so.

Creating the Main Console

Now that we have everything in place, let's get started:

1. Start a new script.

   You can do this from the **Utilities** panel by clicking the **MAXScript** button and then the **New Script** button.

2. Type in the first few lines of the script:

   ```
   --------------------
   -- UI Window --
   --------------------
   w=newrolloutfloater "Animation Controls" 300 320
   --Here we create an empty window and store an instance
   --of it in the variable w
   ```

   We begin by using comments to let others know that the following section is used to create the floater window itself. This script is going to grow into quite a long program. Because of this, comments will be one of the most important things in it.

   Our first true line of code uses the variable **w** to store an empty window that has a predefined caption and size. By assigning the floater window to a variable, we simplify the code that will be required later when referring to it.

   Now that we have a window, we need to start attaching the rollouts that will end up housing our UI elements such as buttons and sliders. Without these rollouts, you will not be able to add the various types of elements.

3. Continue with the next few lines of code:

   ```
   ----------------------------
   -- UI Rollout MainConsole --
   ----------------------------
   rollout rll_main "Main Console" width:300 height:280
   ( --begin the rollout
   ----------------------------
   -- UI Controls in rll_main --
   ----------------------------
       groupBox grp2 "Schematic Selection" pos:[5,0] width:262 height:265
       bitmap bmpSchematic fileName:"Alien_map.jpg" pos:[10,15]
       button btnNose "Nose" pos:[110,24] width:48 height:16
       button btnPoleRHand "Relbow" pos:[15,92] width:48 height:16
       button btnPoleLHand "Lelbow" pos:[205,92] width:48 height:16
       button btnRHand "Rhand" pos:[15,175] width:48 height:16
       button btnLHand "Lhand" pos:[210,175] width:48 height:16
       button btnPelvis "pelvis" pos:[110,160] width:48 height:16
       button btnPoleRFoot "Rknee" pos:[75,200] width:48 height:16
       button btnPoleLFoot "Lknee" pos:[150,200] width:48 height:16
       button btnLFoot "Lfoot" pos:[180,232] width:48 height:16
       button btnRFoot "Rfoot" pos:[40,232] width:48 height:16
   ```

   The above code introduces a few new UI elements that you have not seen before, the **groupBox** and **bitmap** commands.

The **groupBox** command creates small frames around various elements that are related in one way or another. The only arguments that a **groupBox** will take are the caption and its dimensions. A small thin line is generated that surrounds the controls. The caption is laid over this line in the upper left corner. In this case, the groupBox displays the caption **Schematic Selection**.

The bitmap control allows us to put a picture (bitmap) into the user interface. By generating the bitmap element first, we can add other elements over it. With this script, we will be adding several buttons on top of it.

The idea is to create a graphical representation of the alien setup and then position selection buttons on top of the bitmap. This will provide a sort of visual map to the user when looking for a particular control handle to select.

The **bitmap** command takes a filename and a position as arguments. At first glance this seems to pose no problems. However, MAXScript might return an error telling you that the bitmap cannot be located. This is because it can't find the path. In our case we have placed the three images into the **Scripts** folder, which is one of the places. 3ds max will look before generating an error, so we should be alright.

With the bitmap now in place, we begin placing buttons. Each button has a position and dimension argument supplied along with the required name and caption. We need to specify a position in order to be able to place the buttons so that they match the bitmap. The dimensions of the buttons are changed from the default to a slightly smaller size. This is a matter of personal preference and is not required.

You may need to adjust your position settings so your buttons are better positioned with your bitmap. You will not be able to do this until after this rollout section is completed and you add the rollout to the floater that is being stored in the variable **w**.

Adding Functionality to Buttons

Now that we have positioned all the buttons on top of the bitmap control, let's add some functionality to them. Each of the buttons will work as a quick selection for the major point helpers and bones. Each event statement will be on a single line.

We do not need to use parentheses to block code when there is only one line to be executed. In this case each button has only one function, which is selecting an object. The format that we will use for each button is:

On [button name] **pressed do select** [object]

What could be simpler?

☀ **TIP** ☀

Remember that there are spaces between the dashes in the names of some objects. If you do not put the spaces in the name and surround the name by quote marks, **3ds max** *will not be able to find the object you are referring to and the script will fail.*

1. Add the following code to the script:

   ```
   -------------------------------
   -- Events for rll_Main --
   -------------------------------
         --Each button selects an object.
         on btnNose pressed do select $'Bone - Nose A'
         on btnPoleRHand pressed do select $'R - Point Elbow Target'
         on btnPoleLHand pressed do select $'L - Point Elbow Target'
         on btnRHand pressed do select $'R - Point Hand Controls'
         on btnLHand pressed do select $'L - Point Hand Controls'
         on btnPelvis pressed do select $'Point Alien Root'
         on btnPoleRFoot pressed do select $'R - Point - Knee Target'
         on btnPoleLFoot pressed do select $'L - Point - Knee Target'
         on btnRFoot pressed do select $'R - Point - Ankle Rotate'
         on btnLFoot pressed do select $'L - Point - Ankle Rotate'
   )     --end the rollout
   ```

   We have used single quotes around each object name. This is only required when an object's name contains spaces. Most of the objects in our scene have names that contain spaces.

   We end the rollout with a closing parenthesis.

   As we've seen before, it is necessary to add the rollout to the window in order for it to be displayed and used.

2. Be sure to put the code at the end of the script after the last parentheses block ends.

   addrollout rll_main w   --here we add the rollouts to the floater
                                     --stored in w

3. Save the script as **Alien UI.ms**, and evaluate the script.

   If you entered everything properly, you should see a new window pop up with the alien picture you took earlier accompanied by various buttons. Clicking each button should result in the corresponding objects being selected.

Adding the Morph Targets Rollout

Next we will add the Morph Target rollout and set up the various controls that will make morph animation a piece of cake. This rollout will contain nine bitmap images and nine spinners, one for each morph target.

The next section of code will begin with another set of comments, again this if only for developers (including yourself) that might read this code in the future.

After the comments, we begin our second rollout in the script followed by its various controls.

1. Add the following code to your script. You will need to add it after the very last closing parenthesis and before the **addrollout** line, which should be the last line of code in your script.

```
-------------------------
--UI Rollout Morph Targets --
-------------------------
rollout rll_morph "Morph Targets"
(
------------------------------------
-- UI Controls in rll_morph --
------------------------------------
    bitmap bmpSchematic1 fileName:"LBlink.jpg" pos:[5,15]
    bitmap bmpSchematic2 fileName:"RBlink.jpg" pos:[93,15]
    bitmap bmpSchematic3 fileName:"Closed.jpg" pos:[181,15]
    spinner spLBlink "L Blink" width:55 pos:[30,85] range:[0,100,0]
    spinner spRBlink "R Blink" width:55 pos:[119,85] range:[0,100,0]
    spinner spClosed "Closed" width:55 pos:[205,85] range:[0,100,0]
    bitmap bmpSchematic4 fileName:"Lmad.jpg" pos:[5,105]
    bitmap bmpSchematic5 fileName:"Rmad.jpg" pos:[93,105]
    bitmap bmpSchematic6 fileName:"Smile.jpg" pos:[181,105]
    spinner spLMad "L Mad" width:55 pos:[30,175] range:[0,100,0]
    spinner spRMad "R Mad" width:55 pos:[119,175] range:[0,100,0]
    spinner spSmile "Smile" width:55 pos:[205,175] range:[0,100,0]
    bitmap bmpSchematic7 fileName:"Sneer.jpg" pos:[5,195]
    bitmap bmpSchematic8 fileName:"Frown.jpg" pos:[93,195]
    bitmap bmpSchematic9 fileName:"Mom.jpg" pos:[181,195]
    spinner spSneer "Sneer" width:55 pos:[30,265] range:[0,100,0]
    spinner spFrown "Frown" width:55 pos:[119,265] range:[0,100,0]
    spinner spMom "MOM" width:55 pos:[205,265] range:[0,100,0]
    button btnUpdate "Activate Morph Camera" align:#right width:262 height:20 pos:[5,285]
```

The above code will create a 3 x 3 group of pictures, each with corresponding spinners located directly beneath each one.

There is a chance that the pictures are laid out differently then the channels in the Morpher, but this does not matter. It is important to lay the pictures out in a sensible manner so that related channels are next to one another.

We will need to be cautious when we write the event code so that each spinner affects the appropriate morph channel.

2. Continue adding the code that follows. This code will handle all of the required event calls.

```
--------------------------
-- Events in rll_morph --
--------------------------
    on spLBlink changed Phovalue do
    (
        $'Alien Body'.morpher[4].value=Phovalue
        -- sets the morpher modifier's 4th channel
    )
    on spRBlink changed Phovalue do
    (
        $'Alien Body'.morpher[2].value=Phovalue
        -- sets the morpher modifier's 2nd channel
    )
    on spClosed changed Phovalue do
    (
        $'Alien Body'.morpher[5].value=Phovalue
        -- sets the morpher modifier's 5th channel
    )
    on spLMad changed Phovalue do
    (
        $'Alien Body'.morpher[1].value=Phovalue
        -- sets the morpher modifier's 1st channel
    )
    on spRMad changed Phovalue do
    (
        $'Alien Body'.morpher[3].value=Phovalue
        -- sets the morpher modifier's 3rd channel
    )
    on spSmile changed Phovalue do
    (
        $'Alien Body'.morpher[9].value=Phovalue
        -- sets the morpher modifier's 9th channel
    )
    on spSneer changed Phovalue do
    (
        $'Alien Body'.morpher[6].value=Phovalue
        -- sets the morpher modifier's 6th channel
    )
```

```
on spFrown changed Phovalue do
(
    $'Alien Body'.morpher[8].value=Phovalue
    -- sets the morpher modifier's 8th channel
)
on spMom changed Phovalue do
(
    $'Alien Body'.morpher[7].value=Phovalue
    -- sets the morpher modifier's 7th channel
)
on btnUpdate pressed do
(
    a=$morphCamera
    viewport.setcamera a
)
)
```

The above code snippet contains an event code for each spinner and one for the **Activate Morph Camera** button at the bottom of the rollout.

The event code for each spinner stores the changed spinner value in a temporary variable called **Phovalue**. This value is then fed to the appropriate morph channel on the Morpher modifier. For example, **morpher[1]** is channel 1. The result is the alien morphing more toward the weight you just assigned.

The event for the button being pressed will cause the current active viewport to switch to the morphCamera that you created earlier. This way you can get a straight-on view while working with the Morpher.

This rollout can now be added to our floater.

3. Add the following line to the very end of your script. It should come after the line: **addrollout rll_main w**.

   addrollout rll_morph w

4. Save, evaluate and test the script.

   Verify that all of the morph targets/channels respond accordingly. If you run into any errors, review your code for any misspellings.

   Once you are satisfied that everything is functioning the way it should, you can move on to the hand controls.

Adding the Hand Controls

Our next rollout will allow for easy hand control. Our approach will be similar to the Main Console where we used a graphic with overlaid buttons. To keep the UI size from growing, we will use only one graphic to represent both hands. Simple radio buttons will allow us to tell the UI which hand we are referring to.

This rollout will also contain a feature for storing poses. Once we create a pose with one of the hands that we may need to use again, so we can simply store it and grab it again later. A checkbox will be used to determine if we are storing or recalling a pose.

1. Continue adding the following code after the last closing parenthesis in the editor:

```
-------------------------------------
-- UI rollout hand's control --
-------------------------------------
rollout rll_hands "Hands Control"
(
-------------------------------------
-- UI Controls in rll_hands --
-------------------------------------
bitmap bmpHand fileName:"Alien_hand.jpg" pos:[5,5]
--again the bitmap is created first
--it is important to note the creation order of the spinners - they match
--the creation order of the custom attributes that they will control.
--In this way we can easily update the custom attributes by using an
--index variable in a for loop.
```

The first thing we are going to do in this rollout is add a bitmap, the *Alien_hand.jpg* picture you created at the beginning of this chapter. It is created first so that all of the other controls added will be on top. Again, we are looking for a way of organizing all of the controls (spinners) so that they can be quickly identified with the assistance of this picture. Don't forget, if you put your image in a different folder than scripts, you will need to provide a path and this path is case-sensitive.

Generally the creation sequence of controls does not really matter unless you are planning to put one on top of the other as with the bitmap we used earlier. But in this section of code, the creation sequence of each spinner is critical. We will create the spinners in the same order as the custom attributes that were added to control the finger and thumb rotations. In this way, we will be able to use a single index value to loop through the controls whenever they need to be updated.

TIP

You might need to adjust the position of the controls so that the spinners better match your bitmap.

Updated? Don't forget, we are using only a single graphic of the hand and will need to switch back and forth between the two hands. When we switch from one hand, we will need to have code that will read all of the attribute values for that hand and assign those values to the spinners. When we switch back to the other hand, the script will need to repeat the updating action. If we did not take the time to update the spinners when we switched back and forth between the two hands, you could easily set a motion by accident.

All the spinners are positioned so that they appear at the corresponding place on the bitmap and their ranges match the custom attributes they are driving.

2. Continue adding the following code:

```
spinner spIndexBase "" width:40 height:1 pos:[80,100] range:[0,10,0]
spinner spIndexTip "" width:40 height:1 pos:[80,50] range:[0,10,0]
spinner spMiddleBase "" width:40 height:1 pos:[135,100] range:[0,10,0]
spinner spMiddleTip "" width:40 height:1 pos:[135,50] range:[0,10,0]
spinner spPinkyBase "" width:40 height:1 pos:[185,100] range:[0,10,0]
spinner spPinkyTip "" width:40 height:1 pos:[185,50] range:[0,10,0]
spinner spPinkyCup "" range:[-5,5,0] pos:[185,170] width:40 height:1
spinner spThumbSpread "" range:[-5,5,0] pos:[10,70] width:40 height:1
spinner spThumbMiddle "" width:40 height:1 pos:[10,125] range:[0,10,0]
spinner spThumbTip "" width:40 height:1 pos:[10,100] range:[0,10,0]
spinner spWristUpDown "Up/Down:" width:50 height:1 pos:[130,200]
    range:[-10,10,0]   -- this goes on the same line as previous line
spinner spWristSide "Side:" width:50 height:1 pos:[130,170] range:[10,10,0]
spinner spWristRotate "Rotate:" width:50 height:1 pos:[130,230]
    range:[-10,10,0]   -- this goes on the same line as previous line
spinner spFingerSpread "Spread:" width:50 height:1 pos:[130,140]
    range:[-5,10,0]   -- this goes on the same line as previous line
```

Next, we will create the radio buttons that will allow the user to toggle between the left and the right hand. A radio button takes a name and a caption argument just like most other UI elements. However, instead of creating one radio button at a time (like other controls), you create them in groups. The number of elements in the 'labels' array defines the number of radio buttons that get created. In this case all we need is a left and right radio button. We name the radio button set **LRToggle** and position it near the top of the rollout.

3. Add the following code:

radiobuttons LRToggle labels:#("left", "right") pos:[10,10]

The only UI elements missing are the buttons that will allow us to store and recall a pose and the checkbox that will determine which mode the buttons are functioning in.

4. Continue with the following code:

    checkbox chkbxStore "Store" pos:[10,268]
    button btnPose1 "Pose 1" pos:[70,265]
    button btnPose2 "Pose 2" pos:[120,265]
    button btnPose3 "Pose 3" pos:[170,265]

5. Now that we have all of the controls in place, it is time to start adding the required event code so that each control will have functionality.

    on spPinkyCup changed rotValue do
    (
        toggleString = whichSide LRToggle
        obj=execute (toggleString+"Point Hand Controls'")
        obj.pinky_cup=rotValue
)

    Since each spinner can represent one hand or the other, the first thing we need to do is determine which hand is being referred to. We will need to do this often in both the hand control and feet control sections so it is best to create a function that assists with this task. By creating a function, we will reduce the amount of redundant code.

    In this case, we are going to create a function called **whichSide**. This function will be setup to receive a single argument; a UI Control. In both hand and foot controls we will use radio buttons to determine which side of the character we are referring to. We will simply send these radio button groups to the function and let the function figure out whom we are talking about.

    The function will return a string value of **L** or **R** depending on the state of the radio buttons. The number 1 will always be left and 2 will always be right. This returned value is then stored in a variable called **toggleString**. After **toggleString** is assigned, we add it to the string **'Point Hand Controls'**, creating the name of the correct point helper.

    After determining the correct point helper object that contains the hand control attributes, we store it in the variable **obj**. The code at this time will be the same for each of the spinners. Using dot notation we can now add the respective finger control. In this case we add the text **pinky_cup**.

    At this point, you can continue adding all of the remaining events for the spinners. You will see that the code is exactly the same with the exception of the spinner name and the custom attribute that we are referring to.

6. Continue adding the following events:

   ```
   on spThumbSpread changed rotValue do
   (
       toggleString = whichSide LRToggle
       obj=execute (toggleString+"Point Hand Controls'")
       obj.thumb_base=rotValue
   )
   on spWristSide changed rotValue do
   (
       toggleString = whichSide LRToggle
       obj=execute (toggleString+"Point Hand Controls'")
       obj.wrist_side_to_side=rotValue
   )
   on spWristUpDown changed rotValue do
   (
       toggleString = whichSide LRToggle
       obj=execute (toggleString+"Point Hand Controls'")
       obj.wrist_up_down=rotValue
   )
   on spWristRotate changed rotValue do
   (
       toggleString = whichSide LRToggle
       obj=execute (toggleString+"Point Hand Controls'")
       obj.wrist_rotation=rotValue
   )
   on spThumbTip changed rotValue do
   (
       toggleString = whichSide LRToggle
       obj=execute (toggleString+"Point Hand Controls'")
       obj.thumb_middle=rotValue
   )
   on spThumbMiddle changed rotValue do
   (
       toggleString = whichSide LRToggle
       obj=execute (toggleString+"Point Hand Controls'")
       obj.thumb_spread=rotValue
   )
   ```

```
on spIndexBase changed rotValue do
(
    toggleString = whichSide LRToggle
    obj=execute (toggleString+"Point Hand Controls'")
    obj.index_knuckle=rotValue
)
on spIndexTip changed rotValue do
(
    toggleString = whichSide LRToggle
    obj=execute (toggleString+"Point Hand Controls'")
    obj.index_middle=rotValue
)
on spMiddleTip changed rotValue do
(
    toggleString = whichSide LRToggle
    obj=execute (toggleString+"Point Hand Controls'")
    obj.middle_middle=rotValue
)
on spMiddleBase changed rotValue do
(
    toggleString = whichSide LRToggle
    obj=execute (toggleString+"Point Hand Controls'")
    obj.middle_knuckle=rotValue
)
on spPinkyTip changed rotValue do
(
    toggleString = whichSide LRToggle
    obj=execute (toggleString+"Point Hand Controls'")
    obj.pinky_middle=rotValue
)
on spPinkyBase changed rotValue do
(
    toggleString = whichSide LRToggle
    obj=execute (toggleString+"Point Hand Controls'")
    obj.pinky_knuckly=rotValue
)
on spFingerSpread changed rotValue do
(
    toggleString = whichSide LRToggle
    obj=execute (toggleString+"Point Hand Controls'")
    obj.finger_spread=rotValue
)
```

So that everything executes correctly the next time we test our script, we will need to create the **whichSide** function. Note that all of our functions will be added to the very beginning of the script.

7. Add the following code to the very beginning of your script, before all other lines of code:

fn whichSide UIelement =
(
    holder=case UIelement.state of
    (
        1: "$'L - "
        2: "$'R - "
)
)

As mentioned earlier, this function takes one argument. The incoming argument (a radio button group) will be stored in the variable **UIelement**. From there we will test the state of the control.

Instead of using an **if** statement, we will use a **case of** statement. This is another way of expressing an **if** statement. Our decision to use a **case of** statement is a personal preference in this case -- either one would work fine.

The **case of** syntax is as follows:

Case [exp] **of** ([cases])

The **case** command takes an expression and compares it to a number of cases. The cases are listed within the parentheses block and followed by a colon and code. When a match is found, MAXScript will execute whatever comes after the colon.

In our situation, the radio button groups will have a state of either 1 or 2. If the first radio button is active, the function will return the string **"$'L - "**, indicating the left side of the body.

Updating the Spinners

The next thing that we need to create is a function that will update the UI spinners when you swap between left and right.

Enter the following function just below the last function you entered:

fn updateUI currentRll currentObj end=
(
    holder=custAttributes.get currentObj 1
    --The command takes two arguments, the name of the object and
    --the set's index number.
    --Once this is stored in the holder variable we can start our loop.
    for i=2 to end do
    (
        currentRll.controls[i].value=holder[i-1].value
)
)

We will cover this function in just a moment, but first let us write the event that will be responsible for calling it. The event should be triggered by a click on the **Left** or **Right** radio buttons, so we will need to add an event for **LRToggle**.

1. Add the following event code inside the **rll_hands rollout** commands, just below the last spinner event that you added earlier:

on LRToggle changed side do
(
    case side of
    (
        1: updateUI rll_hands $'L - Point Hand Controls' 15
        2: updateUI rll_hands $'R - Point Hand Controls' 15
)
)

The event uses another 'case of' statement to determine which side of the character we are referring to.

With the hand controls rollout, we are going to update all of the spinners so that they match the values in the custom attributes on the Point Hand Controls.

The **updateUI** function is designed to update any rollout so that it can be reused when we start developing the feet controls rollout. In total, three arguments are passed to the function: The current rollout, the given hand control object and a single integer value, which represents the number of custom attributes + 1. The integer value that is sent to the function will simply be used to determine how long the **for** loop will run, each time updating a new UI element in the specified rollout.

Let's take a closer look at the **updateUI** function. As you can see, the **for** loop is started at 2 rather than 1. The reason is that the first UI element in the rollout is a bitmap. We cannot change the creation order of the bitmap in relation to the spinners, since it then would lie on top of the spinners, hence they would not be visible. The custom attributes, on the other hand, start at 1. This means we will have to subtract 1 in order to make the two sets of values match.

MAXScript stores the custom attributes in sets. In the function we store the custom attribute set in a variable called **holder**.

By using this approach, we can easily use this function for the feet rollout by passing the feet rollout, the foot control object, and the correct integer value to the function. As you can see, it is crucial that the UI elements order of creation matches the creation order of the custom attributes of the hand, else we might have gotten the thumbs custom attribute values dropped into the pinky's spinners. We could have used specific names for all of the UI controls and custom attributes but that would not have been an efficient approach.

Storing and Retrieving Poses

The only things left to create on the hands rollout are the pose storage buttons. The setup is rather simple, but leaves you room to expand upon the code.

We are going to take a new approach to this task. Ordinarily we would save each related spinner value in a variable and then recall the variable when the user wished to retrieve a pose. The problem with this approach is that the poses are only good during your current session of **3ds max**. If you close and restart **3ds max**, your stored poses will be gone. Not good.

Fortunately for us, when Discreet wrote **3ds max** they left a slot on all nodes open for third-party developers. When information is stored in this slot, it gets saved when the scene is saved. When the file is reloaded, the information is still there. By taking advantage of this setup, we can easily store and retrieve poses even after the file has been saved, closed, and later reopened.

The commands required to gain access to this special slot are **setAppData** and **getAppData**.

Before we can continue writing code, we must first decide which object we will use to store the extra data. In our case, we have decided to add a dummy object to the scene and store the poses with it. Let's create the dummy object now.

1. Create a dummy helper object anywhere in your scene. Name the object **poseSaver**.
2. You can hide the dummy object if you wish. We only need it to be present so we can information in it.

    Next, we will be creating an event call for each of the three Pose buttons and two functions, one to set a pose and one to retrieve a pose.

3. Add the following code in the **rll_hands rollout**; just below the last event call you created:

```
-- Pose button 1 pressed event
on btnPose1 pressed do
(
    if chkbxstore.state==true then
        (
            setPose rll_hands $poseSaver 0
        )
        else
        (
            getPose rll_hands $poseSaver 0
        )
)
-- Pose button 2 pressed event
on btnPose2 pressed do
(
    if chkbxstore.state==true then
    (
        setPose rll_hands $poseSaver 50
    )
    else
    (
        getPose rll_hands $poseSaver 50
    )
)
-- Pose button 3 pressed event
on btnPose3 pressed do
(
    if chkbxstore.state==true then
    (
        setPose rll_hands $poseSaver 100
    )
    else
    (
        getPose rll_hands $poseSaver 100
    )
)
)           -- this is the final closing parenthesis for the hands rollout
```

Now, when any of the three **Pose** buttons are pressed, the state of the checkbox **chkbxStore** is checked. If the checkbox is checked then the state will return true letting us know that the user has chosen to store a pose, and the **setPose** function is called. We pass the following arguments:

rll_hands - The current rollout

$poseSaver - The dummy object where the pose will be stored

0, 50, 100 - Offsets for the **setAppData** command for poses 1, 2, and 3 respectively

If the checkbox returns false, then the given pose is loaded. The exact same arguments are passed, but instead of calling **setPose** we call **getPose**.

The **setPose** function will store the current hand pose to the dummy object **poseSaver**.

4. Add the **setPose** function to the top of your script with the other functions:

```
fn setPose currentRll valHolder btnINC=
(
    for i=1 to currentRll.controls.count do
    (
        try (
            setAppData valHolder (i+btnINC)
((currentRll.controls[i].value) as string)
        )
        catch (print "not valid UI component")
    )
    --then we store them at the correct char on
)
```

In order to understand how this function works, you must understand how the **setAppData** command works. As mentioned earlier, this command allows you to store a string in an object using an index value to determine where the value should be written. A **for** loop is used to make stepping through each UI control within the rollout a breeze.

We are able to determine the number of controls in the rollout by checking the parameter **currentRll.controls.count**. With each control, we attempt to perform the **setAppData** command by embedding it in a try/catch error handler. We have to do this because we are trying to extract a value (numerical value) from each control. Since we are walking through every control in the hands rollout, the loop will encounter the bitmap control, along with a few others that do not have a value parameter associated with them. If we did not use the try/catch code, the script would fail the moment it tried to look for a value on a control that does not have one.

If the **setAppData** code does execute, then the current control's value is written to the **poseSaver** dummy. Where it is written is very important. Each of the pose buttons send a different index value that will be used as the starting point of the data to be written: **Pose 1** sends 0, **Pose 2** sends 50, and **Pose 3** sends 100. The actual numbers are not important as long as they do not overlap one another.

As the data is being written to **poseSaver**, it is written at the index number that was sent plus the offset. This offset is simply the current iteration number through the loop. We do this so that each control value is not written on top of the last control value.

The **getPose** function is very similar to the **setPose** function.

5. Add the **getPose** function in the functions section of your script:

```
fn getPose currentRll valHolder btnINC =
(
    for i=1 to currentRll.controls.count do
    (
        try
        (
            holder=(getAppdata valHolder (i+btnINC))
            currentRll.controls[i].value=(execute holder)
            currentRll.controls[i].changed (execute holder)
        )
        catch(print "not valid UI component")
    )
)
```

This function also takes; a rollout, an object that holds the stored data, and a pointer for where to start reading the data. Again we use the try and catch to allow the script to pass through empty index ID's.

In this function, we dynamically store the values in the variable holder. Remember that the value was stored as a string, so in order to read it back as a value we use the execute command. We drop the value into the current spinner and force its event call by using **changed** and the value. In this way we force the alien's pose to be updated.

6. So that this rollout is added to our current floater, we need to add the next line of code to the very bottom of your script:

addrollout rll_hands w

You should now be able to test the script for any problems. Evaluate it and try all of the different UI controls that you've added. If anything goes wrong, recheck all of your spelling and punctuation for errors. After you have verified that everything functions correctly, move on to the feet controls rollout.

Adding the Feet Controls

The final rollout is for the foot controls. This rollout is very similar to the hand control rollout where we will be using a bitmap image of the foot with overlaying controls.

As always, the bitmap control must be executed first so that the other controls will exist on top of it. We will also use a radio button control just like we did in the hands rollout so that we can toggle back and forth between the left and right foot.

Buttons will be strategically placed so that we can easily select key pivot points on the alien's foot. Finally, sliders are used to allow quick access to the foot custom attributes that were set up in **Chapter 7**.

1. Add the following code after the **rll_hand rollout**:

    ```
    ---------------------------------------
    -- UI rollout feet control --
    ---------------------------------------
    rollout rll_feet "Feet Controls"
    (
    ---------------------------
    -- Controls in rll_Feet --
    ---------------------------
        bitmap bmpHand fileName:"Alien_foot.jpg" pos:[5,5]
        radiobuttons LRToggle labels:#("left", "right") pos:[70,10]
        button btnAnkle "" pos:[35,52] width:10 height:10
        button btnHeel "" pos:[22,87] width:10 height:10
        button btnBall "" pos:[92,89] width:10 height:10
        button btnToe "" pos:[130,89] width:10 height:10
        slider slRoll "Foot Roll" pos:[0,110] width:160 range:[-5,10,0]
        slider slBig "Big toe" pos:[160,10] width:110 range:[-5,5,0]
        slider slMiddle "Middle toe" pos:[160,60] width:110 range:[-5,5,0]
        slider slLittle "Little toe" pos:[160,110] width:110 range:[-5,5,0]
    ---------------------------
    -- Events in rll_Feet --
    ---------------------------
        on LRToggle changed side do
        (
            case side of
            (
                1: updateUI rll_feet $'L - Point - Ankle Rotate' 4
                2: updateUI rll_feet $'R - Point - Ankle Rotate' 4
            )
        )
    ```

The above code defines the foot control's UI elements along with the first event. The **LRToggle** event is nearly identical to the event code that we used in the hands toggle between left and right.

The event uses the 'case of' statement as before. In total three arguments are passed to the updateUI function: the current rollout, the given foot control object, and then a single integer value. As you saw earlier, the integer value is used to determine how long the 'for loop' will run. The updateUI function is described in the hand controls section above and there is no difference in the way we are using it now.

2. Add the following events for all of the sliders:

   on slBig changed rotValue do
   (
       toggleString = toeRotation LRToggle
       obj=execute (toggleString+"Point - Ankle Rotate")
       obj.bigToe=rotValue
)
   on slMiddle changed rotValue do
   (
       toggleString = whichSide LRToggle
       obj=execute (toggleString+"Point - Ankle Rotate")
       obj.middleToe=rotValue
)
   on slLittle changed rotValue do
   (
       toggleString = whichSide LRToggle
       obj=execute (toggleString+"Point - Ankle Rotate")
       obj.littleToe=rotValue
)
   on slRoll changed rotValue do
   (
       toggleString = whichSide LRToggle
       obj=execute (toggleString+"Point - Ankle Rotate")
       obj.footRoll=rotValue
)

   Notice that the events are identical to the ones that we used for the hand spinners. We simply make a function call to the whichSide function, assemble a string that represents the correct Point - Ankle Rotate object and then update the appropriate custom attribute with the sliders changed value.

3. Now we need to add the events that will handle the button presses:

   on btnAnkle pressed do
   (
       toggleString = toeRotation LRToggle
       obj=execute (toggleString+"Point - Ankle Rotate")
       select obj
)

   on btnHeel pressed do
   (
       toggleString = whichSide LRToggle
       obj=execute (toggleString+"Point - Heel Rotate")
       select obj
)

   on btnBall pressed do
   (
       toggleString = whichSide LRToggle
       obj=execute (toggleString+"Point - Ball Rotate")
       select obj
)

   on btnToe pressed do
   (
       toggleString = whichSide LRToggle
       obj=execute (toggleString+"Point - Toe Rotate")
       select obj
)
)   --closing parenthesis for foot controls rollout

   These events use the same approach as seen above in determining the correct object. A call to the whichSide function is placed with the returned value being added to the appropriate string representing the correct object. Then the object is selected.

4. Finally, you will need to add the feet rollout to our floater window. Add the following line of code to the very end of your script:

   addrollout rll_feet w

   Save your script, evaluate it, and test it for problems. If you do find that something is not functioning correctly, go back through the code and locate the problem. Most of the time it is some sort of punctuation problem.

   You're all done with scripting!

5. Save your script as *Alien UI.ms*.
6. Save the scene as *Alien - Setup with UI.max*.

SUMMARY

In this chapter, you have learned how to:

- Capture and place icons for a custom user interface
- Create a UI for quickly selecting body parts
- Save and retrieve poses so they can be used repeatedly
- Set up a UI for working with morph targets

This UI will speed you up tremendously when animating the alien. It may seem like a lot of code to write, but the payoff will be great.

Mastering MAXScript can take time, but as long as you have the patience to continue working with it, there's no limit to what you can do.

Camera Tracking

This chapter will show you how to integrate CG elements into prerecorded video, giving the appearance that the CG objects truly existed in the scene when the video was shot.

This is accomplished by *tracking* a 3D camera. Tracking is the process of making a 3D camera's animation match the motion of a real camera used to shoot live footage. In this way, the 3D elements can later be placed against the live background for seamless compositing.

TECHNIQUES IN THIS CHAPTER

In this chapter, you will learn how to:

- Place and measure tracking points in the live footage.
- In **3ds max**, create and place helper point objects for use in tracking.
- Match a 3D camera's motion to live footage.

TUTORIALS IN THIS CHAPTER

The tutorials in this chapter will take you through the entire process of animating a 3D camera to match the camera motion in live footage.

- **Tutorial 10.1 CamPoint Setup** will show you how to place helper objects called CamPoints for use in the tracking process.

- **Tutorial 10.2 Camera Tracking** takes you through the process of using the Camera Tracker utility with your live footage.

For the tutorials in this chapter, you can use the live footage provided on the CD, or practice with your own digitized footage.

CAMERA MATCHING

Matching a software camera to a real world camera is not as hard as it sounds, but there is a lot of pre-planning that must be done on the front end. If you do not follow all the appropriate steps, camera matching becomes extremely difficult.

Before getting into the specifics, let's take a quick look at the overall process of integration.

CHOOSING A LOCATION

To start things off, you will need to pick a location that is appropriate for the shot. Don't worry if you can't find a place that is 100% perfect as this is typically impossible. In our shot, we need an alien terrain. When was the last time you saw one of those? Unless we can hitch a ride to the moon, we'll have to work with earthly landscapes.

Fortunately, post production applications nowadays can easily change the color and contrast in footage and paint out specific elements, so we just need to find a land area with shapes and features similar to what we want.

With our scene, the closest thing that we could find to alien terrain was a nearby rock quarry. Of course there were problems with our location. We've got trees in the background, the environment is the wrong color, and the sun was in the wrong location (they were blasting at 8am when the sun was in the perfect spot. Rats!). We will fix these minor problems in our post production session later on.

SHOOTING THE LOCATION

After the location has been picked out, you need to spend some time setting up specific tracking points and recording measurements between them. After all the scene setup is done, you are ready to shoot the video.

You will need to make sure that the video you shoot matches the storyboards that were approved earlier. It is best to do a few practice runs first, then play them back and make sure the motion is what you need. When you begin the final shoot, it is very wise to shoot the scene three of four times just in case something goes wrong in one of them.

In our work with motion tracking, we have found that the footage is best shot with a handheld camera. Although conventional video wisdom says you should use a tripod whenever possible, a handheld camera gives the tracking function more varied data to work with, and thus actually gives fewer tracking errors than footage shot with a tripod.

After the shoot, you return to your studio where you digitize or save the video on your editing system. From there you convert those files into a series of TIF or TGA images that can be loaded into the background of your **3ds max** scene. At this point the footage is strictly a 2D element in the background of the scene.

Lucky for you, we have done all that work already. This chapter picks up the process at the point where you're ready to match the 3D camera to the live footage.

MATCHING THE 3D CAMERA TO THE LIVE FOOTAGE

Next, you will use **3ds max** utilities to match a 3D camera to the real world camera on one frame. Matching a 3D camera simply means having it placed in the scene so that its perspective is the same as the original real world camera. The focal length is also adjusted to match the original one. With a good camera match in place, an object in the scene would look as if it were actually sitting in there in that environment.

TRACKING THE CAMERA

Next comes tracking. *Tracking* means having the 3D camera animate over time so it matches a moving real world camera throughout the entire length of the animation. If the camera is matched and tracked correctly, geometry created in the scene and rendered through this camera will appear to actually exist in the scene.

Once these steps are complete, you will composite the tracked camera, live footage and alien animation in Combustion. This step is covered in a later chapter. Right now, we're only concerned with tracking the camera with the live footage.

CAMERA MATCHING UTILITIES

There are several software packages available that will both match and track a 3D digital camera such as the **3ds max** camera to a real world camera. The problem is that these packages can get quite expensive. Fortunately for us, **3ds max** includes these capabilities so you don't have to purchase any additional software. For these tasks, we can use **3ds max**'s Camera Match and Camera Tracker utilities.

The Camera Match utility matches a 3D camera in a still shot, while the Camera Tracker utility is designed to match a 3D camera to a moving real world camera. Together, these two utilities will allow us to integrate our alien and his toys into the scene later on. Seeing as the footage was shot with a moving camera, we will focus on the Camera Tracking utility in this chapter.

REFERENCE POINTS

Before recording any video, you must prep your location. The camera matching/tracking utilities require fixed reference points that don't move during the shoot in order to do the calculations needed to place the **3ds max** camera appropriately. When performing camera tracking later on, these points in the footage are referred to as *features*.

PLACING FIXED POINTS

At your location, the reference points you choose should be relatively small yet easy to find. A reference point can be anything from a small paint chip on the side of a wall to a rock of a particular size and shape. There should be a considerable contrast between the colors of the point and its immediate surroundings so it is easy to spot in the scene.

For the tracker to work properly, a minimum of six points is required. Two of the six points must exist on a different plane than the others. For example, let's say that you have six golf balls on the floor to use as tracking points. You would have to take two of these golf balls and move them up on a rock, log, or something to prevent all the CamPoints from being on the same plane.

Do not group all of the points in the center of the shot. This will not provide enough information to the tracker to do a good match. It is best if you position the points all over the scene, some close, some far, some high, and some low. The more spread out they are, the better the match.

Figure 10.1 *Tennis and golf balls placed as features*

For our footage, we chose to use tennis and golf balls placed at various locations around the area. These markers will be painted out of the footage in post production. Note that some of our markers were perched on rocks to provide sufficient tracking data along the Z axis.

MEASURING DISTANCES

The tracking utility is interested not only in the locations of the points themselves, but in the distances between them. The first point you place will be your main reference point, which is considered to be at 0,0,0.

At your location, this point can be anywhere in the view, but it is best if the main reference point is close to the center. In this way, all other points will be at a substantial distance from the main reference point.

All other points are measured from this point, and are expressed as XYZ locations based on the unit scale. The choice of what you call the X and Y axes at your location are somewhat arbitrary, but Z should always be straight up and down.

If you are using feet and inches, for example, you might measure a point at the location as being 20' away from the origin point on the X axis, 15'2" on the Y axis and 2' on the Z axis. This point's location will be expressed as 20', 15'2", 4'.

If your location points are not too far apart, you can use a tape measure to measure the distances between them. If they are far apart, you can use a piece of string. Simply place the string so it spans the distance between the main reference point and another reference point along an axis, then measure the string.

You should, of course, write everything down, and measure all points two or three times just to be sure.

Accurate measurements are crucial! Just in case you did not get that last sentence, let us say it again: Accurate measurements are vital, essential, and really important. If you do not take the time to get perfect measurements, then your tracking experience will not be an enjoyable one. You will end up with a jerky camera or so many errors that the camera will not even track. An extra hour or two spent verifying your measurements will save you several hours or even days of work when tracking the camera later on.

THE RIGHT TOOLS FOR THE JOB

We are shooting an outside scene on a landscape that is not perfectly flat. This complicates the measuring process. To make things easier, here is a list of items we recommend you take with you:

- Measuring tape
- Clear packing tape or scotch tape
- Roll of string
- Stakes
- A level
- A square
- An assistant with lots of patience

You can use the stakes to tie off lengths of string for measuring. A square is recommended so you can tie off lines at 90 degrees from one another.

If you are unable to use stakes, have your assistant hold one of the string ends directly above a point while you hold the other end, then cut the string and then measure it. Do whatever is necessary to ensure the measurements are as accurate as possible.

After you have recorded all of the correct measurements, remove any stakes and string, taking care not to disturb any of the points. Then you can shoot the video and digitize it for camera matching.

3D SCENE POINTS

After you bring the digitized footage into **3ds max** and set it up as a background, you will need to set up helper objects called CamPoints. These are point objects created especially for tracking points in live footage. A point object is a tiny object with no size or shape, essentially just a point point used just for tracking purposes.

The CamPoints are placed at locations that correspond precisely to the visible points in the digitized footage. You simply move the CamPoint in the scene so it sits right at the spot where the reference point appears in the digitized image.

After that, the tracking process can begin. This process is described in detail over the course of this chapter.

> **TUTORIAL 10.1 CAMPOINT SETUP**
>
> Now that you have everything videoed and you've written down accurate measurement data, we need to load the files into **3ds max**. In our case we have saved each frame of the video in a separate file.
>
> In order for **3ds max** to understand these files as a sequence, we must generate an IFL (image file list) file. This is a simple text file that contains the list of every file that is in the sequence, one after the other. We will use this IFL file to load the video footage into the environment and tracker.
>
> You could create the IFL file by typing in each filename one by one in a text editor, but it's easier to use the IFL Manager utility. This utility automatically creates an *.ifl* file of sequentially numbered files just by picking the first one.
>
> In order to use the footage, you must first save it on your hard disk. To do this, copy the file **low-res footage.zip** from the **Movies/Background Plates** folder on the CD to your hard disk. Unzip the contents of this file into a separate folder.

Creating the IFL File

1. On the **Utilities** panel, click the **More** button and choose **IFL Manager**.
2. On the IFL Manager rollout, click the **Select** button.
3. On the Browse Images for Input dialog, select the file *quar0001.jpg* from the folder containing the unzipped footage files.

   This is the first file in our sequence.
4. Make sure the **Sequence** checkbox is unchecked.
5. Click **Open**.

   The parameters are now filled in on the **Utilities** panel.

6. Click the **Create** button. On the Create IFL File dialog, enter the file name *Tracker* and click **Save**.

   The IFL file has now been created and saved.

Loading the IFL into the Rendering Environment

Next we are going to load the IFL file into the rendering environment so when you render a scene you will be able to see the background video plates.

1. From the menu, choose **Rendering/Environment**.

2. In the Environment dialog, click the button currently labeled **None** (underneath Environment Map).

   The Material/Map Browser appears.

Figure 10.2 *Material/Map Browser*

3. Choose **Bitmap** and click **OK**

4. On the Select Bitmap Image File dialog, locate and select the file *Tracker.ifl*. Click **Open**.

5. Close the Environment dialog.

6. Activate the Perspective viewport.

7. On the Main Toolbar, click **Quick Render (Production)**.

   If you have set everything up correctly, the frame will render with first frame from the recorded video as the background.

Figure 10.3 *Rendered background image*

8. Close the rendered window.

Loading the IFL into the Viewport

1. Verify that the Perspective viewport is still active.
2. From the menu, choose *Views/Viewport Background*.
3. On the Viewport Background dialog, check the **Use Environment Background** and **Display Background** checkboxes. Click **OK** to close the dialog.

You should now see the background image in your Perspective viewport.

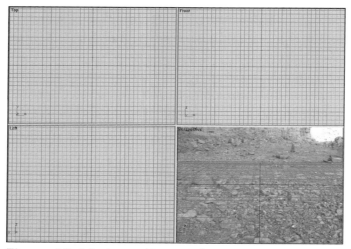

> ☼ **TIP** ☼
>
> Don't forget to load the background into both the rendering environment and the viewport, as we just did here.

Figure 10.4 *Background image in Perspective viewport*

Creating CamPoints

The camera match utility requires a minimum of 6 tracking points and two of these points cannot be on the same plane. The more points you have the more accurate your track will be. We used a total of 12 reference points in our scene. You can see them in the viewport as tennis balls and golf balls. Some of the golf balls are hard to see in the compressed *.jpg* files provided on the CD.

1. On the **Create** panel, click **Helpers**. Choose **Camera Match** from the pulldown menu, and click **CamPoint**.

2. Expand the Keyboard Entry rollout.

3. Verify that X, Y, and Z are all set to 0 and click **Create**.

    A CamPoint has been created in the viewport at 0,0,0. This will be our main reference point.

4. Go to the **Modify** panel. On the Name and Color rollout, verify that the object name is **CamPoint01**. If it is not, change the name.

5. Click the **Create** button 11 more times so that there are a total of 12 CamPoints in the scene.

    Don't be concerned that all the points are at 0 0 0. We are going to fix that in just a moment.

6. Right-click in the viewport to exit the CamPoint creation tool.

Setting the Units

Before we set the position of each CamPoint, we need to fist change the units setting in max. We need to match the max units to the unit type that we recorded when we were taking our measurements.

1. From the menu, choose **Customize/Units Setup**.

    The Units Setup dialog appears.

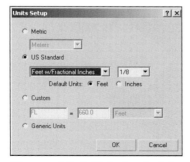

Figure 10.5 *Units Setup dialog*

2. Select **US Standard**.

3. Set **Units** to **Feet w/ Fractional Inches**, and click **OK** to accept the settings.

Positioning the CamPoints

Earlier, we described the process of getting the XYZ values for each point in the physical location.

1. Press **<H>** to open the Select by Name dialog and select **CamPoint02**.
2. On the Move Transform Type-In dialog, enter the following Absolute:World coordinates:

   X 9
   Y 46
   Z 0

   Leave the dialog open when you have finished.

3. Continue selecting each CamPoint helper and entering their position coordinates from the table below.

Tracker	X	Y	Z
CamPoint01	0	0	0
CamPoint02	9	46	0
CamPoint03	8' 2 4/8"	85' 0"	10' 0"
CamPoint04	14' 0"	199' 0"	17' 0"
CamPoint05	41' 6 1/8"	-31' 11"	0
CamPoint06	63' 9 4/8"	61' 3 7/8"	0
CamPoint07	74' 11 1/8"	-57' 4 1/8"	0
CamPoint08	87' 3 6/8"	125' 0"	26' 4 2/8"
CamPoint09	141' 1"	-74' 7 3/8"	0
CamPoint10	145' 1 3/8"	-70' 3 4/8"	17' 4 6/8"
CamPoint11	123' 9 1/8"	66' 0"	0
CamPoint12	138' 3 5/8"	139' 0"	23' 0"

4. Save the scene to the file *Set - Tracked Camera.max*.

TUTORIAL 10.2 CAMERA TRACKING

Now that the CamPoints are all set up, we can use the Camera Tracker utility to animate the 3D camera in such a way that it will move with the recorded footage.

Setting up the Camera Tracker Utility

The first thing we are going to need to do is load the movie into the camera tracker utility. To do this we will load the IFL file that we created earlier.

1. On the **Utilities** panel, click **More**. Select the **Camera Tracker** utility from the Utilities dialog and click **OK**.
2. On the **Utilities** panel, click button under Movie file button, currently labeled **None**.
3. On the Browse Images for Input dialog, choose *Tracker.ifl* and click **Open**.

A new window appears showing the first frame of the digitized footage.

Figure 10.6 *First frame of digitized footage*

Creating a Motion Tracker

In the next section we will perform a 2D track of all the points in our scene. This involves creating and placing tracking gizmos (trackers), and placing each one over the corresponding place in the footage display.

The goal is to have each tracker track the movement of its related video point. In other words, as the feature shifts around the screen due to real-world camera movement, the related tracker should always stay on top of it. This gives the tracking utility the information it needs to animate the 3D camera appropriately.

It is very important that each tracker added to the scene be connected with the appropriate CamPoint. This association makes it possible for the camera tracking utility to calculate the camera match.

The following figure provides a guide to the locations of the CamPoints you placed earlier. You can use it as a guide to help you place your trackers.

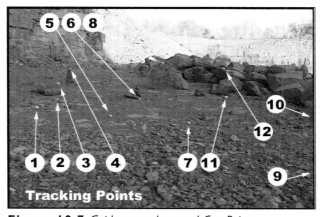

Figure 10.7 *Guide to trackers and CamPoints*

1. In the Motion Trackers rollout, click **New Tracker**.

   A tracker gizmo is placed in the scene.

Figure 10.8 *First tracker in the scene*

A tracking gizmo appears as a box within a box. The innermost box has a set of crosshairs running through it. You can move the gizmo anywhere on the screen by clicking and dragging inside either of the boxes. You can also resize the inner box or the outer box by clicking on one of the box corners and dragging it larger or smaller.

The inner box, known as the *feature bounds box*, has the task of trying to stay centered on an assigned point (feature) at all times. The outer box, the *motion search bounds box*, provides a search area on a frame-by-frame basis.

Together, the two boxes perform the tracking process as follows:

- The tracker takes a snapshot of what's within the feature bounds box.
- On the next frame, it looks within the search bounds box to find a match to the snapshot taken in the previous frame.
- If the tracker can't find a match, errors are generated.

This is why it is so important to make sure to use points that have a lot of contrast against the surrounding objects — the tracker needs to be able to recognize the feature when it sees it.

Be careful not to get carried away with the size of your motion search bounds box. The larger the box, the longer the tracking operation will take. Typically the outer box is twice the size of the inner box.

2. Position the gizmo so it is directly over the first tennis ball (point 1).

3. Once you have the gizmo lined up, adjust the size of the inner and outer boxes so that the inner box closely surrounds the tennis ball and the outer box is roughly twice the size of the inner box.

TIP

With some movie footage, it might be difficult to see the gizmo clearly. If you are having a hard time seeing the gizmo, check the Fade Display checkbox on the Movie rollout.

TIP

You can use the shortcut keys <I> to zoom in and <O> to zoom out of the movie window.

Associating a Tracker with a CamPoint

1. On the Tracker Setup section, click the **Scene object** button and press **<H>** on the keyboard. Select **CamPoint01** and click **Pick**.

   CamPoint01 is now associated with Tracker1.

2. On the Movie rollout, set the **Show frame** value to 100.

   This shows frame 100 for demonstration purposes only.

3. In the Motion Trackers rollout, click the **Set stop** button.

   In the listbox above, you can see that CamPoint01 has a start frame of 0 and an end frame of 100. It also has an **X** to its left side indicating that the tracker is active. Only active trackers are tracked.

   So why are active and inactive trackers important? A tracker can only track while its tracking point (feature) is in the viewport. The moment the feature exits the viewport, the tracker will return an error. To avoid this, we will need to make the tracker inactive at the frame where the feature exits the screen.

   Once a feature leaves the screen and the tracker has been made inactive, a new tracker will have to be used if the feature point reenters the viewport. In other words, a tracker has only one life, and the moment it goes inactive, its life is over permanently. A new tracker must be born at the time the feature comes back into view.

4. Set the **Show frame** parameter back to 0.

Tracking the First Point

Now that we have setup our first tracker, let's see if we can track it. There are two ways of tracking footage, manually or with batch tracking. Batch tracking is easy — just click the **Complete Tracking** button and the utility will do the rest. The manual approach give you far more control.

Let's try out the tracking procedure.

1. Expand the Movie Stepper rollout.

   On this rollout we have the option of tracking a single frame, 10 frames, or all the frames at a time. We will start by tracking a few individual frames.

2. Click the button next to **Feature tracking** to turn it **On**.

   Tracking will only be performed if **Feature tracking** is on. If it is off, the step buttons will still step you through the frames but no tracking will take place.

3. Zoom in on the tracker.

   We are zooming in so that it will be easy to see what is happening after we start tracking. If you can't zoom in with the **<I>** key, this means that your tracker has been deselected. Simply select it and zoom in.

4. Click **Step forward one frame or keyframe button** 10-20 times.

   As you click, watch the Movie Display. You should see it updating as the inner box tries to stay centered around the tennis ball.

   Let's try tracking 10 frames at a time.

5. Click the **Step forward 10 frames** button. You will see the tracker try to maintain a good track on the tennis ball. The inner box should stay centered around the ball and a dotted line showing the path of the track should be visible on the screen.

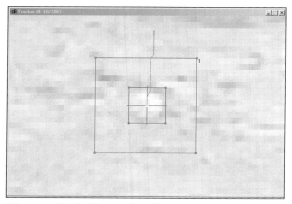

Figure 10.9 *The tracker doing its thing*

6. Click the **Step to last active frame for selected tracker** and watch the tracker track the remaining frames.

   The tracking will stop at frame 100 because this is where we told the tracker to become inactive.

7. Turn the **Feature tracking** button **Off**.

Checking an Incomplete Track

At this point you should have had a good track with no problems. You can thank the bright yellow tennis ball for that. But what if there were problems? Where would we look to see what the errors were? In this section, we are going to create an error.

1. In the Movie Stepper rollout, click the **Step to first active frame for selected tracker**.

2. Click **Clear Tracking To End**.

   This will delete the current track for tracker #1.

3. Change the sizes of the tracker's inner and outer boxes to make them extremely small.

4. Turn **Feature tracking** back on.

5. Click **Step to last active frame for selected track** and watch as the tracker tries to track its feature point.

6. Once the track has completed, turn off **Feature tracking**.

7. Expand the Batch Track rollout and look at the Tracking Status section.

   You should see Incompletes and Errors both with a value of 1.

8. Click the **Check status** button.

9. In the Tracking error review window, you will find a list of errors.

   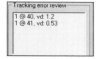

   You can see that this past track was not successful. We have received some errors that we must deal with.

Handling Errors

As the 2D track is executed, you will most likely encounter errors. These errors are displayed in the tracking error review window.

You should always click the **Check status** button after completing a track to see if there are any errors. The tracker utility will tell you how many trackers are incomplete and how many errors occurred during the track.

The syntax of an error is:

[tracker#] @ [frame], [error] [error value]

A quick way of fixing errors is by going to the given frame and repositioning the reference feature and the search area. Then use the manual tracking control under the movie stepper rollout. After you are done fixing the error, click the check status button again. The error should go away from the list.

When interpreting the error messages, keep in mind that the camera tracker works primarily with colors and contrast. When you set a tracker over a point on the footage, the Camera Tracking utility looks for that pixel and its surrounding area on subsequent frames. If the lighting changes significantly over the course of a few frames, for example, the Camera Tracking utility will have trouble finding the same spot on the footage.

With this in mind, you can fix the error by adjusting the error thresholds. There are three thresholds, each with an abbreviation that appears in the error display:

- Match Error (Me)
- Variance Delta (Vd)
- Jump Delta (Jd)

Match Error is a measure of the difference between best match in the current frame to the match at the previous keyframe. The error shown is a percentage of the maximum possible image difference. Good matches are usually below 0.05 percent.

Variance Delta measures whether the target and match have similar color ranges. This threshold can compensate for mistakes made by the **Match Error** measure.

Jump Delta sets the maximum by which a tracker can move to find a pixel. The average amount by which a pixel has moved is calculated for the previous five frames. If the pixel then moves by more than this average plus the Jump Delta, you'll get an error. If your camera motion includes some sudden accelerations, you will need to increase the Jump Delta to prevent errors from occurring.

All of these thresholds can be adjusted to make the tracking utility more lenient. However, keep in mind that the higher the threshold, the more likely it is that the resulting camera motion will be bumpy.

1. Change the first tracker's inner and outer boxes back to their original sizes.
2. Turn on **Feature tracking** and track the scene again.
3. Click **Check status** to see errors.
4. Adjust thresholds as necessary and repeat the tracking process. See if you can get rid of some of the errors in this way.

Completing the Tracking Process

Using the techniques described earlier, you will now create all of the needed trackers to track the scene.

Preplanning can be a big help here. Before creating more trackers, find out everything you can about each feature in the movie. For example, ask the following questions about each one:

- At what frame does the tracker need to become active?
- At what frame does the tracking feature exit the screen, and therefore need to be made inactive?
- Does the point reenter the scene? If so, it will require a new tracker.

1. On a piece of paper, note any entry/exit frame numbers and other pertinent information for the remaining trackers.
2. Set up the rest of the trackers by placing them and associating them with CamPoints.
3. Track the scene by clicking **Complete Tracking** on the Batch Track rollout.
4. Deal with errors one tracker at a time until you have eliminated all of them.

Once you have completed the tracking process, it is time to move on to 3D match moving.

Using 3D Match Moving

3D match moving is the process of matching a **3ds max** camera to the motion of the real world camera that was used to film the scene. A combination of the CamPoints and their distances combined with the trackers and their tracked motion is what makes this process possible.

1. Create a free camera in your scene. Name the camera **Camera01 - Tracked**.
2. On the Match Move rollout, click the **Camera** button (currently labeled: None) and select the camera that you just added to your scene.

3. Determine which actions to match.

   You have several options to choose from. In our case we only changed the FOV and pan of the camera, but more options were required when we performed the match. It is recommended that you start with all of the options and start removing them as you work your way toward a perfect match.

4. Set the range of the movie.

   You may chose to work on the entire sequence at once or split up the match moving process over several smaller ranges. The Animation Start setting allows you to offset the **3ds max** time in comparison to when the footage was loaded.

   Before clicking the Match Move button, you can choose to animate the display without generating any keyframes during the first run. In this case you will be able to see if your match holds up or falls apart without having to deal with bad keyframes afterwards.

6. Click **Match Move** to set keyframes for the camera based on the tracking data.

Testing the Match

It's an easy matter to test the animated camera with a simple primitive in the scene.

1. Create a teapot or other simple piece of geometry in the scene.

   Watching the scene through the matched camera view, move and/or resize the object so it looks as if it is really sitting in the scene on the current frame.

2. Move the time slider to various frames and see if the object's perspective looks correct with the rest of the scene.

   This is a quick test to give you an overall idea of whether the camera is matched.

3. If everything looks good, render all frames to an AVI file and watch the animation to check if the camera is truly tracked correctly. Check to see if the object slides around or appears anchored to the ground.

4. If there is sliding or jerking, then the camera must be tracked again. Make adjustments to the Camera Tracker utility and perform the **Match Move** again.

   If your camera is tracking correctly, congratulations!

☆ **TIP** ☆

*To fix jerky movements, try a smoothing filter such as a **Pan** filter with smooth amount of 1.*

Saving the Tracker Data

When **Auto Load/Save settings** is checked on the Movie rollout, the Camera Tracker utility generates a *.mot* file with all the camera data when a Match Move is performed. The name of the *.mot* file is displayed below the Load and Save buttons on the rollout. The *.mot* file is saved in the same folder as your IFL file. The *.mot* file is stored independently of the *.max* file, and contains only tracker data but no 3D scene elements.

1. Note the name of the *.mot* file for later use.
2. Save the scene to the file **Set - Tracked Camera.max**.

WHAT TO DO IF THE CAMERA WON'T TRACK

You might run into a situation where the camera simply refuses to track correctly despite all your efforts, and you don't have time to fine-tune keyframes.

If you have Combustion, you can use it instead to track the footage. To use this approach, you have to render your animated scene as a long shot showing all the action on one still camera. We are not suggesting this approach for our alien commercial project, but as an alternative for your own work in the future should you need it.

1. Load your 3D scene and animate the objects as needed.
2. Place a free camera in the scene and display the footage in the camera view.
3. Position the camera so the CG elements match up with the background plates on the first frame of the animation. Hide the background plates.
4. Dolly the camera away from the objects until the entire action area fits within the camera view. Move the time slider to make sure all the action stays within the camera view at all times. Do not animate the camera.
5. Render the animation to a high resolution such as 987x666. With a higher resolution, you can zoom into the rendered image in Combustion without degradation.
6. In Combustion, visually adjust the scale and placement of the rendered animation to match the footage on the first frame of the animation. Use Combustion's tracking tools to track the entire composite.

SUMMARY

In this chapter, you have learned an important set of skills that will serve you well when compositing animation with live footage. You now know how to:

- Place and measure tracking features at the shoot location
- Place CamPoints for tracking
- Use the Camera Tracker to match the 3D camera to the footage

You will use the tracked camera in *Chapter 12 Animation and Rendering* to render the alien animation so it will composite seamlessly with the live footage.

Texturing and Lighting

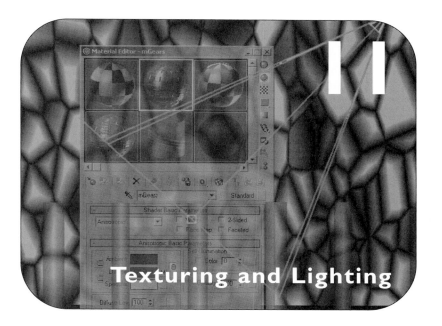

In this chapter, we will focus on the lighting and materials required to make our models appear real when placed against our background plates.

TECHNIQUES IN THIS CHAPTER
In this chapter, you will learn:

- How to use the Material Editor
- How to create custom materials
- Methods for applying multiple bitmaps or materials to one object
- The best setup for soft, diffuse lighting
- How to set up separate lights for controlling overall illumination and shadows
- How to make an object transparent while receiving shadows

TUTORIALS IN THIS CHAPTER

The tutorials in this chapter show you how to create and apply materials and to light the set for our scene.

- **Tutorial 11.1 Texturing the Alien**, **Tutorial 11.2 Texturing the Rover** and **Tutorial 11.3 Texturing the Battery** use the Material Editor to create custom materials for these objects from scratch.

- **Tutorial 11.4 Lighting the Set** shows you how to set up a working light solution for our project.

- **Tutorial 11.5 Shadows for the Ground** creates a transparent ground plane with specially placed rock objects that will catch shadows from our animated objects. The shadows will be used later in compositing.

Two tutorials in the *Tutorials* folder on the CD, *CD Tutorial 1* and *CD Tutorial 2*, are used with this chapter. These tutorials show you how to use Photoshop to make custom textures. The resulting maps are included on the CD in the *Maps* folder.

TEXTURING

The art of giving objects color, patterns, bumpiness, shininess and other surface qualities is called *texturing*.

Texturing is usually done late in the production phase, often in conjunction with lighting. Texturing defines the way the scene will look when struck by light, so it makes sense to do these two tasks together.

Different light and texture setups produce different types and qualities of highlights on objects (also called *hotspots* or *specular highlights*). It takes practice to be able to identify the best material type for the job, but in time it becomes a snap.

MATERIALS

In **3ds max**, texturing is accomplished with *materials*. A material is comprised of color and pattern definitions. In a material, you specify the color and patterns for the overall color, and the highlights, bumpiness and transparency, among other attributes. The material is then applied to an object. Materials are created and applied from the Material Editor.

It is important to note that you can only see the final effect of a material by rendering the scene. What you see in viewports is just a simple representation that might or might not resemble the material in the final rendering. It's important not to panic and start making wild adjustments to materials just because they look strange in viewports. Always check the rendering to see what the material really looks like in the scene.

MAPS

The building block for a material is the *map*. A map is a pattern of colors, either defined by a file (such as a bitmap) or through some sort of computer-based procedure (as with smoke and other random patterns).

A material has several *map channels*, where a different map can be assigned to each different material attribute. For example, you can use one map to define the overall color of the material, another to define the transparent/opaque areas, and yet another for bumps. With these map channels, you can create just about any material for any occasion.

The use of materials and maps will become clearer as you go through this chapter.

THE MATERIAL EDITOR

All material creation takes place in the Material Editor, a special window with all the tools you need. You can open the Material Editor in a number of ways:

- Click the **Material Editor** button on the Main Toolbar
- Choose **Rendering/Material Editor** from the menu
- Press the **<M>** key on the keyboard

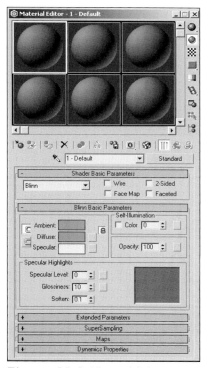

Figure 11.1 *Material Editor*

The Material Editor consists of three main areas:

- Sample slots at the top where a representation of each material is shown on a primitive, such as a sphere
- A toolbar below and to the right of the sample slots for assigning and naming materials, and other functions
- A rollout area for assigning colors, maps and settings

The Material Editor is a modeless window, meaning it can remain on the screen while you select objects or perform other tasks in viewports. Any changes to settings in the Material Editor are updated automatically as you make them.

In the next few tutorials, we'll take you through the process of creating materials for the scene elements and explain each new concept as it comes along.

We'll start with the materials for the alien.

TUTORIAL 11.1 TEXTURING THE ALIEN

The alien consists of three objects, the two eyes and the body. We will create separate materials for the eyes and body.

Creating the Basic Eye Color

1. Open the **Alien - Setup with UI.max** file that you created in **Chapter 9**, or load the file **Alien - Setup with UI.max** from the **Scenes/Alien** folder on the CD.

Figure 11.2 *Alien file*

2. Hide all the bones and point helpers in the scene.

3. Open the **Material Editor** using any of the methods described on the previous page.

   By default, the top left sample slot is active, as indicated by the white border around the slot.

4. On the toolbar, change the name of the material from 1 - Default to **mEye**.

TIP
To change a material name, highlight the old name, type in the new one and press <Enter>.

5. On the Shader Basic Parameters rollout, make sure the shader type is set to **Blinn**.

6. Click on the **Diffuse** color swatch. A color selector appears. Change the color to a pure white with the following settings:

   R 255
   G 255
   B 255

   The Diffuse color sets the overall color for the material. The term *diffuse* comes from the study of light in physics.

7. Click the **Specular** color swatch and also change it to a pure white.

TIP
You don't have to close the Color Selector before clicking on the Specular color swatch.

   The Specular color sets the color of any highlights on the material. Highlights appear only on shiny materials. The level of shininess is set by the Specular Level and Glossiness parameters.

8. Change the **Specular Level** parameter to 54.

   This sets the intensity of the highlight for the material.

9. Change the **Glossiness** parameter to 39.

   Glossiness sets the size of the highlight for the material. Higher values make a smaller highlight.

   The graph to the right of these two parameters gives you an idea of the size and intensity of the highlight in graphical form.

   This material will work as the base for the eye. Now we need to make an eyeball as part of the material. For this, we will use a map to define the color and size of the eyeball.

Assigning a Map for the Eye Definition

If all you ever wanted to do was assign plain colors to objects, you could do without maps. Usually, however, you will want different kinds of colors and patterns in materials, and for this you need maps.

Here we will assign a map to the Diffuse Color channel of the material. The Diffuse color channel defines the overall color of the material.

1. Expand the Maps rollout.

2. Click on the button labeled **None** across from the **Diffuse Color** channel.

The Material/Map Browser appears.

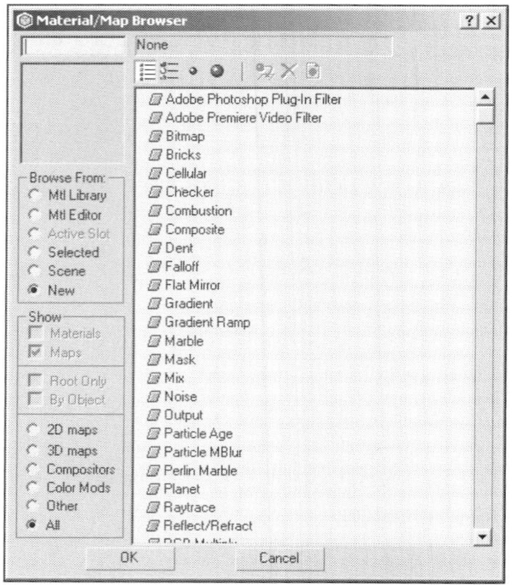

Figure 11.3 *Material/Map Browser*

This is where you pick the type of map to assign to the selected channel.

3. Choose the **Gradient Ramp** map and click **OK**.

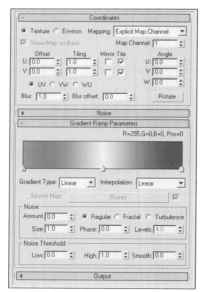

Figure 11.4 *Gradient Ramp map rollouts*

The Gradient Ramp map allows you to create a blend of colors. Note that the rollout area on the Material Editor has changed to display the parameters for the Gradient Ramp map.

> ☼ **TIP** ☼
>
> *A new rollout appears because you are now at the map level of the material.*

Adjusting the Gradient Ramp

For the most part, we will be working with the parameters on the Gradient Ramp rollout. The gradient shown on the rollout works with flags, the small green icons at the bottom of the gradient. The flags set specific colors along the gradient. By default, there are three flags on the gradient, one at each end and one in the middle.

You can add flags by clicking and dragging on the gradient. To change the color of a flag, double-click it and change the color on the Color Selector. To move a flag, click and drag on it. To remove a flag from the gradient altogether, right-click it and choose **Delete** from the pop-up menu.

The goal is to create a cartoon-style pupil and cornea for the alien's eyes with this map.

1. Click between the center and rightmost flag to add another flag.
2. Right-click the new flag and choose **Edit Properties** from the pop-up menu.
3. Set **Position** to 89.
4. Click the color swatch and change the color to **RGB 0,3,5** (near-black).
5. Use the same method to change the remaining flags with the following parameters. Note that you can simply click on each flag to display its parameters in the Flag Properties dialog.

   Flag #1: **Position**   0
             RGB   255,255,255 (white)

   Flag #2: **Position**   86
             RGB   255,255,255 (white)

   Flag #4: **Position**   100
             RGB   0,0,27 (very dark blue)

This creates a ramp from white to black, with a gray gradient between the second and third flags.

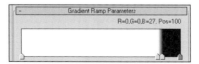

Figure 11.5 *Resulting gradient for eye*

6. In the Coordinates rollout, choose the **VW** option to orient the map properly for the alien's eye.

Take a look at the sample slot. Looks rather like an eyeball, doesn't it?

Assigning the Material to the Alien's Eyes

By now, you're probably itching to see what the material looks like on the alien's eyes. We will keep you waiting no longer.

1. In the scene, select both the alien's eyes.

2. On the Material Editor toolbar, click **Assign Material to Selection**.
3. Click the **Show Map in Viewport** button to make the map show up on the alien's eyes in the scene.

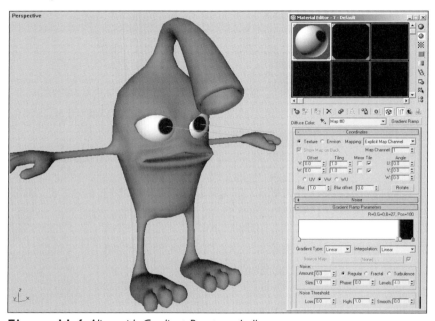

Figure 11.6 *Alien with Gradient Ramp eyeballs*

Copying the Map

We will need the Gradient Eye map later for another use, so we'll copy it to another slot where it will be easy to grab later on.

☀ **TIP** ☀
Maps within a material can be named separately.

1. Change the name of the map from Map #1 to **Gradient Eye**.

When you assigned the Gradient Ramp map, you went down to a lower level of the material, where the Gradient Ramp parameters were displayed. Every time you assign a map to a map channel, you will be taken to a lower level of the material where you can set parameters for the map. To get back to the root level where the original rollouts are displayed, simply click Go to Parent.

2. Click **Go to Parent** on the Material Editor toolbar.

Now we can easily grab and copy the Gradient Eye map to another sample slot.

3. On the Maps rollout, click and drag the **Gradient Eye** map to an empty sample slot at the top of the Material Editor. In the Instance (Copy) Map dialog, choose the **Instance** option and click **OK**.

This will save the map separately so it will be easy to grab for later use.

When the map is copied, it appears as a flat image in the sample slot. This is because a map doesn't have the 3D attributes that a material has such as shininess, bumpiness and so on. It is simply a flat pattern of colors.

Setting up a Reflection Map

You might think the eyeball looks just fine the way it is, but any real eyeball generates a lot of reflections. Eyeballs as we know them on this planet are always wet so they reflect the world around them. To make the eyeball look more realistic in the scene, we will assign a map to the Reflection map channel.

When a map is used in the Reflection map channel, **3ds max** creates a giant invisible sphere around the scene and puts the map on it, then figures out what part of the map is reflected in the object based the viewing angle.

The alien will be in the landscape, so we will set up a reflection map that utilizes the landscape scene. The landscape will only reflect in the whites of the eyes. We have saved the first frame of the landscape footage to a file on the CD so you can easily use it.

This might seem unnecessary for such a subtle effect, but it is these subtleties that add realism to the scene.

1. Click the **mEye** material sample slot again to select it.
2. On the Maps rollout, click the **None** button across from the **Reflection** map channel.
3. On the Material/Map Browser dialog, choose **Mask**.
4. On the Mask Parameters rollout, click the **Map** button and choose **Bitmap** from the Material/Map Browser.
5. Select the bitmap **Eye_Reflection.tif** from the **Maps** folder on the CD.

   This bitmap holds the first frame of the landscape footage. Setting the map up in this way will cause the first landscape image to be used as the reflection map.
6. Set the **Blur Offset** parameter to 0.01.

   Before going on, let's talk a little about where we're at in the material.

   When you assigned the Mask map, you went down one level from the root level of the material. On the Mask map, you then assigned submap to the Map channel, and went down another level. This means you are now two levels down from the root level.

The **Go to Parent** button goes up one level at a time. You could click it once to get back to the Mask map, or twice to get up to the root level. Don't do this yet as we still have some work to do on the Mask map.

Reusing a Map

Now we are going to reuse the map Gradient Eye that we separated earlier.

1. Click the **Go to Parent** button once to return to the Mask map level.
2. Drag the **Gradient Eye** map from its separate material slot to the **Mask** map. On the Instance (Copy) map dialog, choose **Instance** and click **OK**.

   The Gradient Eye map we just copied will now act as a mask for the reflection. The white areas of the Gradient Eye map will reflect while the black areas will not. This will cause the whites of the eyes to reflect the landscape scene in a subtle way.

3. Click the **Mask** map button, now labeled **Gradient Eye**.
4. On the Coordinates rollout, click the **Texture** radio button.

Setting the Map Amount

So far, every time you have assigned a map on the Maps rollout, you have used the map at 100% intensity. There is a way to cause a map to only affect the material partially.

If you look at the eye material in the sample slot, you can see that the reflection is a little too strong. We want to reduce the effect of the reflection map so the reflection will not overwhelm the eye.

1. Click **Go to Parent** twice to return to the root level of the material.
2. On the Maps rollout, set the **Amount** to 50 for the **Reflection** map channel.

   The reflection is now more subtle.

   The eye material is complete. Now we will move on to create the material for the alien's body.

Creating the Base Body Material

> **☼ TIP ☼**
>
> *Shader types are usually named after the person(s) who developed the science behind them. For example, the Blinn shader is named after graphics mathematician James Blinn.*

1. Select an empty sample slot and name the material **mAlien**.
2. Assign the material to the alien's body.

   Every material uses a shader type to determine the shading method used to calculate the material. Different shader types produce different kinds of diffuse color and highlight effects.

   The shader type is set with the pulldown at the left of the Shader Basic Parameters rollout. The default shader type is Blinn.

3. Verify that the shader type is set to **Blinn**.
4. Set the **Specular Level** to 97.
5. Set **Glossiness** to 25.

This will produce a smooth bright highlight on the alien's body. We will be making his skin a little bumpy, and these settings will catch highlights off the bumps.

Defining the Alien Skin Maps

The alien body material will use three map channels:

- **Diffuse** for overall color, using a Speckle map for spots
- **Bump** to add a bumpy texture, using a Cellular map for a skin texture
- **Specular Level** for highlights, using the same map as the Bump map channel to enhance the highlights at the bump edges

The Speckle and Cellular maps are *procedural* maps, which means they are randomly generated by the program rather than by a bitmap or a set pattern.

Procedural maps can go in infinitely in any direction without ever showing a noticeable repeating pattern.

You can also assign a Diffuse Color map by clicking the small box next to the Diffuse color swatch on the Blinn Basic Parameters rollout.

1. On the Maps rollout, click on the **Diffuse Color** map channel. On the Material/Map Browser, select **Speckle**.

The speckle map randomly places blobs of color on a background color.

2. Set the following parameters on the Speckle Parameters rollout:

   Color #1   RGB 51,128,255 (blue)
   Color #2   RGB 185,211,255 (pale blue)
   Size   200

   On the Coordinates rollout, set the **Tiling** for **X**, **Y**, and **Z** to 0.3.

3. Click **Go to Parent** to return to the material's root level.

4. On the Maps rollout, click the **Bump** map channel and select the **Cellular** from the Material/Map Browser dialog.

The Cellular map makes an irregular pattern of cells or chips. It is ideal for simulating the etched pattern of human or reptile skin.

To see what the Cellular map itself looks like, click the Show End Result button on the Material Editor toolbar to turn it off. Be sure to turn it back on again so you can see the entire material.

5. In the Cellular Parameters rollout, set the following parameters:

   Type   Chips
   Fractal   (checked)
   Size   1.0
   Spread   0.5
   Bump Smoothing   0
   Adaptive   (checked)

6. Click the **Go to Parent** button to return to the material's root level.

7. On the Maps rollout, set the **Bump Amount** to 10.

We will use the same map as a Specular Color map. A Specular Color map sets the colors of the highlight over an object. When the same map is used as a Bump and Specular Color map, highlights at the edges of bumps are enhanced.

8. On the Maps rollout, click and drag the **Bump** map channel to the **Specular Color** map channel. Choose **Instance** when prompted for a clone method.

By doing this, any adjustments made to either map will change the other.

All the alien materials are now complete. If you render the scene, you should see a bumpy, shiny, speckled bluish skin on the alien.

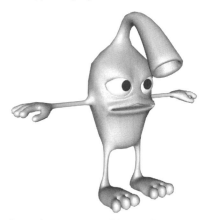

9. Save the scene as *Alien - Setup and Textured.max*.

TUTORIAL 11.2 TEXTURING THE ROVER

To keep your system working fast, work in wireframe mode as much as possible and hide the rover body when you're not working with it.

Texturing the rover will be a more involved process than the alien because of the number of separate objects and the combination of maps needed.

The bitmap needed to texture the metallic rover body is included on the CD. You can also create this bitmap yourself in Photoshop 6 by following *CD Tutorial 1* from the *Tutorials* folder on the CD.

The rover body will be assigned three separate materials that will be merged into a single material. The three materials are the overall rover material, a material for the windshield and another for the windshield rim.

Converting the Rover to an Editable Mesh

1. Load the file *Rover - Setup.Max* that you created in *Chapter 6*. You can also load the file *Rover Setup.max* from the *Rover* folder on the CD.

The rover body is currently an Editable Poly with its smoothness determined by NURMS subdivisions. When assigning multiple materials to one object, it is easier to work with an Editable Mesh. If we kept the model as an Editable Poly, we would have to apply numerous Mesh Select and Material ID modifiers to the stack to apply submaterials to parts of the rover body, which would make the process more complicated and perhaps slow down the system.

2. Select the **Rover Body**.

TIP
Converting the object to an Editable Mesh will make no visible change to the object.

3. On the **Modify** panel, click on the **Editable Poly** modifier on the modifier stack.

4. Verify that **Use NURMS Subdivision** is checked and that Display **Iterations** is set to 2.

5. Right click on the modifier stack. On the pop-up menu, select **Convert to: Editable Mesh**.

Creating the Rover Body Material

Next we will start creating the general body material. The goal is to achieve a silver/golden brushed metal look.

1. Open the **Material Editor**.

2. Select an empty sample slot, and name it **mBody**.

3. On the Shader Basic Parameters rollout, set the shader type to **Phong**.

4. Click the lock button to the left of the Diffuse and Ambient colors to unlock the two colors.

When these colors are unlocked, you can have different Diffuse and Ambient colors. The Ambient color is used when the object is in shadow.

5. Change the colors on the Phong Basic Parameters rollout to the following:

Diffuse   **RGB 42,37,31** (dark brown)
Ambient   **RGB 55,55,55** (dark gray)
Specular **RGB 252,232,221** (pale pink)

6. Set the **Specular Level** to 18 and the **Glossiness** to 10.

Setting up the Diffuse Color Map

If you want to create the bitmap for this step yourself in Photoshop 6, do **CD Tutorial 1** from the **Tutorials** folder on the CD before continuing.

1. Expand the Maps rollout and click on the **Diffuse Color** channel button. Choose **Bitmap** from the Material/Map Browser dialog and click **OK**

2. On the Select Bitmap Image File dialog, choose the **BaseColor.tif** file you created with **CD Tutorial 1**, or choose it from the **Maps** folder on the CD.

3. Click **Go to Parent** to return to the root level of the material.

Setting up the Ambient Map

1. On the Maps rollout, click the lock icon between the Ambient Color and Diffuse Color map channels to turn it off and unlock the two map channels.

When the lock is off, you can assign different maps to the Diffuse and Ambient Color channels. Here we will use the same map, but with slightly different settings.

TIP

Setting the Amount value lower than 100% causes the map to provide a percentage of the color and pattern, while the Ambient and Diffuse colors on the Phong Basic Parameters rollout provide the remainder of the color.

2. Click and drag the **Diffuse Color** channel to the **Ambient Color** channel and choose **Copy** from the dialog.

3. Click the **Ambient Color** map channel button to access the bitmap parameters for this map channel.

4. On the Output rollout, set the **RGB Offset** to -0.35.

   This will darken the bitmap so the ambient shade will not be the same as the diffuse color, which will enhance the depth of the colors in the final rendering.

5. Click **Go to Parent** to return back to the root level of the material.

6. On the Maps rollout, set the **Amount** for both the **Ambient Color** and **Diffuse Color** to 30.

Setting up the Bump and Specular Level Maps

TIP

Be sure to change the HSV values and not the RGB values for this step.

1. On the Maps rollout, click the **Bump** channel button. Choose the **Speckle** map.

2. Set the following parameters for the Speckle map:

   Color # 1 HSV 0,0,138 (medium gray)
   Color #2   HSV 0,0,175 (light gray)
   Size   5

3. Click **Go to Parent** to return back to the root level of the material.

4. On the Maps rollout, set the **Bump Amount** to 12.

5. Click and drag the **Bump** channel map to the **Specular Level** channel map and choose **Copy** from the dialog.

6. Click the **Specular Level** channel button.

7. On the Speckle Parameters rollout, change the **Size** parameter to 100.

8. Click **Go to Parent** to return to the root level of the material.

Setting up the Reflection Map Channel

TIP

A Noise map makes an irregular pattern of soft blobs.

1. On the Maps rollout, click the **Reflection** map channel, and assign a **Noise** map to it. Set the following parameters for the Noise map:

   Color #1 RGB 168,138,111 (light brown)
   Color #2 RGB 88,74,67 (dull brown)

2. Click **Go to Parent** to return back to the root level of the material.

3. On the Maps rollout, set the **Reflection Amount** to 15.

4. Assign the material to the **Rover Body**.

5. Apply a **UVW Map** modifier to the rover body and set the Mapping type to **Box**.

6. Save the scene with with the filename *RoverTex01.max*.

Creating the Rim Material

Next you will work with the polygon sub-object level to select the windshield and windshield rim polygons. These polygons will receive a material different from the one applied to the remainder of the rover body. To make this step easier, we will isolate the rover body.

1. Make sure the **Rover Body** is selected.
2. **<Ctrl>-right-click** in any viewport to display the Quad menu, and choose *Isolate Tool* from the menu.

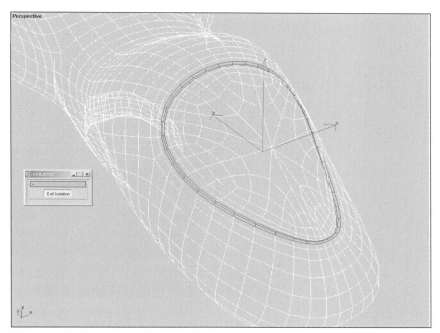

TIP

Isolating the rover body will make it much easier to select polygons in the steps that follow.

Figure 11.7 *Polygon selection defining the windshield rim*

3. On the **Modify** panel, access the **Polygon** sub-object level.
4. Select a rim around the windshield two polygons thick.
5. Once you have completed the polygon selection, drag an empty sample slot from the Material Editor onto any one of the selected faces.

   This applies the material just to the selected polygons, and makes the new material a submaterial of the general rover material. You will get to see how this material is represented in the Material Editor in just a few moments.

6. Name the new material **mRim**.
7. Set the shader type to **Phong**.
8. Set both the **Ambient** and **Diffuse** color swatches to pure black.
9. Set the **Specular Level** to 25 and the **Glossiness** to 60.

Creating the Windshield Material

1. Verify that you are still at the **Polygon** sub-object level.
2. Select all the polygons that make up the windshield, excluding the windshield rim faces.

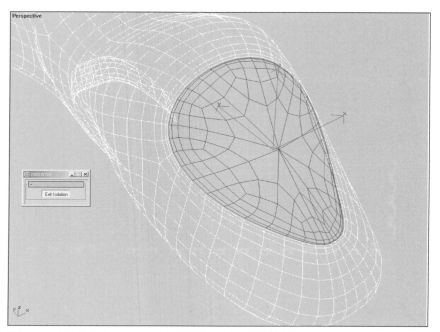

Figure 11.8 *Polygon selection defining the windshield*

3. From the Material Editor, click and drag an empty sample slot onto the selected polygons.
4. Name the new material **mWindshield**.
5. Set the shader type to **Phong**.
6. Set the **Specular Level** to 71 and the **Glossiness** to 53.
7. Set the **Diffuse** color swatch to an orange color with the RGB settings **255,122,4** (orange).

8. On the Phong Basic Parameters rollout, unlock the Ambient and Diffuse colors.
9. Set the **Ambient** color swatch to a darker orange color with the RGB colors **117,56,2** (dark orange).

Setting up the Windshield Reflection Map

1. Click the **Reflection** map channel button and select **Bitmap** from the Material/Map Browser.

2. In the Select Bitmap Image File dialog, navigate to the **3dsmax4\Maps\Backgrounds** folder and select **Sunset90.jpg**.

   This orange sunset image, which comes with **3ds max**, works well as a reflection map for shiny objects in red or orange environments.

3. Click the **Go to Parent** button.
4. Set the **Reflection Amount** to 50.

Viewing the New Multi/Sub-Object Material

When you assign a material to a selection of polygons on an Editable Mesh, the Material Editor generates a special kind of material called Multi/Sub-Object. This type of material is made up of two or more submaterial. The materials are assigned to polygons based on their Material ID numbers. Material #1 goes to polygons with Material ID 1, Material #2 is applied to polygons with Material ID 2, etc.

When you assign materials to polygons in the way we just did, the Material Editor automatically generates Material IDs for the object and sets up the Multi/Sub-Object material. You could use the Material modifier to set up material IDs for selected polygons, but the method we used is much faster.

Let's take a look at the material that was generated.

1. Click **Exit Isolation** to bring the rest of the objects back into the scene.
2. Select an empty sample slot in the Material Editor.

3. On the Material Editor toolbar, click **Pick Material from Object**, and click on the rover body in any viewport.

The empty sample slot is now filled with the Multi/Sub-Object material for the rover. On the Basic Parameters rollout, you can see that the number of materials has already been set to three. The material names are listed for each submaterial, and the Multi/Sub-Object material has already been assigned to the object.

> **TIP**
> A Multi/Sub-Object material displays in a sample slot as a checkered sphere with the checkers showing alternating submaterials.

4. Name the new Multi/Sub-Object material **mRoverMaster**.
5. Save the scene with the filename **RoverTex02.max**.

Before we continue, let's make sure you understand what we did here. We created a Multi-Sub/Object material automatically by following this sequence of steps:

- Convert the object to an Editable Mesh.
- Select polygons, and assign a material to the polygons. Do this for one or more areas of the object.

At this point, the material has been created. We can see it by doing one more step:

- Select an empty sample slot, click **Pick Material from Object**, and click the Editable Mesh object.

At the beginning of this tutorial we knew we were going to perform this procedure, which is why we converted the rover body to an Editable Mesh. You can create a Multi/Sub-Object material in this way only with an Editable Mesh.

Make sure you understand this process as you will need it in later steps.

Converting and Isolating the Reactor

The reactor and engine are part of a single piece of geometry for which we will also need another Multi/Sub-Object material. We will take the same approach we used for the rover body, first converting it to an Editable Mesh and then creating the Multi/Sub-Object material through polygon selection.

1. Select **Reactor - Engine**.
2. On the **Modify** panel, right-click on the modifier stack and select **Convert To: Editable Mesh** from the menu.
3. **<Ctrl>**-right-click and select the *Isolate Tool* from the Quad menu.

Creating the Submaterials

1. On the **Modify** panel, access the **Polygon** sub-object level.
2. Select all the polygons on the reactor, the bubble-shaped section of the object.

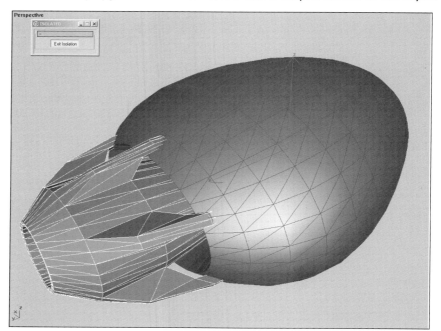

Figure 11.10 *Reactor polygons selected*

3. On the Material Editor, click and drag an empty sample slot onto the selected polygons.
4. Name the new material **mReactor**.
5. From the menu, choose *Edit/Select Invert*.

   This will cause the selection to reverse and the engine polygons to be selected.

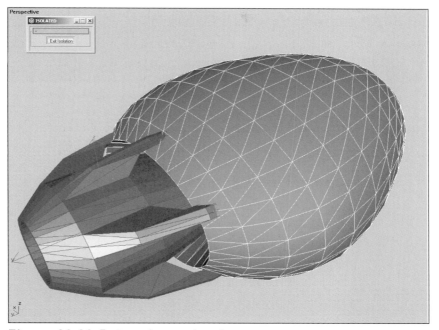

Figure 11.11 *Engine polygons selected*

6. In the Material Editor, click and drag a different empty sample slot onto the selected polygons.
7. Name this material **mEngine**.
8. From the menu, choose *Edit/Select Invert* to invert the selection back to the reactor.

Creating the Base Material for the Reactor

For the reactor we will use an Anisotropic shader, which creates the ellipse-shaped highlights common to many metals.

1. In the Material Editor, select the **mReactor** material.
2. Set the shader type to **Anisotropic**.

3. Unlock the Diffuse and Ambient color swatches.
4. Set the following parameters for the material:

Ambient   RGB 50,50,50 (very dark gray)
Diffuse   RGB 72,72,72 (dark gray)
Specular Level 130
Glossiness   17
Anisotropy   66
Orientation 0

Setting up the Reactor's Diffuse Map

1. On the Maps rollout, assign a **Noise** map to the **Diffuse Color** map channel.
2. Under the Noise Parameters rollout set the following:

   Noise Type Fractal
   Size   15

3. Click **Go to Parent** to return to the root of the material.
4. On the Maps rollout, set the **Diffuse Color** map **Amount** to 50.

Setting up the First Bump Mix Channel

1. Assign a **Mix** map to the **Bump** map channel.

   A Mix map allows you to blend two maps. Each of the maps you will use can also have submaps.

2. Under Mix Parameters, click the button under Maps to the right of the Color #1 swatch. This button is currently labeled **None**.

3. On the Material/Map Browser, select **Checker**.

4. Set the following parameters for the Checker map:

   On the Coordinates rollout:
   U Offset   5
   U Tiling   0
   V Offset   0
   V Tiling   8

   On the Checker Parameters rollout:
   Color #1   RGB 116,116,116 (gray)

> ☼ **TIP** ☼
> *Assigning two levels of checker maps in this way creates the appearance of etched metal.*

5. Click the **Color #2** map channel and choose another **Checker** map. To create smaller indentations in our material, set the following in the Coordinates rollout:

   U Offset   0
   U Tiling   10
   V Offset   5
   V Tiling   5

6. Click the **Go to Parent** button twice to return to the **Mix** map level.

Setting up the Second Mix Channel

1. Click the **Color #2** map channel and assign it a **Checker** map. In the Coordinates rollout, set the following parameters:

   U Offset 5
   U Tiling 0
   V Offset 0
   V Tiling 25

2. Click **Go to Parent** to move up one level on the material.

3. Click the **Mix Amount** map channel and choose a **Gradient** map from the dialog.

4. Set the **Color #2** to a pure black (RGB 0,0,0) and set its **Position** to 0.2.

5. Click **Go to Parent** twice.

6. On the Maps rollout, set the **Bump Amount** to 7.

Setting up the Reflection Map

1. Assign a **Noise** map to the **Reflection** map channel.

2. On the Noise Parameters rollout, set the following parameters:

   Color #1   RGB 255,121,0 (bright orange)
   Color #2   RGB 255,215,193 (pale orange)

3. Click **Go to Parent** to return to the root of the material.

4. Set the **Reflection Amount** to 15.

Using the Material/Map Navigator

We've got a lot of maps and submaps here. Let's take a moment to see what we've got. We'll use the Material/Map Navigator to display a hierarchical listing of all the maps and submaps in the material.

1. On the Material Editor toolbar, click **Material/Map Navigator**.

The Material/Map Navigator appears, showing all the submaps for the currently selected material.

Figure 11.12 *Material/Map Navigator for reactor material*

2. Compare your Material/Map Navigator with the one shown in the previous figure to ensure you have set up the material correctly. The map numbers can be different, but the map types should be the same.

Assigning Mapping Coordinates to the Reactor

Now that we have the material set up, we need to assign mapping coordinates so the material will know which way to place itself on the geometry.

1. The polygons making up the reactor (bubble-shaped area) should still be selected. If they are not, select them now.
2. On the **Modify** panel, apply a **UVW Map** modifier.
3. On the modifier stack, expand the **UVW Map** modifier and click on the **Gizmo** sub-object.
4. On the Parameters rollout, click the **Cylindrical** selection.
5. In the Alignment section, click **Fit**.
6. Move or rotate the Perspective viewport to get a good view of the reactor.
7. Perform a quick render of the Perspective viewport.

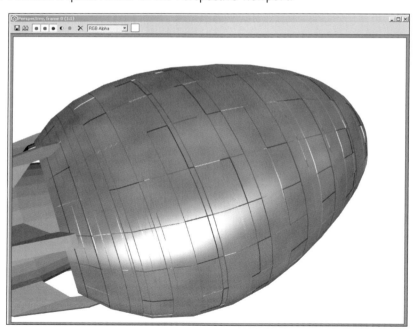

Figure 11.13 *Reactor texture placement*

The material for the reactor is now complete.

8. Save the scene with the filename *RoverTex03.max*.

Creating the Engine Material

Next, you will create the material for the engine, called **mEngine**. This material will also have a metallic appearance.

Now that you have the hang of setting up materials in the Material Editor, we will simply provide you with settings and you can set up the materials.

Some maps are assigned submaps. The submaps are indented so you can see the levels. After entering the parameters for each map, be sure to click **Go to Parent** to return to the Maps rollout to set the **Amount** value for each map channel where indicated.

Material name	mEngine
Shader Type	Anisotropic
Ambient/Diffuse	(unlocked)
Ambient color	RGB 50,50,50 (very dark gray)
Diffuse color	RGB 72,72,72 (dark gray)
Specular Level	130
Glossiness	17
Anisotropy	66
Orientation	0

Diffuse Color map: Noise
    Noise Type Fractal
    Size   15
Amount   50 (on Maps rollout)

Specular Color map: Noise
    Noise Type Fractal
    Size   25
Amount   100 (on Maps rollout)

The nested Noise and Checker maps will produce a bumpy metallic surface.

Bump map: Mix

    Color#1 map: Checker
        U Offset   0
        U Tiling   50
        V Offset   0.5
        V Tiling   0
        Color #2 map: Checker
            U Offset   0.5
            U Tiling   0
            V Offset   0
            V Tiling   25
    Color #2 map: Noise
    Mix Amount   65.7
Bump Amount   25 (on Maps rollout)

Reflection map: Noise
   Color #1   RGB 255,121,0 (orange)
   Color #2   255,215,193 (pale pink)
Amount   15 (on Maps rollout)

Check your work by clicking the **Material/Map Navigator** button on the Material Editor toolbar and comparing your map hierarchy to the one shown below.

Figure 11.14 *Material/Map Navigator for engine material*

Assigning the Final Mapping Coordinates

You should still have the reactor's polygons selected.

1. On the modifier stack, click on the **UVW Map** modifier.
2. Apply a **Mesh Select** modifier to the stack.
3. Expand the **Mesh Select** modifier and access the **Polygon** sub-object level.

   This will automatically select the previous selection, which is all the polygons for the reactor.
4. Invert the selection by choosing *Edit/Select Invert* from the menu.
5. Apply a **UVW Map** modifier to the selection.
6. In the Parameters rollout, click **Cylindrical** as the mapping type.
7. Click the **Fit** button.

Rendering a Test and Collapsing the Stack

1. Do a quick render and verify that the materials look right.
2. Make sure that the **UVW Mapping** modifier is highlighted on the modifier stack. Right-click on it and select *Collapse All*. If asked if you would like to continue, answer **Yes**.

   By collapsing the stack, we keep all the mapping coordinates but they are no longer adjustable.
3. Save the scene with the filename *RoverTex04.max*.

Creating the Track Material

Now we will create the material for the two tracks.

1. Select an empty material slot and name it **mTracks**.
2. Drag and drop the material from the Material Editor onto each of the rover's tracks.
3. Set the following parameters for the material:

   Shader Type   Blinn
   Diffuse   Near black
   Ambient   Near black
   Specular Level 32
   Glossiness   17

   Diffuse Color map: Noise
       Noise Type Fractal
       Low   0.3
       Color #1   RGB 47,47,47 (very dark gray)
       Color #2   RGB 115,115,115 (medium gray)
   Amount   25 (on Maps rollout)

Creating the Track Wheel Material

Now let's create the map for the wheels within the treads. This material will resemble the material used for the reactor, with a metallic finish that makes the wheels look like gears.

1. Select an empty sample slot and name the material **mGears**. Assign it to the four track wheels.

2. Unlock the Ambient and Diffuse colors.
3. Set the following shader parameters:

   Shader Type   Anisotropic
   Ambient color RGB 50,50,50 (very dark gray)
   Diffuse color RGB 72,72,72 (dark gray)
   Specular Level 130
   Glossiness   17
   Anisotropy   66
   Orientation   0

4. Set up the following maps:

   Diffuse Color map: Noise
       Noise Type Fractal
       Size   10

Bump map: Mix
    Color #1 map: Checker
        U Offset   0.5
        U Tiling   0
        V Offset   0
        V Tiling   8
        Color #1   **RGB 116,116,116** (gray)
    Color #2 map: Checker
        U Offset   0
        U Tiling   10
        V Offset   0.5
        V Tiling   5
    Mix Amount map: Gradient
        Color #2 Position   0.1
Amount   **6** (on Maps rollout)

Reflection map: Noise
    Color #1   **RGB 5,3,31** (dark blue)
    Color #2   **RGB 92,90,99** (dull purple)
Amount   **15 (on Maps rollout)**

5. Compare with the Material/Map Navigator below.

Figure 11.15 *Material/Map Navigator for gears material*

In modeling the rover track wheels in Chapter 4, you were left to your own skills to model the track wheels. For this reason, the material we use for our track wheels here might not work for yours without some adjustment.

If necessary, you can apply mapping coordinates to individual polygon selections such as the hub or rims. To do this, apply a Mesh Select modifier to the wheel, select polygons and then apply a UVW Map modifier. A Planar type map applied to polygons can solve some material problems.

Creating the Light Housing Material

Next we will create the material for the light housing. Again we will be building a Multi/Sub-object material. We will construct another similar metal material for the housing and a blue self-illuminating material for the lights themselves.

1. Convert each of the four lights to an Editable Mesh.
2. Select an empty slot in the material editor and name it **mLightHousing**.
3. Set the shader type to **Phong** on the pulldown menu.

4. Unlock the Diffuse and Ambient colors on the Phong Basic Parameters rollout.
5. Set the following parameters for the shader.

   Ambient color HSV 0,0,75 (dark gray)
   Diffuse color   HSV 0,0,168 (light gray)
   Specular Level 18
   Glossiness   10

6. Set up the following maps:

   Diffuse Color map: Noise
       Noise Type Fractal
       Size   100

   Bump map: Speckle
       Color #1   **HSV 0,0,138** (medium gray)
       Color #2   **HSV 0,0,175** (light gray)
       Size   5
   Bump Amount 12

   Specular Level map: Copy map from Bump channel

   This completes the material for the light housing.

Creating the Base Light Color

We are going to create the blue light material before we assign these two materials to the actual geometry.

1. Select an empty material and name it **mLightColor**.
2. Set the shader type to **Blinn**.

3. On the Blinn Basic Parameters rollout, lock the Diffuse and Ambient colors together.
4. Set the following shader parameters:

   Specular Level 0
   Glossiness   0
   Diffuse color   RGB 32,0,194 (bright blue)

TIP
Self-illumination causes an object to look as it if is lit up from the inside.

5. On the Blinn Basic Parameters rollout, check the **Color** checkbox in the Self-Illumination section.

6. Drag and drop the **Diffuse** color swatch to the **Self-Illumination** color to make a **Copy** of it.

Setting up the Diffuse Color Map

1. Set up the following map for the material:

   Diffuse Color map: Gradient Ramp
       Gradient Type Radial
       Interpolation Linear

2. Create a fourth color flag by clicking anywhere on the gradient strip.

3. Set the following values for the color flags. The color flag numbers below represent the flags from left to right, not the actual number that is assigned to them.

   Flag #1: **Position**   0
                 RGB   0,0,255 (bright green)

   Flag #2: **Position**   56
                 RGB   69,161,185 (blue)

   Flag #3: **Position**   90
                 RGB   244,249,251 (pale blue)

   Flag #4: **Position**   100
                 RGB   255,255,255 (white)

4. Click **Go to Parent**.

Setting up Self-Illumination

1. On the Maps rollout, drag and drop the **Diffuse Color** map onto the **Self-Illumination** channel to make a **Copy**.

2. Click the **Self-Illumination** channel button to edit its **Gradient Ramp** map.

3. Make the following modifications to the color flags:

   Flag #2 **Position**   42
                 RGB   3,7,250 (bright green)

   Flag #3: **Position**   87
                 RGB   59,139,159 (blue-green)

   Flag #4: **Position**   100
                 RGB   200,200,200 (light gray)

4. Click **Go to Parent**.

5. On the Material Editor toolbar, change the **Material Effects Channel** from 0 to 1. This will allow us to glow this particular material in **Combustion** later on.

Assigning the Two Materials to the Lights

With the techniques used earlier, assign the two materials to the appropriate polygons on all four lights. After you have applied the materials, apply a **UVW Map** modifier to each polygon selection with **Planar** mapping. Name the Multi/Sub-Object material **mLightAssembly**.

Creating the Base Fuel Tank Material

1. Select an empty material slot, name it **mFuelTank**, and drag and drop it onto the rover's fuel tank.

2. Unlock the Ambient and Diffuse colors, and set the following shader parameters:

   Shader type Phong
   Diffuse color RGB 55,55,55 (dark gray)
   Ambient color RGB 42,37,31 (dark brown)
   Specular color RGB 252,232,221 (pale orange)
   Specular Level 18
   Glossiness 10

3. Set up maps for the material as follows:

   Diffuse Color map: Bitmap
      Bitmap BaseColor.tif (from *CD Tutorial 1* or *Maps* folder on CD)
      Texture (selected)
   Diffuse Amount 30

   Unlock **Ambient Color** and **Diffuse Color** channels

   Ambient map: Copy from Diffuse map as a Copy
      RGB Offset 3.5
   Ambient Amount 30

   Bump map: Speckle
      Color #1 HSV 0,0,138 (medium gray)
      Color #2 HSV 0,0,175 (light gray)
      Size 5
   Bump Amount 12

   Specular Level map: Copy from Bump map as an Instance

   Reflection map: Gradient
      Color #2 RGB 0,0,0 (black)
      Color #3 RGB 0,0,0 (black)
      Color #1 map: Noise
         Color #1 RGB 168,138,111 (medium brown)
         Color #2 RGB 88,74,67 (brown)

4. Assign a **Gradient Ramp** map to the Gradient map's **Color #2** map channel. Set the **W Angle** to 130. Delete all flags except for the two on the ends, and change the two remaining flags to black.

5. Add a series of flags at the center,. Change their colors to make three yellow lines.

6. Click **Show Map in Viewport** to show the map on the fuel tank.

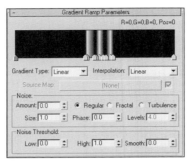

Figure 11.16 *Gradient Ramp Parameters*

7. Click **Go to Parent** twice, and set the **Reflection Amount** to 15.
8. Drag and drop the **Reflection** map onto the **Self-Illumination** map channel and make a **Copy** of it.
9. Compare with the Material/Map Navigator below.

Figure 11.17 *Material/Map Navigator for fuel tank material*

Applying Mapping Coordinates

1. Apply a **UVW Map** modifier to the fuel tank.
2. Set the type to **Planar** and click the **X** axis for the alignment. Depending on how you made your fuel tank, a different axis might be required in step 4.
3. Click the **Fit** button.
4. Adjust the **UVW Mapping** modifier or **Gradient** map parameters so the yellow lines run along the fuel tank.

Creating the Inner Wheel Material

1. In the Material Editor, locate and select the **mRoverMaster** Multi/Sub-Object material. If you can't find it, select any slot and use **Pick Material from Object** to pick it from the rover body.

2. Under the Multi/Sub-Object Basic Parameters, locate the **mBody** Standard material and drag it to an empty sample slot to make a **Copy**. Name the new material **mWheelCap**.

3. Locate the material **mTracks** (pick it from the tracks if necessary) and copy it to another slot as a Copy. Name this material **mWheelRubber**.

4. Select one of the inner wheels (not the track wheels) and convert it to an Editable Mesh.

5. Select the polygons that make up the wheel cap and assign the material **mWheelCap** to them.

6. Invert the selection on the wheel objects to select the rubber part of the wheel, and assign the material **mWheelRubber** to them.

7. Repeat the last three steps on the other three wheels.

8. Apply a **UVW Map** modifier to each wheel.

Texturing the Struts, Springs and Track Supports

The struts, springs and track supports are less visible than the other materials found on the rover so we encourage you to test your skill level in creating these materials. Try to create materials that make the struts and springs blend well with the other materials on the rover.

When you have finished texturing the entire rover, save the scene with the filename *Rover - Textured.max*.

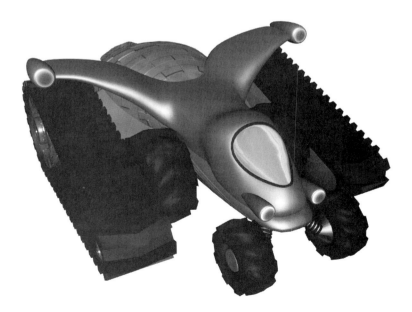

THE BATTERY MATERIAL

The last texturing task on the rover model is the battery. This material will be a little on the complex side, with two decals in one material.

There is a good chance that the battery you created is different from the one we'll be using in the next tutorial. To get this material to look right, you might need to texture our battery first using the following tutorial. After you see how the material is supposed to look, you can make it work with your own battery. You can find our battery on the CD in the *Battery* folder, with the filename *Battery Model.max*.

BATTERY TEXTURES

The battery requires three separate maps to define its color: an overall color map, a plus sign and a minus sign. All three textures need to be placed differently on the object. We also want to be able to adjust the mapping coordinates for each one separately so we can move one without affecting the others.

Placing two or more Diffuse maps on the same object is accomplished with the use of a *map channel ID*. Each bitmap can be associated with a map channel ID, and then two or more UVW Map modifiers can be applied to the object, each set with a map channel ID that corresponds to one of the bitmaps. In this way, you can adjust the UVW Map modifier that pertains to the bitmap with the same map channel ID, and adjust mapping coordinates just for that bitmap.

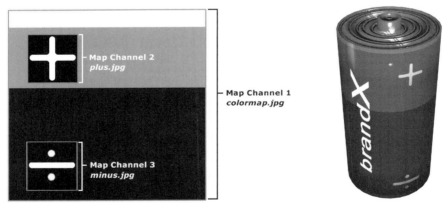

Figure 11.18 *Map channel IDs, and resulting battery texture*

Sample slots in the Material Editor don't always display map channels accurately. Information from the UVW Map modifier is needed to display the maps in their correct locations, and the Material Editor sample slots don't have access to this information. For this reason, don't rely on sample slots for accurate material display when using map channel IDs.

If you want to create the textures for the battery yourself in Photoshop 6, do the tutorial in *CD Tutorial 2* in the *Tutorials* folder on the CD. These textures are also provided in the *Maps* folder on the CD.

TUTORIAL 11.3 TEXTURING THE BATTERY

We have included the battery in a separate file so you can load it alone or merge it into your scene.

Creating the Base Material

1. Load the file **Battery Model.max** from the **Battery** folder on the CD. Alternately, you can load **Rover - Textured.max** and merge the entire contents of **Battery Model.max** into the scene. The battery object is named **Battery**.

2. Select an empty sample slot and name it **mBattery**. Assign it to the battery object.

3. Set the shader type to **Phong**.

 4. Unlock the Ambient and Diffuse colors.

5. Set the shader parameters to the following:

   Ambient color RGB 42,37,31 (dark brown)
   Diffuse color   RGB 55,55,55 (dark gray)
   Specular Level 109
   Glossiness   10

Setting up the Colors and Maps

1. Assign a **Bitmap** map to the **Diffuse** Color map channel, and choose the map **ColorMap.jpg** that you made with **CD Tutorial 2** in the **Tutorials** folder. You can also load this map from the **Maps** folder on the CD.

 2. Click the **Show Map in Viewport** button to make the map appear on the battery in shaded viewports.

 3. Click **Go to Parent** to return to the root of the material.

 4. Unlock the Ambient Color and Diffuse Color channels.

5. Drag and drop the **Diffuse Color** channel to the **Ambient Color** channel and make a **Copy** of it.

6. Set both the **Ambient Color Amount** and **Diffuse Color Amount** to 30.

Assigning Mapping Coordinates to the Battery

1. Select the battery object.

 2. On the **Modify** panel, apply a **UVW Map** modifier.

3. Set the type to **Cylindrical** and choose **Z** as the alignment axis.

4. On the modifier stack, expand the **UVW Map** modifier and select the **Gizmo** sub-object.

 5. Click **Select and Non-uniform Scale**. In the Perspective viewport, scale the gizmo on the Z axis to about 110%.

6. In the Perspective viewport, use **Select and Move** to move the gizmo on the Z axis until the white color at the top just covers the top rim of the battery. If necessary, scale the gizmo more on the Z axis so the black color extends to the bottom of the battery.

Figure 11.19 *UVW Mapping gizmo moved up*

7. Click, then right-click the **UVW Map** modifier and select **Rename** from the pop-up menu. Rename the modifier to **baseCyl_ch1**.

When you applied the UVW Map modifier to the battery, by default the modifier's Map Channel parameter is set to 1. This parameter sets the map channel ID. Renaming the modifier to reflect this setting will make it easier for us to find the modifier after we have applied more UVW Map modifiers to the stack with different map channel IDs.

Adding the Plus Map

As of right now, the Diffuse Colors and Ambient Color maps are the same. We need to change the Diffuse Color map so we can layer on the new plus and minus images. We will leave the Ambient Color map as the battery base color, and use a Mix map as the Diffuse map so we can mix the base color map and plus map.

We will use a map for the Mix Amount. When a Mix Amount map is used, the map under Color #1 shows through the white areas of the map, while the map under Color #2 shows through the black areas.

1. On the Maps rollout, click the **Diffuse Color** channel button.
2. Click the **Bitmap** button (located to the right, just under the toolbar). On the Material/Map Browser, select **Mix**. In the Replace Map dialog, select **Keep old map as sub-map** and click **OK**.
3. Click drag the **Color #1** map to the **Color #2** map and choose **Swap** from the dialog.
4. Assign a **Bitmap** to the **Color #1** map channel and select the *Plus.jpg* file you created with *CD Tutorial 2*, or select *Plus.jpg* from the *Maps* folder on the CD.
5. In the Coordinates rollout, make sure **Texture** is selected, and set the **Mapping** pulldown to **Explicit Map Channel**. Uncheck the **Tile** checkbox for both **U** and **V**.
6. Set the **Map Channel** parameter to 2.

Figure 11.20 *Coordinates rollout for Bitmap*

7. Expand the Output rollout, and change the **Output Amount** to 3.0.

   This brightens up the map so it will show up better on the rendered object.

8. Click **Show Map in Viewport**.

Setting up the Mix Amount

Now we need to tell these two maps how to mix. When a map is used for the Mix Amount, the white areas of the map show the Color #1 map while the black areas show Color #2. We will use *Plus.jpg* as the Mix Amount map to show the plus sign over the base color.

1. Click **Go to Parent** to step up to the **Mix** map base level.
2. Drag the **Color #1** map channel to the **Mix Amount** map channel to make a **Copy** of it.

3. Click **Show End Result** on the Material Editor toolbar to turn it off and see what the map looks like at this level.

There seems to be a problem here. The wrong part of the base color is showing through. We want to invert the black and white colors so the base color will show outside the plus boundary while the plus sign's white color shows inside the plus boundary.

You have two choices for inverting the way the Mix Amount map affects the two maps. You can swap the Color #1 and Color #2 maps, or you can enable the Invert option for the Mix Amount map. We are going to use the second option.

4. Click the **Mix Amount** map to move to its level on the material.
5. On the Output rollout, check **Invert**.

The plus sign should now display in white over the base color in the sample slot.

6. Click **Go to Parent** to return to the **Mix** map base level.

Applying Mapping Coordinates for the Plus Sign

We now need to apply a new UVW Map modifier with its Map Channel set to 2 so we can control the placement of the plus sign.

1. Verify that the battery is still selected. On the **Modify** Panel, apply another **UVW Map** modifier.
2. Set the type to **Planar** and use **Y** as the axis of alignment.
3. Set the **Map Channel** to 2.

   This corresponds to the Map Channel value on the Coordinates rollout for the **Plus.jpg** bitmaps.
4. Rename this modifier to **topLogo_ch2**.
5. Expand **topLogo_ch2** in the modifier stack and select **Gizmo**.

☀ **TIP** ☀

In the Material Editor sample slot, the plus sign is still as big as the color map even though you just adjusted its mapping coordinates in the viewport. UVW Map info is not passed on to sample slots, so the battery material will always look a little funky in the Material Editor though it looks fine in renderings. When using map channel IDs, use viewports or renderings, not sample slots, to check the material.

Figure 11.21 *Battery in Perspective viewport*

6. Use both scaling and moving to fit the gizmo in such a way that the plus symbol appears near the top of the battery, as shown in the previous figure.

7. When you are done, click on the **topLogo_ch2** listing in the modifier stack to exit the Gizmo sub-object level.

Adding the Minus Image

Now we need to make room for the minus sign. To do this, we will use the same technique we used to set up a plus sign Mix map. We will end up with a Mix map that consists of two Mix maps, one for the plus and one for the minus.

1. Click the **Mix** button just below the toolbar, above the Mix parameters rollout. On the Material/Map Browser, choose **Mix**, and choose **Keep old map as submap**.

2. Copy the Mix map on the **Color #1** map channel to **Color #2** with the **Copy** option.

3. Click the **Color #2** map channel, then click the **Color #1** channel. Replace the bitmap with the file **Minus.jpg** that you created in **CD Tutorial 2**, or load it from the **Maps** folder on the CD.

4. In the Coordinates rollout, set the **Map Channel** to 3.

5. Click **Show Map in Viewport**. This will replace the map currently shown on the battery with the minus map.

6. Click **Go to Parent** to go up one level, then drag the Mix map from the **Color #1** map channel to the **Mix Amount** map channel as a **Copy**.

7. Click the **Mix Amount** map and check **Invert** on the Output rollout.

8. Apply another **UVW Map** modifier to the battery. As before, set its type to **Planar** and alignment on the **Y** axis. Set its **Map Channel** parameter to 3 and move the gizmo so the minus sign appears at the bottom of the battery. Name this new modifier **bottomLogo_ch3**.

Mixing the Plus and Minus

So far you have seen how to add a decal to an object by using the Mix map and the black/white image as a Mix Amount map. We will now use an image that's white at the top and black at the bottom to mix the plus and minus signs' Mix maps. We will use a Gradient Ramp for the Mix Amount map.

1. Click **Go to Parent** twice.

Make sure that you are at the top level **Mix** map, the first one you assigned to the **Diffuse Color** channel. If you are not sure, click **Go to Parent** to go all the way to the top root level of the material, then click across from **Diffuse Color** on the Maps rollout to access the **Mix** map.

2. Assign a **Gradient Ramp** map to the **Mix Amount** channel.

TIP

Using map channel ID 1 will make the gradient cover the entire battery.

3. In the Coordinates rollout, click the **Texture** radio button and set the Mapping pulldown to **Explicit Map Channel**.
4. Set the **Map Channel** to 1.
5. Set the **W Angle** parameter to 90.

   This rotates the ramp 90 degrees around the W axis, which points straight out from the center of the object.

6. Click between the second and third flags to create a new flag in the gradient.
7. Set the color and position for each flag as follows:

   Flag #1: **Position**   0
              RGB   0,0,0 (black)

   Flag #2: **Position**   50
              RGB   0,0,0 (black)

   Flag #3: **Position**   51
              RGB   255,255,255 (white)

   Flag #4: **Position**   100
              RGB   255,255,255 (white)

8. Click **Go to Parent** twice to return to the material root level.
9. Compare your material the Material/Map Navigator shown below.

Figure 11.22 *Material/Map Navigator for battery plus and minus maps*

At this point, you should be able to see both the plus and minus on the battery when you render it. Do a test rendering of the battery to see how it is shaping up.

Creating a Mix Map for the Bump Channel

We'll use the same technique to create a bump map with the plus and minus.

1. Set up a **Mix** map for the **Bump** map channel with **Plus.jpg** as the **Color #1** map and **Minus.jpg** as the **Color #2** map. For both maps, check the **Tile** checkbox for both the **U** and **V** on the Coordinates rollout. Set the **Map Channel** for **Plus.jpg** to 2 and **Minus.jpg** to 3.

2. Use the same **Gradient Ramp** map as the **Mix Amount** map.
3. Set the **Bump Amount** to 100.

Creating the Glossiness Map

1. On the Maps rollout, assign a **Gradient Ramp** map to the **Glossiness** channel.
2. Set the **Mapping** type to **Explicit Map Channel** and the **Map Channel** to 1.
3. Add color flags and adjust their parameters as follows:

Flag #1: **Position**   0
        RGB   **0,0,0** (black)

Flag #2: **Position**   23
        RGB   **0,0,0** (black)

Flag #3: **Position**   30
        RGB   **255,255,255** (white)

Flag #4: **Position**   32
        RGB   **255,255,255** (white)

Flag #5: **Position**   39
        RGB   **0,0,0** (black)

Flag #6: **Position**   56
        RGB   **0,0,0** (black)

Flag #7: **Position**   60
        RGB   **255,255,255** (white)

Flag #8: **Position**   67
        RGB   **255,255,255** (white)

Flag #9: **Position**   74
        RGB   **0,0,0** (black)

Flag #10:**Position**   100
        RGB   **0,0,0** (black)

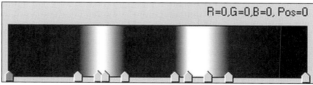

Figure 11.23 *The resulting Gradient Ramp color strip*

This will generate highlights running along the sides of the battery.

4. On the Output rollout, check the **Invert** checkbox.

   We found that inverting the gradient made the battery look its best.

5. Click **Show Map in Viewport** to display the highlight in viewports.

6. Click **Go to Parent** to return to the root of the material.

Assigning the Self-Illumination Map

Now we need to get the brand name text on the battery. Rather than add another map channel for the Diffuse map, we will use the text as a Self-Illumination map. This will cause the base colors to appear where the text lies over them.

> ☀ **TIP** ☀
>
> Using a Self-Illumination map ensures the letters will be bright and easy to read in the rendered scene.

1. In the Phong Basic Parameters rollout, check the **Color** checkbox in the Self-Illumination section.

2. On the Maps rollout, assign a **Bitmap** map to the **Self-Illumination** channel.

3. Navigate to and select the **BrandX.tif** file that you created in Photoshop with **CD Tutorial 2**, or use the file **BrandX.jpg** from the **Maps** folder on the CD.

4. Set the **Mapping** type to **Explicit Map Channel** and set the **Map Channel** to 1. Set additional parameters as follows:

V angle	18
W angle	-90
U Tiling	1.2
V Tiling	6.2

   Uncheck **Tile** for **U** and **V** directions

5. Click the **Show Map in Viewport** to display the text in shaded viewports.

6. Click **Go to Parent** to return to the material root.

7. Check your work against the Material/Map Navigator.

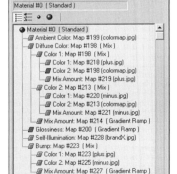

Figure 11.24 *Material/Map Navigator for battery body*

Texturing the Battery Top and Bottom

Use techniques you've learned to texture the top and bottom of the battery. The battery end will barely be visible inside the rover, so this is just to test your skills.

When you have finished, save the file as **Battery - Textured.max**, then open **Rover - Textured.max** and merge the textured battery into it. If you are replacing another battery that is already in the scene, make sure you delete the old untextured battery. Save the scene again as **Rover - Textured.max**.

Texturing the Remote Control

The final object that needs to be textured is the remote control. You are on your own for this one! Open the remote control file and see if you can create a realistic looking material for both the remote control body and the antenna. After you are done, save the file as **Remote Control - Textured.max**.

LIGHTING

Lighting is one of the trickiest areas in the production process. Good lighting can bring a scene to life by giving it depth and warmth while bad lighting can quickly kill a scene by making it look flat and washed out.

We must match our lighting to real world, natural outdoor lighting. This could quickly become a complex job but fortunately we have a few tricks up our sleeves that will provide excellent results.

TYPES OF LIGHTS

Before we get into lighting our scene, let's do a quick review of the types of lights that are available along with a description of what makes each one special.

Target and Free Spot Lights

A spot light creates a focused beam of light. The rays are therefore non-parallel, much like a flashlight or theater spotlight.

The target spot consist of two objects: A source icon and a target icon. The source icon always points toward the target icon. This makes it easy to have a spotlight follow objects, as you can simply link the target icon to the object you want the spot to continuously illuminate.

The free spot is like the target spot, but without a target. In order to aim the light, you can either rotate it in your viewport or navigate a viewport while looking through it.

Every spot light has a hotspot and falloff angle associated with it. The light is at its full intensity inside the hotspot area, then gradually "falls off" to zero at the falloff angle.

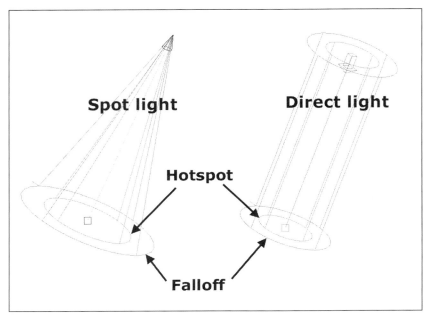

Figure 11.25 *Hotspot and falloff angles*

Target and Free Direct Lights

A direct light emits parallel rays of light. Direct lights are generally used to represent light sources that are long distances away, such as the sun. The light is coming from such a far distance away that the rays of light appear parallel to one another, which casts the parallel shadows that people intuitively associate with outdoor scenes. Compare with target lights, which cast non-parallel shadows.

A target direct light is aimed just as the target spot is, via its target object. The free direct light is like the target direct light, but without a target.

Direct lights also have hotspot and falloff values for setting the size of the area affected by the light.

Omni

Omni lights (short for *omni-directional*) emit light from a single point in all directions. Omni lights are great for overall illumination, such as fill lights in a traditional three-point lighting scenario. Shadows cast by omni lights are non-parallel, just like spot lights.

SHADOWS

Shadows will play an important part in the final animation. Shadows are used to create the intuitive connection a viewer makes between the 3D animation and the background plates. If the shadows cast by the alien and his toys match the shadows in the video footage, we will be able to create the illusion that the alien is actually in the environment.

There are two types of shadows in **3ds max**, shadow mapped and raytraced.

Shadow Mapped Shadows

Shadow mapped shadows use a map, or image, projected from the light source to determine where the shadows should fall. The larger the map size, the sharper the shadow. Smaller map sizes can be used to create soft shadows.

Shadow mapped shadows are controlled by three parameters: Size, Sample Range, and Bias. Size sets the size of the map. The Sample Range determines how many times the shadowed area is calculated. You can get very soft shadows with a low Size and a high Sample Range. Conversely, a high Size and low Sample Range produces sharp shadows. Keep in mind that the higher you set these values, the longer it will take to render the shadows.

The Bias parameter is used to compensate for shadow calculation errors, which can be especially obvious on tall, thin objects. Sometimes, **3ds max** will place a shadow in such a way that it is detached from the object even though the object is sitting right on the surface on which the shadow is cast. You can compensate for this error by adjusting the Bias parameter to bring the shadow closer to the object or push it farther away. The Absolute Bias checkbox can be used to keep the Bias from being calculated on each frame of an animation, which prevents shadows from dancing and jittering.

Raytraced Shadows

Raytraced shadows are produced by bouncing a ray of light around the scene and producing shadows from these calculations. Raytraced shadows always render sharply. Unlike shadow mapped shadows, raytraced shadows can generate semi-transparent shadows from transparent objects. Raytraced shadows are generally more accurate than shadow mapped shadows, but they also take longer to render.

There are only two settings to adjust when using raytraced shadows: Bias and Max Quadtree Depth. As with shadow mapped shadows, the Bias parameters offsets the shadow position from the object. The Quadtree Depth, in general, sets the number of times the light rays are bounced around the scene. You can decrease render times by decreasing this value.

3D LIGHT LIMITATIONS

By default, lights added to a scene do not cast shadows -- they penetrate every object they strike to illuminate the objects behind. They also do not attenuate (lessen in effect) over distance. These attributes are opposite the real-world effects of lights. This means that you will usually have to adjust light parameters to get a realistic look.

THE SET

The first step in lighting and texturing the scene is to create a basic set, a stage for our animation. The set will consist of a plane and the lights needed for our scene lighting to match the background plates.

A simple plane will serve to mark the floor or ground for our set, and will be the base reference point for such things as object placement, the rover movements, and the alien. The plane will also receive the shadows cast by our scene objects. The plane itself will not be visible in the final rendering.

GLOBAL ILLUMINATION

In matching our digital environment to the real world environment that was captured on video, we will use a technique to simulate the global illumination you get outdoors on a sunny, hazy day. This method uses a hemispherical arrangement of free direct lights to illuminate the objects in the scene from all directions, creating a even illumination. This is a great approach when lighting an outdoor scene for a nice ambient feel.

For our dome of lights, we are going to use free directional lights because we want the light rays to remain parallel to one another and not distort our shadows.

TUTORIAL 11.4 LIGHTING THE SET

As we advance through this tutorial, we will explain any new light parameters as they come up. In later tutorials and steps, we will simply tell you the settings.

Before you can do this tutorial, you must have already copied and unzipped the file *low-res footage.zip* from the *Movies/Background Plates* folder on the CD to a folder on your hard disk, and you must have the file which holds the list of background images, *Tracker.ifl*, in the same folder. If you did the tutorials in *Chapter 10*, you have already done this. If you haven't, then do it now.

Let's start off by creating the ground plane for the set.

Creating the Ground Plane

1. Open the file *Set - Tracked Camera.max* file that you created in *Chapter 10*. Alternately, you can load the file *Set - Tracked Camera.max* from the *Scenes/Set* folder on the CD.

If you load this file from the CD, you will have to point the background image to the folder that contains your *Tracker.ifl* file. Choose **Rendering/Environment** from the menu, click the button under **Environment Map**, choose **Bitmap**, and pick the file *Tracker.ifl* from the folder on your hard disk. Click in the camera viewport to make the background appear.

2. Choose **Customize/Units Setup** from the menu. Set units to **Generic** and click **OK**.
3. Select and hide all the CamPoints.
4. Create a plane in the Top viewport.
5. Name the plane **Ground**.
6. Set the **Length** and **Width** parameters to 9000.
7. In the Move Transform Type-In dialog, set the plane's absolute world position to the following:

   X   1280
   Y   1465
   Z   0

8. On the **Display** panel, freeze the new plane.
9. Save the scene as *Set - Lights01.max*.

Creating the First Light

If you're starting at this point in the tutorial, you can load the file *Set - No lights or cameras.max* from the *Set* folder on the CD.

1. In the Left viewport, create a free direct light. Name the light **globalLight01**.
2. Use the Move Transform Type-in dialog to set the light's absolute world position:

   X   -14172
   Y   1465
   Z   100.

3. On the **Hierarchy** Panel, click **Affect Pivot Only**.
4. In the Top viewport, move the pivot to the center of the scene. Use the plane to figure out where the center of the scene will be. We used the following numbers:

   X   1280
   Y   1465 (same as above)
   Z   100 (same as above)

We have moved the pivot to the center of our scene to make it easier to duplicate and place the remainder of the lights.

5. Click **Affect Pivot Only** to turn it off.

6. In the camera viewport, copy and rotate the light 22.5 degrees around its Y axis. In the Clone Options dialog, set the **Object** type to **Instance** and **Number of copies** to 8. Click **OK**.

You have now created your first rib arch of lights.

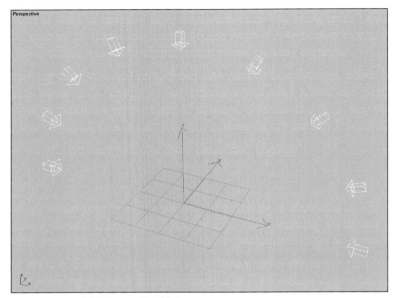

Figure 11.26 *First arch of lights*

7. Save the scene as **Set - Lights02.max**.

Creating the Remaining Light Arches

Next, we will copy and rotate the arch of lights to create a dome of lights.

1. Select the light at the very top of the arch and delete it.

    We have deleted this light so that in the next step we won't create the same light on top of itself as we copy the arch.

2. Select all the remaining lights in the arch.

3. On the Main Toolbar, make sure the transform center is set to **Use Selection Center**. If it is, you should see one transform gizmo at the center of the selection, and not an individual transform gizmo at each light.

4. In the Top viewport, copy and rotate the selected lights -36 degrees around the Z axis. This time, create 4 copies with the **Instance** option.

    This creates copies the original arch four times. When you look in the Top viewport, you should see ten half-arches around the scene.

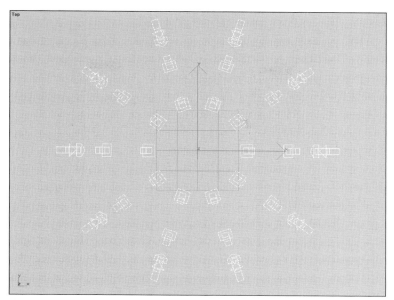

Figure 11.27 *The remaining light arches*

> ☀ **TIP** ☀
>
> *If your arches are not evenly spaced, then you did not rotate by exactly -36 degrees. Undo the rotation and try again until your arches are evenly spaced.*

Next, we need to replace the top light we deleted earlier.

5. Select the light that is positioned directly to the left of the plane's center.

6. In the Top viewport, copy and rotate the selected light around its Y axis 22.5 degrees. Make 1 copy of the light as an **Instance**. This replaces the deleted top light.

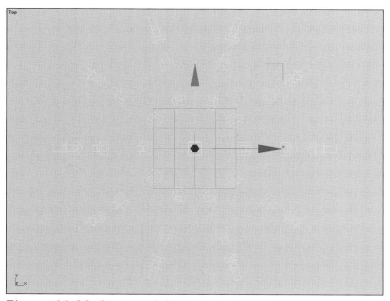

Figure 11.28 *Center top light*

Setting the Dome Parameters

We have made every light an instance so that we only need to modify any one light to make all the other lights update to the same settings.

1. Select any light in the dome.
2. On the **Modify** panel, set the **Multiplier** value to 0.045.

   The Multiplier multiplies the light intensity (its V value). We need a very low setting or the scene will be too washed from so many lights.

3. Set the following RGB values for the lights to make them a pale yellow:

   R 254
   G 240
   B 188

   These settings will give the lights a slight reddish/yellow tint. This will complement the reddish landscape we will generate later on for the alien and his toys.

4. In the Affect Surfaces section, uncheck the **Specular** checkbox.

   When Specular is unchecked, the lights will not generate any highlights where they fall. For this overall illumination setup, we don't want any highlights to appear.

5. On the Directional Parameters rollout, set the **Hotspot** to 9000 and the **Falloff** to 9002.
6. Check the **Show Cone** checkbox.

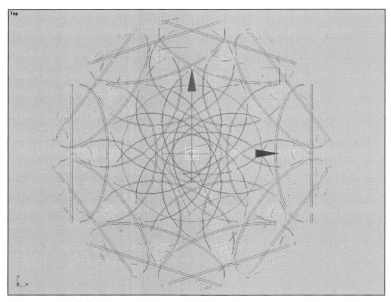

Figure 11.29 *Dome of lights with Show Cone on, or perhaps a Spirograph drawing?*

Checking **Show Cone** causes each light's hotspot and falloff areas to appear in the scene at all times. When unchecked, a light's hotspot and falloff appear only when the light is selected.

You will now be able to see how the lights overlap one another. The idea is to create a dome of lights that have no holes.

7. Look around the dome and check for holes. If you find any, increase the **Falloff** value until there are no holes visible.

Setting the Dome Shadow Parameters

You can also make a light cast shadows by checking Cast Shadows on the General Parameters rollout.

1. In the Shadow Parameters rollout, check the **On** checkbox so the lights will cast shadows.

2. Verify that the pulldown beneath the **On** checkbox is set to **Shadow Map**.

3. Set the **Dens**. parameter to 1.

   This sets the shadow density. Higher values make the shadows darker, while lower values make them less intense.

4. In the Shadow Map Params rollout, set **Bias** to 0.

5. Set the **Size** parameter to 600.

   We need a larger map size to make the shadows more profound while keeping them smooth.

6. Set the **Sample Range** parameter to 6.

7. Save the scene as **Set - Lights03.max**.

Simulating Bounced Light

In real life, outdoor light bounces all over the place. You can easily see this effect on a sunny day by looking under a tree or any other object that casts shadows. If the sun is the only light, then no direct light is hitting the area under the tree. Nonetheless, the area is illuminated enough for you to clearly see the grass or other objects below the tree. This is because of bounced light.

In **3ds max** light rays do not bounce, so you'll have to simulate this effect with the tools you have. A common solution is to place a direct light under the entire scene and point it upward.

To simulate the appearance of light bouncing off of the ground, we will duplicate the top dome light and position it so it points upward from below the scene. We will then set the parameters so that the light gets weaker as it gets farther from the light source. This effect, called *attenuation*, serves to make the light act like real sunlight, where objects farther from the ground get less bounced light.

1. Select the light at the top of the dome.

TIP

This light is a Copy rather than an Instance because it will need to have parameters different from those in the dome.

2. In the Left viewport, copy and rotate the light around its Z axis 180 degrees so that it is pointing up toward all the geometry in the scene. Set the clone type to **Copy** and the light name to **BouncedLight**.

3. Set the following settings for the new light:

   On the General Parameters rollout:
   Cast Shadows checked
   Multiplier 0.2
   Check **Diffuse** and uncheck **Specular**

   On the Directional Parameters rollout:
   Hotspot 4800 (large enough to cover the entire scene)
   Falloff 4802

   On the Attenuation rollout, in the Far Attenuation section:
   Start 15362
   End 17785
   Check the **Use** checkbox

   On the Shadow Parameters rollout:
   Dens. 1

   On the Shadow Map Params rollout:
   Size 256
   Sample Range 4

Testing the Dome Lights

Next, we need a stand-in object to test the lights. Just like in Hollywood, the star (our alien) doesn't need to be present. Instead, he has sent his stand-in, which happens to be a teapot.

1. In the Camera Tracked viewport, create a teapot and position it to sit on top of the ground plane.

2. Make sure the Camera Tracked viewport is active. On the Main Toolbar, click the **Quick Render (Production)** button.

3. Make sure you can see the teapot in the rendering, and that it is softly illuminated with no shadows.

4. When you have seen that the lights are properly set up, group the lights. To do this, select all the dome lights, then choose *Group/Group* from the menu. Name the group **domeLight**.

> ☀ **TIP** ☀
>
> *Teapots make great stand-in objects. They have a variety of curved surfaces so you can test light properties, and they usually charge less than union wage.*

Figure 11.30 *A test rendering*

Creating the Illuminating Key Light

The dome lights and the bounce light have given us some nice overall illumination and very soft shadows. Now we need a few key lights to simulate the sun and create the harder shadows that will make the alien and rover look as though they're actually in the environment.

If we were to use only one key light for both these effects, we would have a problem. A bright key light would cast dark shadows but would wash out the scene. A dim light would give the right amount of illumination, but the shadows would be washed away by the dome lighting.

Instead, we will divide these tasks between two key lights. One light will be responsible for mimicking the direct sunlight but will not cast shadows, while the other light will only cast shadows but provide no illumination. In this way, we can control the shadows independently from the light causing the actual illumination. A pretty slick setup!

1. In the Top viewport, create a Target Direct light and name it **Key Light - Lights**.
2. Position the light in absolute world coordinates to:

   X 3896
   Y -5487
   Z 2239

3. Select the target of the new direct light and set its absolute position to:

   X 1075
   Y 1330
   Z 491

   So how did we determine the position of this light? It takes paying close attention to the background plates. When matching digital lights to real world lights as we are doing here, it is really important to study the lighting captured in the video before you begin. Look for things like shadow direction, specular highlights and dark shaded areas. These can all be helpful in making this determination. Taking notes on the day of the shoot regarding the sun position and brightness can also be helpful.

4. Select **Key Light - Lights** again.

5. On the **Modify** panel, set the following attributes for the light:

   Cast Shadows (unchecked)

   Set the color to a pale yellow:
   R 254
   G 240
   B 188

   Multiplier 0.4
   Diffuse   (checked)
   Specular   (checked)
   Hotspot   3500
   Falloff   3502

TIP
The low Multiplier setting will prevent the light from overlighting the scene.

   Make sure the **Hotspot** is large enough to cover the action area and that there are no signs of fade off in the viewport. This light is our sun, so it has to cover the entire scene.

Creating the Shadow-Casting Key Light

1. Hold down the **<Shift>** key and click on the key light to create a new light. Make the light a copy, and name it **Key Light - Shadows**.

2. Change only the following settings:

   Cast Shadows (checked)

   Multiplier   1.0
   Diffuse   (unchecked)
   Specular   (unchecked)
   Ambient Only   (unchecked)
   Dens.   0.8
   Bias   0.1
   Size   1024
   Sample Range   12

TIP
If the Diffuse, Specular and Ambient qualities of the light will not affect the scene, what does that leave? The shadows only, of course.

Testing the Final Lighting

1. Do another Quick Render of your scene. The new lighting will help give the illusion that the teapot is actually in the scene. It should be better illuminated and there should be a faint shadow behind it.

   The lighting for the scene is now complete.

2. Save the scene with the filename *Set - Lights and Ground Plane.max*.

CATCHING SHADOWS WITH MATERIALS

The Material Editor includes a special type of material designed to render invisibly except for shadows. This type of material, called Matte/Shadow, is ideal for our ground plane. The idea is to render the shadows only, then composite just the shadows over our martian landscape footage.

You will also need to set up small objects as substitutes for rocks in the scene, and assign a Matte/Shadow material to them as well. These objects will receive shadows from the alien and rover, which will add realism to the scene.

Creating and assigning a Matte/Shadow material is a simple process, as illustrated in the next tutorial.

TUTORIAL 11.5 SHADOWS FOR THE GROUND

For this tutorial, you will need the file *Set - Lights and Ground Plane.max* that you created in the last tutorial. If you didn't create this file, you can load the file *Set - Lights and Ground Plane.max* from the *Scenes/Set* folder on the CD.

Before you can do this tutorial, you must have already copied and unzipped the file *low-res footage.zip* from the *Movies/Background Plates* folder on the CD to a folder on your hard disk, and you must have the file *Tracker.ifl* in the same folder. If you did the previous tutorial or the tutorials in **Chapter 10**, you have already done this. If you haven't, then do it now.

If you load the files from the CD, you will have to point the background image to the folder that contains your *Tracker.ifl* file. Choose *Rendering/Environment* from the menu, click the button under **Environment Map**, choose **Bitmap**, and pick the file *Tracker.ifl* from the folder on your hard disk. Click in the camera viewport to make the background display.

Let's get started on our Matte/Shadow material.

1. Open the **Material Editor**.
2. The first empty sample slot should currently be selected. Name this material **mGround**.
3. Click the button labeled **Standard** next to the pulldown where you named the material. On the Material/Map Browser, choose **Matte/Shadow** and click **OK**.

   The options available in the Material Editor have changed from our familiar Standard settings. A Matte/Shadow material has very few parameters.

4. On the Matte/Shadow Basic Parameters rollout, set the following settings:

Opaque Alpha	(unchecked)
Receive Shadows	(checked)
Affect Alpha	(checked)

   Figure 11.31 *Matte/Shadow Basic Parameters rollout*

   Unchecking **Opaque Alpha** will produce an alpha channel with the matte material present.

   Checking **Receive Shadows** will render shadows on the ground plane after the material is applied.

   Checking **Affect Alpha** calculates shadows into the alpha channel.

5. Unfreeze the ground plane if it is frozen.
6. Drag and the drop the material from the sample slot to the ground plane.
7. Render the Camera01 - Tracked viewport.

   The ground plane has become transparent but the teapot shadows appear on the ground. If you have set everything up correctly, the test teapot should appear to be casting shadows on the ground. If the shadow does not look right, make sure you have set the parameters for the lights and the Matte/Shadow material correctly.

Figure 11.32 *Rendered scene with Matte/Shadow material*

> **☼ TIP ☼**
>
> *If the teapot does not look as if it's really part of the scene, don't panic! This could be caused by numerous things, such as the lights needing adjustment or the teapot requiring a real material to look right.*

This simple example should give you a basic idea of how the Matte/Shadow material works.

Continue adjusting the light settings until you are happy with the illumination and shadows in the scene.

8. Delete the teapot.

Creating Rock Stand-ins

To complete the set we need to create matte objects in the locations of several of the key rocks. The objects will not render, but will receive shadows from the alien and rover as they pass by them. The shadows will then be passed on to the image behind them, which is the live footage.

Without these shadows, the CG objects would appear as flat cutouts pasted onto the background plates. With good shadow effects we can tie the CG elements to the scene making it appear as if they were really there.

In creating the rock objects, we must first consider where the alien and rover will cast their shadows. The set has already been lit and we know where the key light is coming from. Of course we don't want to take the time to model every single rock in the scene. We will only model the larger rocks to catch the obvious shadows from the alien and rover.

At least four rocks need to be modeled. We used six in our scene. You will make the rocks from simple primitive boxes converted to editable meshes.

The figure below shows the six rocks we set up for our scene. Frame 175 is shown, where all the rocks are in frame.

Figure 11.33 *Various rock substitutes at frame 175*

1. Go to frame 175 or a frame near this frame where you can see all the larger rocks that will receive shadows from the alien or rover.

2. Select the ground plane and click **<Alt-X>** to make it semi-transparent.

    This will make it possible to see the ground plane and the background plate at the same time.

3. With the **Box** standard primitive tool, create a box in the Top viewport over one of the CamPoint helpers. Name the box **RockMatte01**.

4. Position the box to align it to one of the rocks on the background plate.

    Use the CamPoint helpers to help position the rock. Looking at the CamPoints and the background plate, you can see exactly where the rocks should be positioned in 3D space. For instance, the height of the larger rock the alien sits on can be determined by looking at the CamPoint positioned on top of the rock.

    Work in all viewports to ensure the rock is placed correctly in 3D space. Otherwise, it might look alright at the frame you're at, but will slide around unrealistically when you move the time slider.

5. Add a few **Length Segs**, **Width Segs**, and **Height Segs** so that you will have enough detail to match the shape of the rock in the footage. Two or three should do the trick.

6. Convert the rock to an Editable Mesh.

7. Copy the rock object five times to other rock locations.

8. For each rock object, use box modeling techniques to form the geometry into the shape of the rock you are trying to match. The object should completely cover the rock in the background footage when viewed through the Camera - Tracked viewport.

   Avoid cutting the rocks at edges too sharply. This can lead to chopped shadows and a fake look. It is always best to try to round the edges where shadows will be hitting. Chamfering the top edges can accomplish this.

Testing the Rocks

1. Move the time slider and make sure the rock objects continue to cover the appropriate rocks in the background plate. Make any adjustments to the rock positions or sizes as necessary.

2. Assign the Matte/Shadow material **mGround** to the rocks.

3. Render the scene at a few representative frames and test the shadows.

   The shadows cast by the rock objects should blend seamlessly with the background plate. If you can't tell where the rock objects are in the rendered image, then you've done your job right.

   If your shadows appear rough, try applying a **Smooth** modifier to the rocks with the **Auto Smooth** option checked.

   Make sure the scene renders fast and that the viewport interaction/feedback is good. If not, you will need to decrease the resolution of the rocks.

4. Save the scene as *Set - Lights Ground and Rocks.max*.

SUMMARY

In this chapter, you learned:

- How to layer maps to create custom materials in the Material Editor
- How to use procedural maps such as Noise and Speckle
- An effective method for assigning more than one material to the same object
- How to set up dome lighting
- A method for creating separate lights for controlling the sun and shadows
- How to make an object transparent while receiving shadows

The next step is to pull it all together into one scene, which you'll do in the next chapter.

12

Animation and Rendering

This chapter focuses on the development of the final animated scene. Now that you have all the elements you need to put it together, each piece will be merged into one scene, and the animation completed. You will also learn how to render the scene for compositing later on.

TECHNIQUES IN THIS CHAPTER
In this chapter, you will learn how to:

- Merge scenes together in different ways
- Use a low-resolution substitute object to speed up viewport interaction
- Animate using key poses
- Use previews to fine-tune animation
- Render image files especially for compositing
- Render views from different cameras in one rendering session

TUTORIALS IN THIS CHAPTER

In the tutorials in this chapter, you will merge all the scenes together, animate the objects and render images for later compositing.

- *Tutorial 12.1 Putting Together the Final Scene File* shows you how to bring all the scene elements together into one file using a variety of methods.

- *Tutorial 12.2 Animating the Rover and Alien* goes through the process of using the controls set up earlier to animate the rover, set up key poses and fine-tune the animation for the alien.

- *Tutorial 12.3 Rendering the Shots* uses Video Post to render all the cameras in sequence in one rendering session.

BRINGING FILES TOGETHER

By now you have created all the required models, labored over the intricate setups required by both the alien and the rover, and developed lighting and material solutions that give your CG elements the appearance of actually being in the videotaped footage. But before we can begin animating our scene, we must first integrate all of the separate files into one scene.

There are two techniques that can be used to bring the files together: merging and cross-referencing.

MERGING

With merging, copies of the objects in one file are brought into another. There is no connection between the original objects and the new ones in the current scene. If you were to go back and modify anything in the original file, the objects would need to be merged into the scene again.

CROSS REFERENCING

You can also bring objects into a scene as cross-reference (XRef) objects. These objects are instances of the originals which you cannot change in the current scene. You can only modify the objects in the original scene. When they are modified, they are automatically updated in the scene that references them.

You can cross-reference individual objects or entire scenes. When a scene is referenced, the objects in the referenced scene cannot be animated in the current scene. A cross-referenced object, however, can be animated in the current scene.

The cross-reference approach is useful with a large team, with files changing from day to day. As the original file is worked on, another person in the pipeline can reference it and begin working on some other part of the scene. As the referenced file continues to develop, it is constantly updated in the new file.

BRINGING OUR SCENES TOGETHER

Cross referencing is a great method of working, but in our case we cannot do this for every object. The alien UI requires access to the the alien's modifier stack which is not available with XRefs, so we can't use the XRef method for this file. The same is true for the rover, where the scripts we wrote continuously access the rover's modifier stack.

However, we can XRef the set and the remote control. We will use the XRef Scene option to cross-reference the set as we have no need to animate these objects. We will use the XRef Objects option for the remote control so we can animate it in the scene where needed.

Size Does Matter

As we are bring objects into the scene, you will quickly discover that we have a problem with the scale. We didn't take the step of working out a proper scale for each object before starting this project, so when each object comes in it will undoubtedly be too large or too small for the set. Let's look at the objects one at a time to determine how to deal with this problem.

The set cannot be scaled because it's built around the matched camera, so we'll keep the set as our stable size reference. The remote control can easily be scaled because it's one simple grouped object with no rigging or controls. The rover will also be easy to handle because of the **scaleFactor** custom attribute that we added to the rover's main control helper object.

This leaves the alien. Since the alien is already rigged and skinned, there are very specific objects that must be scaled together for the alien to remain intact. We will cover this process in detail during the next tutorial.

TUTORIAL 12.1 PUTTING TOGETHER THE FINAL SCENE FILE

To merge all the scene elements together, we will start fresh with an empty scene.

Referencing the Set

1. Reset **3ds max**.
2. From the menu, select *File/XRef Scene*.

   The XRef Scene dialog appears.

3. Click the **Add** button and select your final set scene file, **Set - Lights Ground and Rocks.max** from **Chapter 11**. Alternately, you can select the file **Set - Lights Ground and Rocks.max** from the **Scenes/Set** folder on the CD.

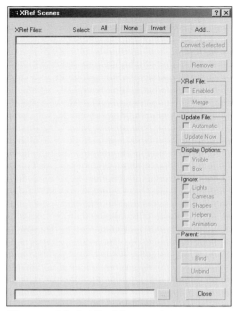

Figure 12.1 *XRef Scenes dialog*

4. Click the **Close** button.

   The set and camera have been brought into the new scene. If you try to select any of the referenced objects, you will find that you cannot. Later, if you find the need to make adjustments on any of the objects in the set, you will need to open the original set file and make the modifications from there.

5. To hide the lights, go to the **Display** panel and check the **Lights** checkbox in the Hide by Category rollout.

6. Save the scene with the filename *Alien Commercial - Final.max*

Displaying the Background Footage

As you may have noticed, the background footage is missing from the viewport. Also, if you do a test render, you will see that everything is black. This is because neither the environment background footage nor the display background are referenced automatically when a scene is merged or externally referenced. Our next step will be to display the environment again in the viewport and rendering.

1. Activate the Perspective viewport and press the **<C>** key to change it to the Camera01 - Tracked viewport.

   Next we will be accessing our live footage. If you did not do the tutorials in *Chapter 10* or *Chapter 11*, copy and unzip the file *low-res footage.zip* and copy the file *Tracker.ifl* file from the *Movies/Background Plates* folder on the CD to your hard disk.

2. From the menu, select **Rendering/Environment**. On the Environment dialog, click the **Environment Map** button in the Background section and apply a **Bitmap** map. Select the **Tracker.ifl** file from the hard disk. Close the Environment dialog.

3. With the Camera01 - Tracked viewport active, choose **Views/Viewport Background** from the menu. On the Viewport Background dialog, check **Use Environment Background** and **Display Background**. Click **OK**.

> ☼ **TIP** ☼
> You will only be able to see a portion of the background footage in the viewport. This is because the ground plane fills up most of the viewport.

The background plate should now be visible in your viewport.

You can turn the background on and off in the viewport by right-clicking the viewport label and choosing **Show Background** from the pop-up menu to turn it on or off. We have elected to work with the background off most of the time.

4. Save the scene.

Merging the Alien

Since we will need access to the alien's modifier stack, we will need to merge the alien file into the scene rather than cross-reference it.

1. From the menu, choose **File/Merge** and navigate to the folder with the file **Alien - Setup and Textured.max** that you created in **Chapter 11**. If you did not create this file, you can choose it from the **Scenes/Alien** folder on the CD. Click **Open**.

The Merge dialog appears.

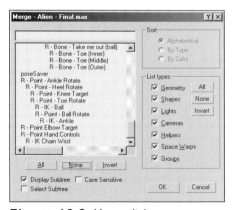

Figure 12.2 Merge dialog

2. On the Merge dialog, click the **All** button at the lower left and click **OK**.

The alien is merged into the scene. The alien comes into the scene at a fraction of its intended size. This is not a problem if you know the correct items to scale. To aid in the selection of these items (in case we need to scale the alien again) we will create a selection set containing the proper objects.

3. Press **<H>** to open the Select by Name dialog and use **<Ctrl>** to select the following objects:

   Eye Control
   L - Point - Ankle Rotate
   L - Point Elbow Target
   L - Point Hand Controls
   Point Alien Root
   R - Point - Ankle Rotate
   R - Point Elbow Target
   R - Point Hand Controls

   Click the **Select** button to select the objects.

4. Click the **Lock Selection** button at the bottom of the screen to lock the selection.

   This will prevent you from accidentally losing any of the objects in the selection for the time being.

5. On the Main Toolbar, type **AlienScale** in the **Named Selection Sets** entry area, and press **<Enter>**.

Figure 12.3 *Named Selection Sets entry area*

The named selection set will make it easier to select these objects later on if necessary.

6. Save the scene.

Scaling and Positioning the Alien

Now that we have the alien in the scene, we need to position him at his starting post and scale him to the proper size.

Moving the alien close to the rock will help you determine the appropriate scale for him.

1. Move to a frame where you can see the rock the alien will be sitting on. This is the largest rock for which you created a matte rock object earlier.

2. In the Top viewport, use **Select and Move** to move the selected points and position the alien over the rock.

   In the next step you will scale the alien. You'll be scaling the alien body and eyes separately. In order to scale them to the same percentage, you will have to note how much you scale the body, then scale the eyes by the same amount.

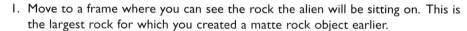
3. Use **Select and Uniform Scale** to scale the points, which will in turn scale the alien. Refer to the storyboards and character sketches if necessary to figure out the appropriate size for the alien. Note the percentage to which you scale the points. If you scale more than once, note the percentage for each scale operation.

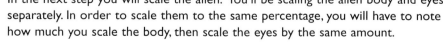
4. Click **Lock Selection** to turn it off.

5. Select one of the eyes, and right-click **Select and Uniform Scale** to display the Scale Transform Type-In dialog. In the Offset column, enter the scale percentage used for the point helpers. If you performed more than one scaling operation, enter each scale percentage in this column, pressing **<Enter>** between each entry.

6. Select the other eye and perform the same process.

   The relative size we used for our alien is shown in the following figure.

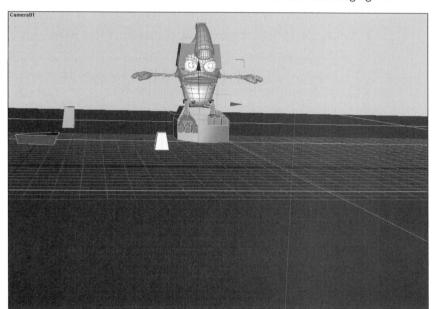

☠ **TIP** ☠

You must scale the alien with all the selected point objects at one time, or you will have problems when you attempt to animate the alien.

Figure 12.4 *Alien sitting on largest rock*

7. Use the **Point Alien Root** helper to position the alien so his bottom is right on the rock.

8. Use the **L - Point - Ankle Rotate** and **R - Point - Ankle Rotate** helpers to position his feet so they are dangling over the front of the rock, as shown in the figure above.

   You can leave the arms outstretched for now as you will position them later in this chapter to hold the remote control.

9. Save the scene.

Merging the Rover

Now it's time to bring in the rover. In order for the scripts to work properly, the rover must be merged. As with the alien, we cannot use the XRef feature because our scripts need access to the rover's modifier stack.

1. From the menu, select **File/Merge**. Choose the file **Rover - Setup and Textured.max** that you created in **Chapter 11**, or choose this file from the **Scenes/Rover** folder on the CD. Click **Open**.
2. On the Merge dialog, click **All** to select all the objects and click **OK**.

   If your textured battery is in a different scene, you should delete the existing battery, merge in the textured battery using the same process, and link the textured battery to **mainControlNode**.

   Next, you'll scale the rover.

3. Select the **mainControlNode**.

   All rover objects have been linked to this dummy object. As this object is moved, everything linked to it will move as well. If we scale this object, everything else on the rover will also scale.

4. Click **Lock Selection** to turn it on.
5. In the Top viewport, move the **mainControlNode** (and thus the rover) near the alien so you will have a reference for how large it needs to be.
6. On the **Modify** panel, set the **scaleFactor** attribute to approximately 1.5.

   This parameter should provide the correct approximate scale. You might want to scale the rover a little larger or smaller depending on the scale of the alien.

7. In the Left viewport, position the rover so it is sitting on the ground plane.
8. Go to frame 0.
9. In the Top viewport, position the rover so it is just outside the Camera - Tracked viewport.

   This will cause the rover to drive into view at the beginning of the animation.

Recording the Rover's Start Position and Rotation

Next you will set the rover's initial position and rotation in the custom attributes you set up earlier so they can be used when you click Reset Position after a motion capture session.

1. Right click the **Select and Move** button to see the Move Transform Type-In dialog. This will give you the current position of the **mainControlNode**.

Animation **549**

2. Note the values for the Absolute: World parameters. Set the **initPosX**, **initPosY**, and **initPosZ** parameters on the **Modify** panel to the same values.

3. Click then right-click **Select and Rotate**, and set the **initRotZ** attribute to the Absolute:World value for the Z parameter on the Rotate Transform Type-In dialog.

4. Click **Lock Selection** to turn it off.

5. Save the scene.

Bringing in the Remote Control

There is no special need to access the remote control's modifier stack, so we can bring it in as a cross-referenced object using the XRef Object feature.

1. From the menu, select **File/XRef Object**. On the XRef Objects dialog, click **Add** at the upper right corner. Choose the file **Remote Control - Textured.max** that you created in **Chapter 11**, or choose it from the **Scenes/Remote Control** folder on the CD.

2. On the dialog that appears, choose the **[Remote Control]** group and click **OK**.

The objects that make up the remote control appear in the window at the bottom of the XRef Objects dialog.

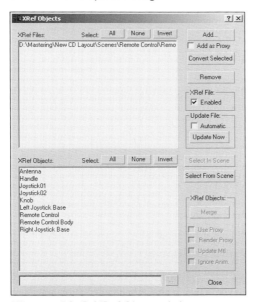

Figure 12.5 *XRef Objects dialog*

3. Use **Select and Uniform Scale** to scale the grouped object **[Remote Control]** to the correct size for the alien. Move the remote control so it is in front of the alien.

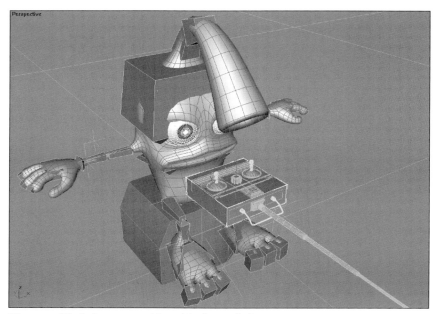

Figure 12.6 *Remote positioned in front of alien*

To maintain quick feedback, we will use a proxy (substitute) for the remote control with less detail than the actual object. The main body of the remote control is roughly in the shape of a box, so we will use a box primitive as the proxy.

4. In the Top viewport, create a box that matches the size of the remote control body. Name the box **Remote Proxy**.

5. Select the grouped object **[Remote Control]** and use **Select and Link** to link it to **Remote Proxy**.

   With the remote control linked to the proxy, you can hide the detailed object if you like. Since the Remote Proxy is now the parent of the remote control, we can move, rotate, and scale the proxy, and all modifications will be mimicked by the detailed remote control. We can later hide the proxy and unhide the detailed remote when we get ready to render.

6. Hide the object **[Remote Control]**.

7. Add a few segments to the **Remote Proxy** box. Convert the box to an Editable Mesh. Pull up a few vertices on the top and front so you'll easily be able to tell where the top/front is later on.

   This will help you keep track of which side is which when you animate the remote control later on.

8. Save the scene.

Positioning the Alien's Hands

1. Hide the alien body, named **Alien Body**.

   This will make feedback and interaction faster in viewports. The alien's bones have been scaled large enough to give a good representation of the alien's geometry. Because of this, the skeleton will do a great job serving as a proxy for the alien.

2. Rotate the **Remote Proxy** so it is angled toward the alien.
3. Position the alien so that its hands hold the remote control.

   While working with the alien's hands and the Remote Proxy, you can unhide the remote control if you like to see how they look together. Just be sure to adjust only the Remote Proxy and not the remote control itself.

   In the following figure, we show our positioned alien with the remote control unhidden so you can see how they look together.

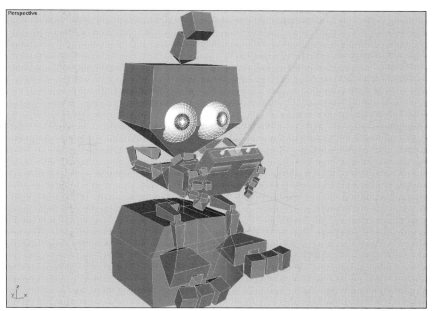

Figure 12.7 *Remote proxy and alien positioned*

4. Save the scene.

Setting up Link Constraints

Our character will be required to interact with two objects during the animation, the remote control and the rover. The first thing we need to decide is which hand will be interacting with each object.

In our animation, the alien holds the remote with both hands, but at one point he holds it with just his left hand, then uses this hand to toss the remote control over his shoulder when he becomes disgusted.

The alien also uses one hand to pick up the rover off the ground to inspect it for trouble. We've decided to use the left hand for this action as well.

It would seem to make sense to link the remote control to the left hand so it will move with the hand. The right hand will have to be animated manually to move along with the remote control and look as though it's holding the other side, but this can be done easily with a some careful keyframing.

The problem arises when the alien goes to throw the remote control. We don't want the remote control to continue to be linked to his left hand at that point as it will get dragged along the ground when the alien gets up and walks over to the rover. Also, we want the rover to be linked to the alien's left hand at the point when he comes into contact with it.

 When you use the **Select and Link** tool, one object is linked to another over the entire animation. The link cannot be selectively broken at any point. Using this tool would cause problems for us as we can't unlink the remote control from the left hand when it gets thrown.

The solution is to use a *link constraint*. This is a type of controller that allows us to link one object to another, then link it to something else at a particular frame.

The remote control will be linked to the left hand at the start of the animation. At the frame where the alien throws the remote control and it leaves his hand, the remote control will be linked to nothing. Then at the point when the alien puts his hand on the rover, the rover will be linked to the left hand.

Linking the Remote Proxy

The link constraint is an animation controller, so it is assigned through the **Motion** panel.

1. Go to frame 0.
2. Select the **Remote Proxy**.
3. On the **Motion** panel, expand the Assign Controller rollout.
4. Highlight the **Transform** track and click the **Assign Controller** button.
5. On the Assign Transform Controller dialog, select **Link Constraint** and click **OK**.

The Link Constraint parameters appear on the **Motion** panel.

6. In the Link Params rollout, click the **Add Link** button. In the camera viewport, click the **L - Bone - Hand** bone, or press **<H>** to pick it from a list.

   The remote control proxy is now linked to the hand on frame 0.

7. Click **Add Link** to turn it off.

8. Save the scene.

   Later, once the frames have been worked out for each shot, we will need to come back to this object and add a **Link to World** entry on the frame where the alien begins to toss the remote control. Right now, we're not exactly sure at what frame that will occur, so we'll leave it for a little later.

Testing the Link

TIP

Be sure to use the point helpers to position the hands and fingers.

1. Zoom in on the remote and hands. If necessary, move the remote control proxy and bend or move the hands and fingers to curl around the remote control and make it look as if the alien is really holding it.

2. Once the proxy is properly aligned with the hands, select **L - Point Hand Controls** and move it around.

   The remote control should move with the left hand while the right hand stays behind.

Figure 12.8 *Left hand positioned*

3. Undo the motion and make any adjustments necessary.

Assign a Link Constraint to the Rover

The alien will interact with the rover near the end of the animation when he reaches out and picks it up. We will need to apply a link constraint to the mainControlNode to make it follow the left hand at that time.

1. Select the **mainControlNode**.

2. On the **Motion** panel, expand the Assign Controller rollout.

3. Highlight the **Transform** track and click the **Assign Controller** button.
4. On the Assign Transform Controller dialog, select **Link Constraint** and click **OK**.
5. On the Link Params rollout, click **Link to World**.

This will leave the rover under its own power for the time being. Linking to the world makes the object behave as if it is not linked to anything at all. Later we will cause the rover to link to the left hand at the appropriate frame.

6. Save the scene.

Assigning the Link Constraint to the mainControlNode breaks the connections made earlier from the scaleFactor attribute to the Scale X, Scale Y and Scale Z tracks. We don't need this connection any more as the rover is already scaled.

If you like, you can test your skills with animation controllers and see if you can rewire the **scaleFactor** attribute so it will once again control the rover's global scale.

Setting up Additional Cameras

All of the objects are now in one scene and sized properly with one another. It is now time to figure out how many cameras we are going to need, where they will be positioned, and at which frames we will look through them.

Figuring out the frame count for each camera requires some review of the storyboards. How many different camera shots do you see?

We have decided we need four cameras to match the storyboard:

- Tracked camera for wide shots
- Close-up camera for alien facial expressions while sitting on the rock
- Over-the-shoulder shot of the Brand X battery in the rover
- Shot of alien shouting "Mom!"

We already have the first camera and need only to set up the other three.

1. In the Front viewport, create a free camera named **Camera - Closeup**.
2. Position the camera slightly lower than the alien, looking up at his head.

The camera should be close enough to clearly show the alien's face and arms, and the remote control.

Figure 12.9 *The Close-up camera*

3. Create another camera and name it **Camera - Mom**.
4. Position this camera so that it is looking down on the alien.

Figure 12.10 *The Mom camera*

This camera will be moved after we have established where the rover will be located when the alien reaches out to pick it up. At this point we are merely trying to get the correct final angle and an approximate location for the camera.

5. Create the last free camera near the Mom camera. Name this camera **Camera - Over Shoulder**.

This camera will be used for the over-the-shoulder shot when we look inside the rover to see what the alien sees, the Brand X batteries. We'll position it more exactly later.

Now that we have the cameras in the scene, we need to determine our frame range and what frames will be viewed through which camera.

First, we know that the commercial must be 30 seconds long. This is what the client has contracted you to do and it's what you must deliver. The project is being created for NTSC television, which runs at 30 frames per second. A total of 900 frames must be delivered when all is completed. We need about 1-1/2 seconds for the logo at the end of the commercial, about 45 frames.

First, let's work out which camera we will be using with which frames.

The frames that we have chosen are listed below:

Camera	Frames
Camera - Tracked	44 - 288
Camera - Closeup	289 - 500
Camera - Tracked	501 - 708
Camera - Over Shoulder	709 - 850
Camera - Mom	851 - 900

We've cut off 43 frames at the front end to allow for 1-1/2 seconds of logo and voiceover.

Setting the Animation Range

1. Click the **Time Configuration** button.

2. On the Time Configuration dialog, set the following parameters:

Frame Rate	NTSC
Start Time	44
End Time	900

3. Click **OK** to save the changes.

4. Save the scene with the filename *Final Animation.max*.

TIME TO ANIMATE

With our cameras set up and our basic link constraints in place, we're ready to animate.

In a sense, this entire book has been leading up to this moment. You wanted to be an animator, right? We hope that by now you realize there's a lot more to good animation practices than keyframing. The actual animation process is going to go very quickly because we have so carefully set up all the controls and UIs we need to make it go smoothly. The setup is what takes all the time, not necessarily the animation itself!

STORYBOARD CHECK

This is a good time for a look at the storyboards. Let's consider what needs to be animated here. The rover drives onto the screen in the first shot, and within a couple of seconds it appears to have some type of trouble. Then the alien walks over to it and...

Wait a second. We won't know where to make the alien walk to until the rover has been animated, so it makes sense to animate the rover first.

ROVER ANIMATION

It's up to you to determine when the rover will begin having difficulty. Showing that the rover is having a hard time is simple enough -- just make it go forward and backward a few times as if it were stuck and trying to break free. After a few seconds, have the rover break free, drive around some more, then repeat the back and forth motion again. At this point the alien will walk over to it, and it doesn't need to move any more.

According to the decisions we made earlier about the camera, the rover will be in the shot for frames 44-288 as it moves around then starts to have trouble, then in frames 501-708 as it has its final trouble and stops.

Recall that the rover animation is performed with motion capture using the keyboard and mouse. You're going to have to practice using this setup a little bit so you'll be prepared when it's time to animate. This step is included in the next tutorial.

ALIEN ANIMATION

Once the rover animation is complete, you will be able to start animating the alien. For this purpose, we built a clever user interface in *Chapter 9* for quickly selecting the various joints on the alien. We will use this user interface to aid in animation.

The animation of the alien falls under the heading of *character animation*, where you'll be breathing life into the character to make him look as if he's actually alive. We do not have the room here to cover all aspects of character animation, and many other very thick books are already available on the subject.

If you've never done character animation before, you can certainly practice on the alien. But if you want to become an expert at it, you are advised to get a few good texts on the subject and practice as much as you can.

To become skilled at character animation, lots and lots of practice is essential. An effective exercise is to make a simple character from linked spheres and boxes and make him dance around a bit. You can learn a lot just from this simple exercise. Try to make the character walk convincingly. You'll need this skill when animating the alien.

In our scene most of the motion is simple, just an alien sitting on a rock interacting with his remote control. The most challenging part of the animation is the alien tossing the remote to the ground, hopping off the rock and walking over to the rover. Rather than going over character animation techniques, we will focus on the basics and good workflow habits.

SAFE FRAMES

While animating our objects, keep in mind that the far edges of your scene will most likely be cut off by the average television set. You do not want to spend hours animating only to discover that important elements are chopped out of the scene.

This chopping occurs because video images are overscanned (stretched) to ensure that no black edges appear around the edge of the screen. The overscanning results in roughly a 10% loss of the image at the edges. In addition, the curvature at the edge of many video and television screens can warp the imagery at the edges.

In **3ds max**, borders can be displayed in viewports to help you keep your animation within the safe area. These borders are called *safe frames*.

To enable safe frames, first choose *Customize/Viewport Configuration* from the menu. Under the **Safe Frames** tab, make sure **Live Action**, **Action Safe** and **Title Safe** are checked. Click **OK** to exit the dialog.

Next, right-click on the label of the viewport you to intend to render and choose *Show Safe Frame* from the pop-up menu.

Enabling safe frames for a viewport changes its aspect ratio to match the aspect ratio set on the Render Scene dialog. You might find that some areas around the edge of the viewport are no longer visible. The viewport now shows exactly what will be in the final rendering. This is called the *live area*.

Enabling safe frames also displays a series of borders in the viewport. These borders represent different "safe" regions for video.

The first border inward from the live area represents the *action safe area*, where any action is likely to be visible. It is important to make sure all of your key areas of action stay within this border, or it might not be visible on some video monitors or television sets.

The next rectangle inward represents the *title safe area*. All titles, logos and text must be within this area, or they might appear warped by the curvature near the edge of a video or television screen.

In the tutorial that follows, set up safe frames for your cameras as you go along to ensure none of your beautiful animation is lost in the final viewing.

TUTORIAL 12.2 ANIMATING THE ROVER AND ALIEN

Before you can animate the rover, you must load the scripts you created in *Chapter 6* and set the initial parameters for the rover.

Loading the Rover Scripts

1. Load the scene **Final Animation.max** that you created in the last tutorial if it is not already loaded.

2. On the **Utilities** Panel, click **MAXScript**. Click **Open Script** and open the file **Rover UI.ms** that you created in **Chapter 6**. Alternately, you can load this file from the **Scripts** folder on the CD.

3. Evaluate the script.

4. Minimize the MAXScript editor.

5. Click **Open Script** again and open the **Rover Callback.ms** script you created in **Chapter 6**, or load this script from the **Scripts** folder on the CD.

6. Evaluate the script.

7. Minimize the MAXScript editor.

8. From the pulldown on the **Utilities** panel, select **Treads Control**.

   Even if this option already appears on the pulldown, select it again to make the rollouts appear.

9. On the Main Console rollout that just appeared, check both **Enable MoCap** and **Enable Script**.

Practicing the Rover Motion

It is important that you spend a few minutes practicing the course of the rover. Decide ahead of time on the basic motion you want, and decide where the rover will come to rest before the alien goes to pick it up.

Of course, if you just want to play around first to get the hang of it, please do so. But in the end you should have a firm idea of where you want the rover to go, how fast it should move, and where it should end up.

Depending on the speed of your computer and video card, you might need to reduce the amount of rover geometry displayed on the screen.

1. On the **Utilities** Panel, click the **Motion Capture** button.

2. Verify that all the controllers are enabled.

3. Click the **Play During Test** checkbox.

4. Activate the Camera - Tracked viewport.

5. Set the **Record Range** for frames 44 to 288.

   Get your fingers ready on the keyboard and mouse. You're ready to go!

6. Click the **Test** button and begin your practice session.

7. Using the arrow keys and the mouse, drive the rover around and make it accelerate or decelerate. Right-click to end the motion capture session.

   The purpose here is not to generate any animation. Right now you're just practicing for the real deal in the next set of steps.

8. To reset the rover back to its starting location, click the **MAXScript** button on the **Utilities** panel and click **Reset Position**.

   Sometimes **3ds max 4** behaves strangely when setting rotation parameters with scripts. If clicking the **Reset Position** button doesn't reset the rover rotation properly, evaluate the *Rover UI.ms* script again. Sometimes this causes the rotation reset process to start working properly.

Animating the Rover

Now it's time to record the motion of the rover. This would be the point where you would ask the client to come in and watch as you drive the rover around. In this way, you could get the client's approval before baking the animation.

1. Follow the same procedure outlined for the practice session, only this time, instead of clicking Test, click **Start**.

   The motion capture will stop on its own when you are actually recording, so there is no need to right click when you're done.

2. After you have captured the motion, click **MAXScript** on the **Utilities** panel and uncheck the **Enable MoCap** and **Enable Script** checkboxes.

3. Click **Bake Animation**.

4. Click the **Play Animation** button at the lower right of the screen to play the animation.

   If you (or the client) are not happy with that take, move the time slider to frame 0, click **Delete Animation**, then click the **Reset Position** button to move the rover back to its starting location. Do the process over again until you're happy with the motion.

   After clicking **Delete Animation** and running another session, if you find the rover moving faster or slower than you expected, check the position and rotation of the **RC_Controller**. All values should be set to 0. If they are not 0, the time slider was not at frame 0 when the animation was deleted, which results in the **RC_Controller** staying wherever it was when you clicked the Delete Animation button.. Simply set the values to 0 and try again.

5. Save the scene.

Using the Alien UI

In **Chapter 9** you took the time to create a user interface to aid in the animation of our alien. You can use this tool to reduce the time it takes to animate the character by using it to select the appropriate controls while animating.

Before beginning, it is important to know which helper animates what. You should use the right helper for the job when animating the alien. Below is a listing of all the objects you will be animating and the helper you should use:

- Point Alien Root controls the location of the entire alien.
- Point Hand Control helpers place the characters hands, which in turn control the way the arms are placed. There are also custom attributes on both of these helpers to control the positioning of the fingers and wrist.
- Point Elbow Target helpers control the direction in which the elbow is pointing. You will work with these helpers when you transform the Point Hand Control helpers. Together, you can achieve almost any arm/hand position.
- Point- Ankle Rotate helpers place the foot as well as control the custom attributes for foot roll and toe rotations.
- Bone - Back B controls the upper body rotation.
- Eye Target controls the direction in which the eyes are looking.
- Remote Proxy controls the remote control.

Let's start by loading the alien UI.

1. On the **Utilities** Panel, click the **MAXScript** button. Open the *Alien UI.ms* script you created in **Chapter 9**, or load it from the *Scripts* folder on the CD.
2. Evaluate the script.
3. Close the MAXScript editor.
4. Verify that the UI is functioning properly by operating some of the controls.
5. Save the scene.

Blocking the Scene with Rough Animation

You are now ready to begin animating. Let's start by *blocking* the scene. This means putting in some very basic animation at key poses.

At this point we are not concerning ourselves with the cameras at all. Later on we'll work out the final camera location, then look at some previews and do some fine-tuning. Right now you just want to get the rough motion happening.

Be sure to save frequently as you go along!

1. Move the time slider to frame 44.
2. Turn on the **Animate** button.
3. Use the Alien UI to assist in blocking the following key poses at the given frames. Don't forget to use the Morpher to make the facial expressions.

> ☼ **TIP** ☼
>
> *The term "blocking" comes from the theater, where the director gives actors the overall movements and timing in each scene before they move on to fine-tuning.*

Frame 44

The alien looks at the rover with a happy expression and holds the remote control as if he were using it.

Frame 150

The alien looks at the remote control. His expression is neutral. The left eye is dropped down a bit by using the angry left eye a little.

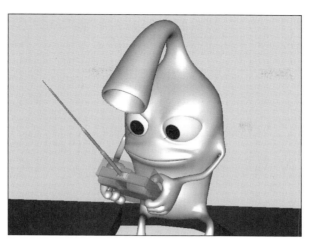

Frame 200

The alien has turned back toward the rover and has a happy expression again.

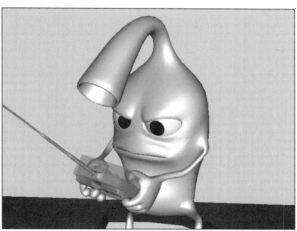

Frame 250

He has turned more toward the rover and now returns to being unhappy.

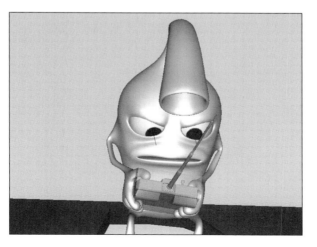

Frame 300

The alien has now turned past the center point of the camera and is studying the remote. His expression blends between being puzzled and aggravated.

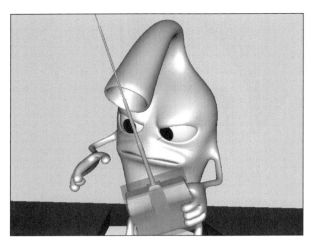

Frame 325

He has now quickly turned his attention back toward the rover while removing his right hand. He is still aggravated.

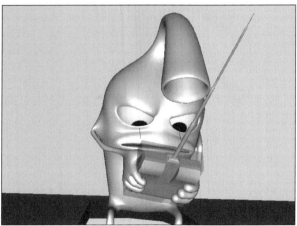

Frame 400

The alien studies the remote trying to figure out what's going wrong.

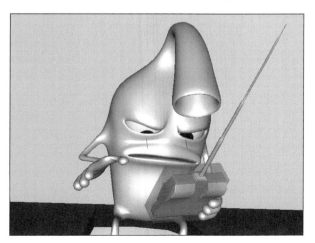

Frame 445

He has drawn his right hand back to strike the remote in the hope of fixing it.

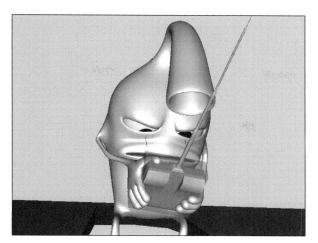

Frame 449

The first hit to the remote.

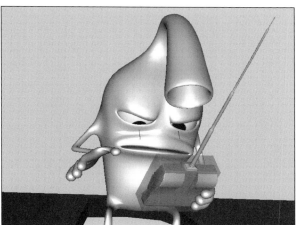

Frame 452

Right hand is drawn again as he prepares to hit the remote a second time.

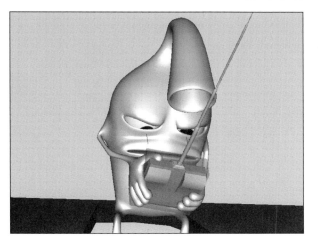

Frame 455

Just like frame 449. He whacks the remote again.

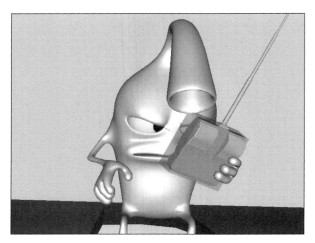

Frame 480

He holds the remote up to eye level using his left hand.

Frame 530

He is now about to toss the remote control over his left shoulder

Frame 538

The alien is just beginning to toss the remote control. To unlink the remote control from his hand, select the remote control, and on the **Motion** panel click **Link to World**.

Animation 567

Frame 550

The remote continues to fly back and drop toward the ground. The Alien begins to lean forward to prepare for his jump off of the rock.

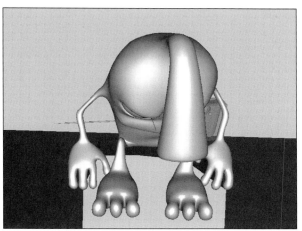

Frame 556

He is now bent completely over, ready to lunge.

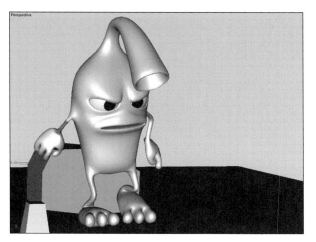

Frame 570

He has landed. His right foot hits the ground first.

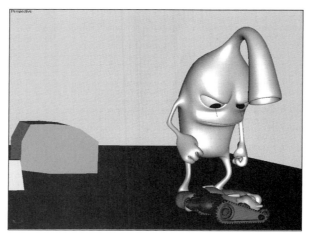

Frame 660

The alien is now standing over the rover, ready to pick it up. Between frames 570 and 660, the alien walks over to the rover. The walking bit is up to you.

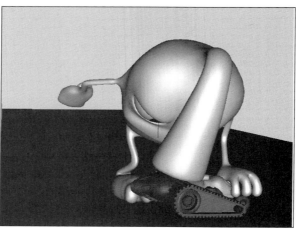

Frame 688

The alien bends over and grabs the rover. At this point, link the rover to the left hand by selecting **mainControlNode**, and on the **Motion** panel, click **Add Link** and choose the alien's left hand bone.

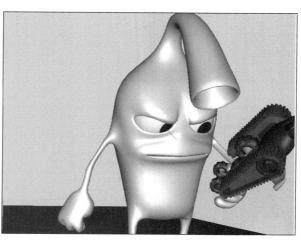

Frame 720

The alien eyes the rover with suspicion.

Animation | 569

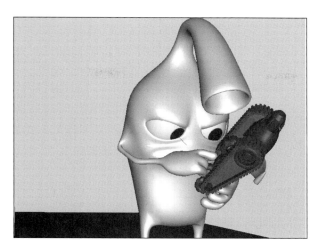

Frame 793

The alien presses the battery door to open it.

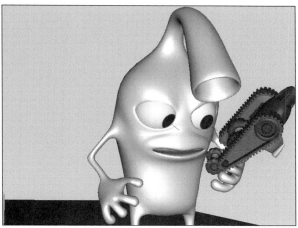

Frame 825

He drops his hand and develops an expression of shock as he discovers that the battery is a generic brand.

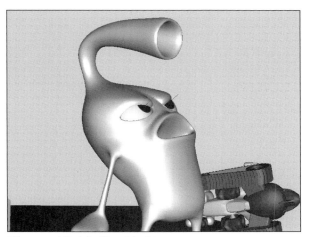

Frames 880-900

The alien screams for his mom.

Figure 12.11 *Alien blocking poses*

Posing the Remaining Shots

After you have blocked in all of the primary poses, it's time to set the positions of the Over Shoulder and Mom cameras.

1. Turn the **Animate** button off.
2. Move the time slider to frame 709.
3. Change one of the viewports to show the view from **Camera - Over Shoulder**.
4. Position the camera so you are looking over the alien's right shoulder and at the bottom side of the car.

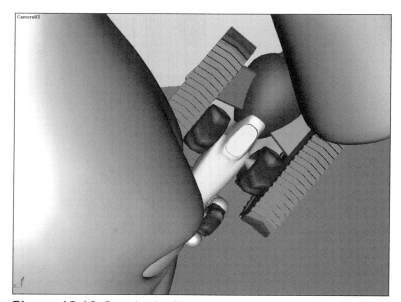

Figure 12.12 *Over the shoulder camera placement*

5. Move to frame 851.
6. Set up a viewport to show the view from **Camera - Mom**.
7. Position the camera so it looks down at the alien from his front.
8. Save the scene.

Figure 12.13 *The Mom camera shot*

Creating a Preview

Previews are an important part of the animation process. A *preview* is a quick rendering of the animation without full textures or lighting. Its main purpose is to show you the motion. We'll create a preview here to see how the alien looks when he's animated.

1. From the menu, choose **Rendering/Make Preview**. The Make Preview dialog appears.

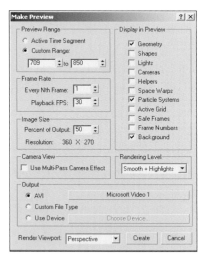

Figure 12.14 *Make Preview dialog*

TIP

Check the Frame Numbers option to display the frame number at the upper corner of the preview. This can be very helpful in spotting exactly where a change needs to be made.

2. On the Make Preview dialog, set the range of frames you want to preview.

3. Set the **Percent of Output** to 50 or 100 percent, whichever is needed to give you a preview with roughly 320 X 240 dimensions.

   A preview of this size is quick to generate, but is sometimes too small to show subtle movements. You can always make a larger preview later.

4. Check or uncheck options in the **Display in Preview** section.

5. Make sure that the **Render Viewport** option is set to the correct camera.

6. Click **Create**.

   Wait a few moments while the preview is created. When the preview is finished, watch it carefully several times. Which areas would you like to improve?

   Go back and change any keyframes that need work. Be sure to put in the walk after the alien jumps off the rock. Be daring, be inventive, and have a good time!

Fine-Tuning the Animation

Now you can start fine-tuning. This process can take hours, days, and sometimes weeks depending on the complexity of the animation.

Our animation is fairly simple so it shouldn't take long to adjust. If you're new to character animation it might take a while, but you'll learn a lot along the way.

Start from the beginning and work in 20 to 30 frame sections before moving on. You will spend most of your time inside Track View adjusting function curves so that the character gives off the most realistic motions. Your job will be to make the alien appear to have weight and be alive. Below are some tips to help you achieve the best possible animation:

- Use function curves in Track View to help you out, but avoid tangent types that result in linear or near linear curves. Straight lines between keys mean robotic or mechanical motion. Most natural motion goes in a slight curve rather than a straight line, so set your curves accordingly.

- Most curves should ease-in and ease-out of their respective keyframes.

- The custom tangent type can give you a great deal of control over your curves.

- The character should "anticipate" large motions with small ones in the opposite direction.

- Exaggerate the facial expressions. This is expected with animation.

- Save often. There's nothing worse than losing a few hours of hard animation work!

- Have patience. Animation is a delicate process that takes time to learn.

When you have finished fine-tuning the animation, save the scene with the filename **Final Animation.max**.

RENDERING

With the completion of our scene, we are ready to start rendering. The process of rendering is a very important step that requires some knowledge of the settings and options that are available.

If you are unfamiliar with the rendering process, we highly recommend that you read the information in *CD Tutorial 3* in the *Tutorials* folder on the CD. This tutorial contains much valuable information about the renderer and how to use it.

If you're interested in learning how to set up network rendering to save rendering time, see *CD Tutorial 4* in the *Tutorials* folder on the CD.

RENDERING FOR POST PRODUCTION

We intend to paint and composite our scene later on in **Combustion**, Discreet's painting/editing/compositing package designed to work hand-in-hand with **3ds max 4**. We use **Combustion** in *Chapter 13*.

One of the benefits of working with **Combustion** is that **3ds max 4** can output various layers of information from the animation for use in compositing with **Combustion**.

One of the ways 3ds max 4 and **Combustion** work together is through a file format called *Rich Pixel Format* (RPF). An RPF file can hold an object ID, material ID and occlusion data (which objects are behind which). The file format also holds UV coordinates and motion data.

RPF files can be rendered directly from **3ds max 4** then loaded directly into **Combustion**. This is the method we will use to create our rendered images for compositing in **Combustion**.

VIDEO POST

To render our commercial, we are going to assume that you have access to a single computer for rendering. If you do happen to have two or more machines networked together, we urge you to render this scene with the network. See *CD Tutorial 4* in the *Tutorials* folder on the CD for information on how to accomplish this.

Our scene consists of four cameras covering five different shots. We could set up the first range of frames to render in the Render Scene dialog, wait for the rendering to finish, set up the next range of frames and so on, but this would take too much of our time.

Instead, we will use Video Post to render all five shots in sequence without us having to come back and start each shot rendering. Video Post is a utility in **3ds max** that allows you to set up various events to take place. An event can be a rendering of a viewport, or an effect on the scene such as a glow. Video Post allows you to put all these events in sequence and adjust them individually as you like. You can also set an effect to take place over a specific number of frames.

It's always nice to submit the job and take a break. And besides, you need the time to go and line up another job. This one's almost finished!

TUTORIAL 12.3 RENDERING THE SHOTS

In this tutorial, we will not be explaining all there is to know about rendering. Instead, we will cover the steps for rendering elements to RPF files for later use with **Combustion** in **Chapter 13**. See **CD Tutorial 3** in the **Tutorials** folder on the CD if you want to learn more about rendering.

Preparing for Rendering

1. Locate an area on your hard disk or network where you have at least 1.2GB of free space. On this hard disk, create a folder called **Rendered Images** to hold all the images that we will render.
2. Under this folder, create five folders named Shot1 through Shot5.

    We will place the rendered images from each shot in separate folders.

3. Load the file **Final Animation.max** that you created in the last tutorial. Alternately, you can load this file from the **Scenes/Final** folder on the CD.
4. Unhide or hide objects in the scene as appropriate. For example, you should hide all the alien bones and the Remote Proxy, and unhide the alien body and the remote control itself.
5. Render a few test frames to make sure all the right objects are hidden or unhidden.
6. From the menu, choose **Rendering/Environment**. On the Environment dialog, uncheck the **Use Map** checkbox. Close the **Environment** dialog.

    We don't need to render the background any more as this will be composited in **Combustion** later on. Turning it off ensures we won't render it by accident.

7. Render another test frame. The background should not render in the scene.
8. Save the scene.

Adding the First Event to Video Post

We will use the Video Post utility to organize our rendering and to add effects to the scene.

1. From the menu, choose **Rendering/Video Post**.

    The Video Post window appears. This window has a queue at the left for adding events. On the right, the frame range for each event is displayed.

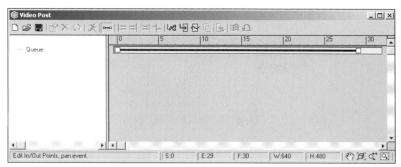

Figure 12.15 *Video Post window*

2. To clear the window and start a new sequence, click **New Sequence** on the Video Post toolbar and click **Yes** on the dialog that appears.

   Next, we will add a scene event to the queue. This type of event renders the 3D scene.

3. On the Video Post toolbar, click **Add Scene Event**.

   The Add Scene Event dialog appears.

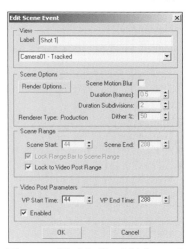

Figure 12.16 *The Add Scene Event dialog*

4. On the Add Scene Event dialog, enter **Shot 1** for the **Label**, and choose **Camera01-Tracked** from the pulldown menu.

5. Set the **VP Start Time** to 44 and the **VP End Time** to 288.

6. On the Add Scene Event dialog, click **Render Options**.

   This displays a dialog with many of the options found on the Render Scene dialog.

   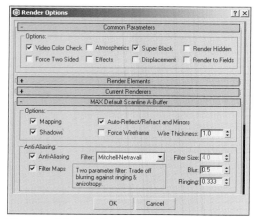

   Figure 12.17 *The Render Options dialog*

7. Set the following parameters:

   On the Common Parameters rollout:
   Video Color Check   (checked)
   Super Black   (checked)

   On the MAX Default Scanline A-Buffer rollout:
   Filter   **Mitchell-Netravali**
   Blur   **0.5**

   Click **OK** to save the settings.

   These settings will be remembered and used by all of the other scene events that we add to the queue, so we won't need to change this dialog each time we add a scene event.

8. Click **OK** to close the Add Scene Event dialog and put the scene event in the queue.

Adding an Output Event

1. If **Shot 1** is highlighted in the queue, click anywhere in the list box to deselect it.

2. From the Video Post toolbar, click **Add Image Output Event**.

An output event saves whatever has been generated in the queue up to that point.

Figure 12.18 *Add Image Output Event dialog.*

3. On the Add Image Output Event dialog, change the **Label** to **Shot 1 Output**.
4. Set the **VP Start Time** to 44 and the **VP End Time** to 288.
5. Click **Files** and navigate to your *Rendered Images/Shot1* folder. Enter the filename **Shot1** and choose the **RPF** file format for output. Click **Save**.

The RPF Image File Format dialog appears. Here you set the information to be stored in the RPF file.

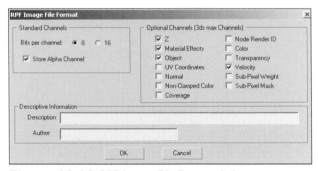

Figure 12.19 *RPF Image File Format dialog.*

6. In the RPF Image File Format dialog, set the following options:

Bits per channel	8
Store Alpha Channel	(checked)
Z	(checked)
Material Effects	(checked)
Object	(checked)
Velocity	(checked)

   Click **OK** to save the settings and exit the dialog.

   The Video Post window should now have two events in the queue. Your queue should look similar to the following figure:

   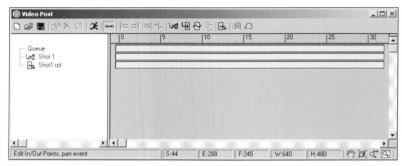

   Figure 12.20 *Video Post window*

Adding the Remaining Events

Use the steps above as a guide for entering the remaining events. Be sure to click the white area of the queue to deselect any highlight events before adding more events.

Add the following events to the queue:

1. **Add Scene Event**

   Label: Shot 2
   Camera - Closeup
   VP Start Frame   289
   VP End Frame   500

2. **Add Image Output Event**

    Label: Shot 2 Output
    VP Start Frame   289
    VP End Frame   500

Folder	*Rendering Images/Shot2*
Filename	Shot2
File format	RPF

Bits per channel	8
Store Alpha Channel	(checked)
Z	(checked)
Material Effects	(checked)
Object	(checked)
Velocity	(checked)

3. **Add Scene Event**

    Label: Shot 3
    Camera - Tracked
    VP Start Frame   501
    VP End Frame   708

 4. **Add Image Output Event**

    Label: Shot 3 Output
    VP Start Frame   501
    VP End Frame   708

Folder	*Rendering Images/Shot3*
Filename	Shot3
File format	RPF

Bits per channel	8
Store Alpha Channel	(checked)
Z	(checked)
Material Effects	(checked)
Object	(checked)
Velocity	(checked)

5. **Add Scene Event**

    Label: Shot 4
    Camera - Over Shoulder
    VP Start Frame   709
    VP End Frame   850

6. **Add Image Output Event**

   Label: Shot 5 Output
   VP Start Frame   709
   VP End Frame   850

Folder	*Rendering Images/Shot3*
Filename	Shot3
File format	RPF

Bits per channel	8
Store Alpha Channel	(checked)
Z	(checked)
Material Effects	(checked)
Object	(checked)
Velocity	(checked)

7. **Add Scene Event**

   Label: Shot 5
   Camera - Mom
   VP Start Frame   851
   VP End Frame   900

8. **Add Image Output Event**

   Label: Shot 5 Output
   VP Start Frame   851
   VP End Frame   900

Folder	*Rendering Images/Shot3*
Filename	Shot3
File format	RPF

Bits per channel	8
Store Alpha Channel	(checked)
Z	(checked)
Material Effects	(checked)
Object	(checked)
Velocity	(checked)

9. Save the scene.

> ☼ **TIP** ☼
>
> *When you save the scene, the Video Post queue and all its settings are saved along with it.*

Executing the Sequence

Now we have input and output events, so we can *execute*, or render, the sequence in the queue.

This event will start the rendering process. It's a good idea to close all applications you don't need to give the computer as much memory as possible to work with. You can still use other programs while the rendering is going on, but keep in mind that this will steal memory from the rendering process.

 1. From the **Video Post** toolbar, click **Execute Sequence**.

The Execute Video Post dialog appears.

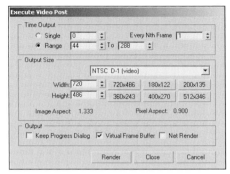

Figure 12.21 *Execute Video Post dialog*

2. On the Execute Video Post dialog, set the following settings:

   Range   44-900
   Output Size   **NTSC D-1 (video)**

   Make sure the **Range** option is actually selected.

   The NTSC D-1 (video) setting will set the proper resolution and image aspect ratio for NTSC broadcast video. Most video output systems use this resolution and aspect ratio.

3. If you want to render over a network, check the **Net Render** checkbox.

   You can only perform network rendering if your system is currently set up to do so with the procedure outlined in *CD Tutorial 4* in the *Tutorials* folder on the CD. If you haven't set up for network rendering, make sure this checkbox is unchecked.

   Are you ready?

The background plate should not render in the scene. The background will be composited with the foreground animation in the next chapter.

3. Click **Render**.

   The rendering process begins. Kick back and relax -- it's going to be a while.

   If you have **Combustion** but haven't yet learned how to use it, this is a good time to do so. Take a look at the next chapter, or use the **Combustion** manual to get a start.

SUMMARY

In this chapter, you learned how to:

- Use cross-referenced scenes and objects
- Use a low-resolution proxy (substitute) to speed up your work
- Animate using blocked key poses
- Create quick previews to help you time and fine-tune the animation
- Render RPF files for use with Combustion
- Use Video Post to render four different cameras in one rendering session

Now you're ready for the last step: compositing. Go on to *Chapter 13* to learn how to put together the background plates and the rendered animation.

Post Production with Combustion

Combustion is a cost-effective, fully featured 3D compositing software package from Discreet. With **Combustion** we have the ability to edit raw video and still images quickly and easily. **Combustion** improves workflow by giving fast feedback of effects and animation as they are applied to our footage by making extensive use of your system's memory. The edited imagery can be output to files ready for film, television, or even streaming web content.

TECHNIQUES IN THIS CHAPTER

In this chapter, you will learn to use **Combustion** to:

- Composite live footage with rendered animation
- Change colors on live footage and rendered animation
- Paint out unwanted objects in live footage

TUTORIALS IN THIS CHAPTER

The tutorials in this chapter will show you how to complete our commercial project with **Combustion**.

- *Tutorial 13.1 Footage Import* shows you how to start a Combustion workspace and import footage.

- *Tutorial 13.2 Color Correction* covers the changing of colors and painting out of unwanted objects in the live footage.

- *Tutorial 13.3 Tracking the Selections* shows you how to apply your color correction and paint operations over the entire sequence of live footage.

- *Tutorial 13.4 Painting the Footage* walks you through the process of painting out scene markers and other unwanted elements from your background footage, and rendering the final frames.

If you don't have **Combustion**, you can use another editing or post production system to do the tutorials in this chapter if you know your system well.

INTRODUCTION TO COMBUSTION

Before we begin editing our scene, let's take a moment to explore the architecture and major features of **Combustion**.

COMBUSTION FILES

A **Combustion** document is called a *workspace*. For most projects a workspace is created as a composite, which is a blank workspace ready to accept footage items.

Footage items refer to anything imported into the **Combustion** workspace, such as digitized footage, image sequences or still images.

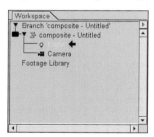

Figure 13.1 *Basic Workspace*

Like 2D paint and compositing packages, **Combustion** works with layers. Each footage item imported into **Combustion** is placed on its own layer and given a listing in the Workspace. Effects such as blur and glow, referred to as *operators*, are placed in the same layer as the footage they affect, and are displayed above the footage item in the Workspace hierarchy.

OPERATORS

Operators are effects or manual changes to footage or imported images. Operators change the way the footage or images are displayed and output, but do not change the actual source files. Operators can be thought of as filters over a camera lens — you see the scene differently, but the actual scene is unchanged.

At any time, operators can be toggled on or off, deleted, or rearranged to change the order in which they are applied.

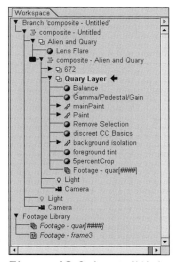

Figure 13.2 *A typical Workspace*

Operators are divided into the following major categories:

- **Color Correctors.** The suite of color correction tools tie **Combustion** to the high-end range of compositing software. **Combustion** allows NT-based access to the Discreet Color Corrector, known from Flame and Inferno. Its major features are the control sets for color basics, histograms, curves, and ranges. These controls can be applied individually to tonal ranges or to the entire gamut of color. Furthermore, the R, G, and B channels may be isolated. We will use this operator to change our rock quarry into another world and to make our alien seem to be under the same light and shadow conditions as the landscape around him. **Combustion** also holds the basic controls for brightness/contrast, color shift, levels, and equalize as well as operators that format clip colors for output to NTSC and PAL video formats.

- **Paint.** The Paint operator is a complete paint program contained within **Combustion**. It uses vector-based curves to define paint strokes, selections, masks, and text. Like similar vector objects found in draw programs, these elements may be moved or edited at any time regardless of their order of creation. Paint components may also be animated and even tracked to follow the motion of footage in our scene. We will be using Paint operators to colorize some of the rocks in our landscape.

- **Selection.** Selection operators are used to define the portions of the footage other operators will affect. Selections can also be accessed through the Paint module of **Combustion**. A good example of using selections would be to fake depth of field.

- **Keying.** Keying operators are used to define masks based on color information. The user defines a color range in the footage, which will then be excluded by the keyer. The most common use of keying is for the isolation of green-screen talent. The fluorescent green is not common in nature so it is easy to isolate with the keyer. The foreground imagery remains and can then be superimposed over new background footage. **Combustion** comes with the award winning Discreet Keyer also from Discreet's high-end applications.

- **Tracking.** The Tracking module matches the movement of a foreground object to the movement of a reference feature that the user defines on a layer. We will use this module for tracking our paint strokes (for tennis ball removal, etc.) to the motion of our footage.

- **Stabilize.** Stabilize operators can be applied to correct rough camera movements like jitters or shakes. A reference feature is defined and tracked to hold its position in the field of view. When footage is stabilized the tracked reference feature is pinned to its point of origin in the viewport. The motion of the footage can be observed in the movement of its borders over time. Together with the Tracking module **Combustion** delivers one of the most proficient tracking solution on today's market.

- **Visual Effects.** The biggest and most diverse collection of operators is the visual effects category. Visual effects include operators that add noise, glow, blur, sharpen, emboss and many more effects. These operators can be used to make effects like film grain, depth of field, and drop shadows.

In addition to operators **Combustion** holds a wide range of filters, which includes 3D post filters made for rich pixel format output (RPF files). The RPF file format, new with **Combustion**, can hold a wide variety of image information useful in compositing.

COMBUSTION USER INTERFACE

The main components of the **Combustion** UI are its command panels and viewports. The command panels use a labeled tab system. These tabbed panels can be grabbed and repositioned in the **Combustion** desktop for a flexible, optimized workflow.

By default, the upper part of the screen contains the viewports, displaying up to four different views at a time. A good workflow is to use two viewports, one to focus on the effect you are currently working with, and the other to display the final composite.

Although the imported footage and images are in 2D, **Combustion** works with these elements as if they were in 3D. For example, tools for zooming, tracking and panning around 2D footage items treat them as though they have depth.

The **Combustion** user interface allows access to all features and can be completely customized.

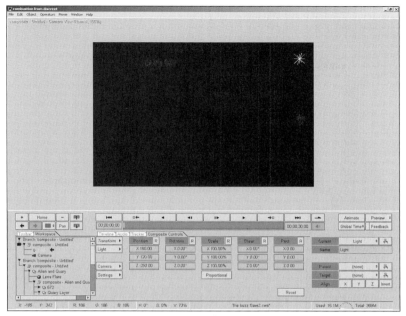

Figure 13.3 *Combustion user interface*

The Workspace panel is located at the bottom left of the screen. This panel is the central control for all your work in **Combustion**.

The Workspace panel displays a hierarchical listing of all of the elements in your workspace, including composite/paint branches, operators (filters and effects), footage items (images, movies, solids), and objects (cameras and lights). On the Workspace panel, these items can be rearranged, toggled on/off, selected for modification, or sent up to the active viewport.

The tab to the left of the Workspace is the Toolbar, which controls what is shown in viewports. Basic manipulation tools can be found here along with viewport navigation tools.

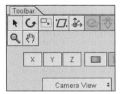

Figure 13.4 *Toolbar*

The main control panel is context sensitive and updates based on the current focus in your workspace. Operator, composite, footage and object attributes may be viewed and changed here.

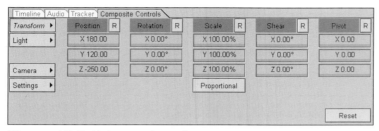

Figure 13.5 *Main control panel*

The Tracker panel is where new Trackers can be initialized and analyzed. Footage can be tracked by position, rotation, and scale, and parameters can be set for tracking and magnification. A preview window displays the pixel information from the footage as our Tracker analyzes it.

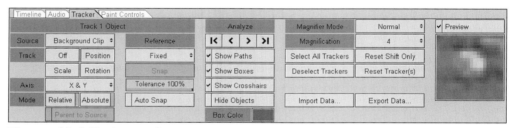

Figure 13.6 *Tracker panel*

Controls for importing sound files are found in the Audio panel. Audio start time and duration can be set on this panel. If audio exists in the scene, **Combustion** can export it to the audio layer of a movie file.

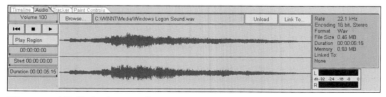

Figure 13.7 *Audio panel*

The Timeline features a time spatial overview of the animation channels in our scene. The list can display different levels of detail, and can be filtered by the user.

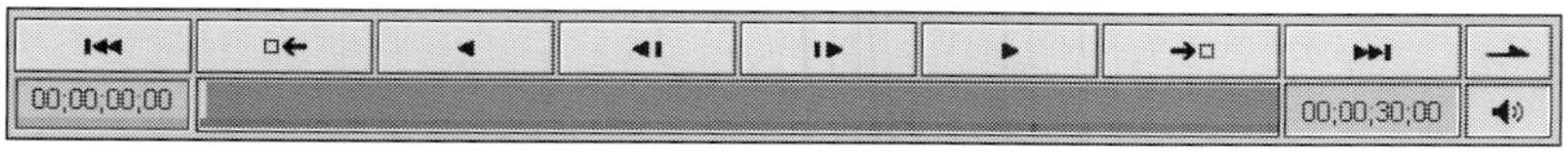

Figure 13.8 *Timeline*

Here we can edit which components are keyed. The attributes keys are visible on the right side of the Timeline panel, and can be visualized and edited in a graph or overview mode.

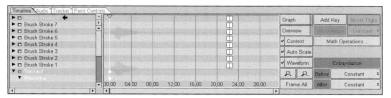

Figure 13.9 *Timeline panel*

Here we have identified some of **Combustion**'s most important tools and their functions. We will use these tools to bring our footage and animation together into a single composition.

RENDERED 3DS MAX FILES FOR COMBUSTION

In *Chapter 12*, we talked about two options for rendering: Rendering elements, or rendering to RPF files.

We chose to render to RPF files. RPFs can hold numerous image channels associated with most post effects such as depth of field, glow, motion blur and so on.

In this chapter, we focus on the RPF files we rendered. If you want to find out more about rendering to elements or RPF files, see *CD Tutorial 3* in the *Tutorials* folder on the CD. The **3ds max** documentation also holds a large body of information on this subject.

OUR RPF FILES

In *Chapter 12*, we rendered the following channels to our RPF file through Video Post:

- Z depth
- Material ID
- Alpha channel
- Velocity

In this chapter, we will work mostly with the velocity and alpha channels. The material ID will be used to create a glow around each of the blue lights on the rover.

We could have rendered depth of field and motion blur information from the **3ds max** scene to RPF files, but these gobble up memory and CPU time. By adding these elements in **Combustion** instead, we can get close to real-time feedback while adjusting the scene, making these settings much easier to adjust.

DEALING WITH SHADOWS

One of the advantages of rendering to elements is that you can save shadow data as a separate file, where with RPF files you cannot. Using rendered elements to keep shadows on a separate layer can save render time in **3ds max**. You can use a very low shadow map size and sample range for minimal render time, then blur the chunky shadows in **Combustion** to create soft, real looking shadows.

We have chosen a different way to deal with shadows. With our RPFs, we can use **Combustion**'s color corrector to separate the highlights from the midtones and the midtones from the shadows, giving us fine control over the shadow tones.

TUTORIAL 13.1 FOOTAGE IMPORT

To start our **Combustion** session we need a workspace, the blank canvas for our project.

The workspace uses references to the footage items rather than the actual footage — they are not loaded and saved as part of the **Combustion** file. In this way, file size remains small for fast feedback, and the raw footage is always left untouched.

Starting a New Workspace

1. From the *File* menu, choose *New*.

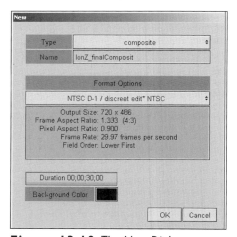

Figure 13.10 *The New Dialog*

2. On the New dialog, set the type to **Composite** and name it **IonZ_finalComposite**.

3. In the **Format** options, set the format to **NTSC-D1**.

   The NTSC-D1 is a high-quality broadcast standard.

4. Set the **Duration** to 30 seconds.

5. Set the background color to black and click **OK**.

> **TIP**
>
> The workspace is set up like a hierarchy. It is always evaluated from the bottom to the top.

You have now created a workspace and a composite branch which appears in your Workspace panel. As we work our way through layers, nests and operators, they will all grow out of this branch.

At any time, you can change the viewport focus by double clicking at any point in the stack. A single click will change the control focus while leaving the viewport at its current point of view.

Importing the Image Sequence

We will start out by generating the Martian landscape for the alien. To do this we will first import the image sequence of the rock quarry.

If you have not already done so, you must copy the file **low-res footage.zip** from the **Movies/Background Plates** folder on the CD, and unzip the file. This will put the quarry background plates on your hard disk so you can use them with **Combustion**.

1. From the *File* menu, choose *Import Footage*.

2. Navigate to the folder that contains the quarry background plates. To optimize your workflow, choose **List View** and set the file browser to **Collapse Sequenced Image** if not already set.

3. Select all the files in the quarry sequence (*quar0001.jpg* through *quar1004.jpg*) and click **OK**.

   The sequence is added to the import queue when it is selected. To add multiple image files or sequences, use **<Ctrl>** or **<Shift>**.

   You can also navigate to other folders and add more files to your selection before clicking **OK**. **Combustion** will remember all the selections you made from various folders, and load the files when you click **OK**.

4. Rename the quarry layer by clicking the layer and entering the new name **Quarry Layer**.

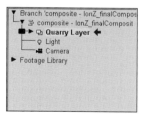

Figure 13.11 *The current Workspace*

The footage is now displayed in your viewport, but as you can see on certain frames, a staircase effect displays. This is because we digitized video frames, where every other line is a field offset by half a frame.

5. Save the file by choosing *File/Save Workspace* from the menu. Save the workspace with the name *AlienComposite01.cws*.

Removing the Interlace Effect

Because this footage was created originally for NTSC video, shot with a DV-CAM, we must adjust the interlacing option on the footage.

1. To clearly see the staircase effect, go to the time code 00:00:07:05. Zoom in on the border between light and shadow on the huge rock in the background.

TIP
If your time is displayed in a format other than time code format (00:00:00:00), you will have to change the format. To do this, choose File/Preferences from the menu to open the Preferences dialog. Change the Display Time As option to Timecode, and click OK to set the format.

Figure 13.12 *Staircase effect*

Here you can clearly see how every other line is offset by half a frame. We will change this before continuing.

TIP
Your time code is now at 00:00:07:05. The frame at this time is excellent for setting up the composite because there are a large number of the scene elements such as rocks and tennis balls in the viewport. From now on, keep the time set to this frame unless you are told otherwise.

2. Click the **Workspace** tab.
3. Select and expand the **Quarry Layer** in your workspace.
4. Select the **Footage** item.

   Footage Controls appear to the right of the Workspace.

5. In the Footage Controls panel, under Source, change the Field Separation from No Fields to **Lower First**.

   The jagged edges will disappear. Combustion takes the necessary measures to display interlaced footage on your computer monitor by unlacing the footage. The length of the footage is unchanged.

CONTRAST AND COLOR ADJUSTMENT

The next step in working with the footage is to make any necessary contrast and color adjustments.

CONTRAST ADJUSTMENT

One of the main purposes of color correction is to improve the balance of light and dark in the image so it has as much depth as possible. To get the best contrast, black should be truly black and white should be truly white. It is rare for digitized video footage to come in with perfect blacks and whites, so color correction almost always needs to be performed.

When looking at an image, it can be hard to tell if certain areas have been washed out and need to be remapped to either white or black. By playing with the levels of an image, you can see how the image should look, or at least how the remapping makes it look.

However, there is a downside to this technique. When using a level operator or a *histogram* (a pictorial representation of black and white levels), you are actually remapping certain shades to black or white and removing the original shade from the image. If too much color information is lost, your image might end up in extreme contrast, with less depth and color information that it had before.

This problem is due to the fact that when footage is shot and digitized, it is converted into a range of 256 levels that can be manipulated digitally. Smooth gradation in brightness is not always possible because the 256 levels cannot cover the entire spectrum of real light distribution. On film and in the real world there are no digital levels; there is enough bandwidth to hold smooth gradations over the entire shade, especially when the footage is shot outdoors.

With video, a bright whitish sky will tend to wash out certain shades -- the white level will be set so high that there will not be enough levels to smoothly transition to the white value. This is where you see hard cuts and faded areas in the footage.

To avoid this problem and obtain more of a film look, we will crop a bit of color information. A good rule of thumb is to crop 5 percent on both ends of the histogram. In this way, you will get more solid black and pure white in your imagery for good contrast, while still giving **Combustion** enough room to generate a smooth gradation over the entire image.

ADJUSTING COLORS

The footage of the quarry will need to be changed to a reddish color. We will accomplish this by selecting various areas and making adjustments to the hue and other values.

The areas in the foreground are the most important, and will be given the most attention. The background areas will be blurred to a large degree and so don't require as much careful adjustment.

TUTORIAL 13.2 COLOR CORRECTION

We will start our color correction session by working with the contrast in the imported footage.

Correcting the Overall Contrast

1. In the workspace, select the **Quarry Layer**.
2. From the menu, choose **Operators/Color Correction/discreet CC Histogram** to add a histogram operator to the quarry footage.
3. Rename this new operator **Overall Histogram**.

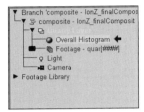

Figure 13.13 *The current Workspace*

4. With the **Master** and **RGB** buttons active, click and drag the **Black** slider to a value of 20.

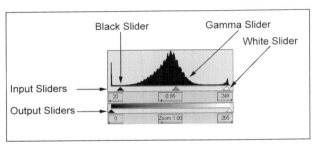

Figure 13.14 *Histogram with Black set to 20*

This action will remap all of the darker shades to black.

5. Move the **Gray** slider to a value of 0.95.

This sets the gamma value. Our new setting will darken all the midtones.

6. Set the **White** slider to a value of 249 to remap lighter shades to white.

Figure 13.15 *The Histogram*

Darkening Washed-Out Highlights

The washed out background needs the most work. Here we remapped the ranges so that the shadows had more influence. The mid-range and the shadow-range should intersect about halfway up through the midtones around a value of 120, just after the midtone peak on the graph.

We will now darken the washed-out highlights.

1. Click the **Highlights** button.
2. Set the **Black** slider to 30.

   TIP

   Values may be entered in decimals from 0 to 1 or whole numbers from 0 to 255.

   This will darken the highlights.
3. Set the **Gray** slider to 0.55.

   The shadows in this image are also too light. Lets fix the shadows.
4. Click the **Shadows** button.
5. Set the **Black** slider to 20.

   Adding black to the shadows gives us more contrast to the image. Be careful using this setting as it can quickly drain color out of your image. Try a setting of 40 to see how the shadow definition disappears and leaves a solid black shade. Set the input slider back to 20 when you have finished experimenting.

Comparing the Images

Let's see how the old and new versions look when compared side by side.

1. Use the Viewport Layout List to set the viewport to a two-side-by-side layout.

Figure 13.16 *Viewport Layout List*

2. Click on in the left viewport.
3. In the Workspace, click the **Footage** item under the Quarry Layer.
4. Click the **Send Up** button.

Here you can see how we have added contrast. The colors appear richer, and the image contains more depth than the original raw footage.

Adjusting the Color Balance

We want to give the rocks and scenery a reddish tint. We will do this by using a *color wheel*.

The color wheel allows you to visually tint shadows, midtones and highlights separately toward any given color, as well as adjusting the master hue of the image. Like the histogram, the color wheel is part of the discreet Color Corrector. It can be found under its color controls, but to save processing time we will add it as a separate operator.

Since the background is so over exposed, we will also modify it further.

1. Select the **Quarry Layer**. From the menu, choose *Operators/Color Correction/discreet CC Color Wheel*.

2. Name this new operator **Overall Tint**.

   The color wheel is great for quickly adjusting the hue of an image. A typical use for a color wheel is to balance colors or remove an unwanted color. For example, if an image seems to be tinted slightly blue or green, you can quickly dub the color, tint toward it, and then use the color wheel to add the opposite color to the image.

3. In the **Contrast** field, enter 90. This will help produce more of a hazy red.

4. You can use your own judgment by activating the **Master** button and clicking inside the color wheel, or use the settings we've provided below:

   - Click **Master**, set **Strength** to 37, set **Hue Tint** to 12
   - Click **Midtones**, set **Strength** to 58, set **Hue Tint** to 14
   - Leave shadows and highlights at default values

   The result should be a reddish quarry, with the exception of the sun-exposed background rocks.

TIP

Adding a color corrector separately will also allow you to change the process order of the different parts of the color corrector. We are not going to use this feature on this project.

Selecting the Background

The background rocks are currently yellow and overexposed, and there is hardly any definition left in the background. We will now use an operator to color the background orange to match the rest of the rocks.

The background area will be blurred later on to add more depth to the footage, so we do not need to spend expensive production time trying to add more detail to the area. We will tint it toward the same red color as the rest of the quarry and try to pull some shadows out of the overexposed rock side.

To isolate the background, we will use a selection operator. This will apply the next operation only to the background, leaving the foreground as it is.

1. From the Viewport Layout List, choose the single viewport option.
2. Select the **Quarry Layer** in your Workspace.
3. From the menu, choose *Operators/Selections/Draw Selection*.
4. Name the new operator **Background Selection**.

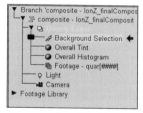

Figure 13.17 *The current Workspace*

Now we will begin drawing a selection around the cliff region.

5. Switch to the **Toolbar** tab, either by clicking the tab or by pulling the tab panel out closer to the viewport to turn it into a floating window.

   Note that the toolbar has now added selection tools because the selection operator was automatically made active when added to the workspace.

6. Click the **Polygon Selection** tool.
7. Begin clicking in the viewport to outline the overexposed rock side (the back canyon wall). Here is a suggested workflow for creating the outlined selection:
   - Press the **<~>** key on the keyboard to maximize the viewport.
   - Use **<Ctrl-+>** to zoom into the viewport.
   - To pan, hold down the **<spacebar>** while you click and drag in the viewport.
   - After you draw your selection, you can go back and modify an edit point by holding the cursor over it. When the mouse pointer changes back to a standard point, click and move it. This procedure does not work on the end point.

TIP

The <~> key is found next to the number 1 key on your keyboard. It is also called the "tilde".

- For further control of the selection outline, you can add Bezier handles. Do this by holding down **<Ctrl>** while dragging an edit point. If you wish to remove a Bezier handle, click the edit point while holding down **<Ctrl>**.

Figure 13.18 *The selected region*

Feathering the Selection

We will now feather the selection to make a soft transition at the selection edges.

1. On the **Selection Controls** tab, click the **Modes** button.
2. In the workspace, select the **Polygon Selection 1** entry.
3. On the **Selection Controls** tab, set the **Feather** parameter to 5.

   This will bleed a little bit onto the rest of the frame, but lower values may produce hard edges.

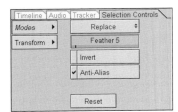

Figure 13.19 *Feather parameter*

4. Make sure that the button above Feather is labeled **Replace**.

5. Name this stroke **Background Isolation**.

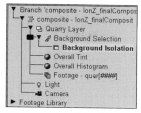

Figure 13.20 *The current Workspace*

Color Correcting the Background Ridge

1. Click the **Quarry Layer** in the Workspace.
2. From the menu, select *Operators/Color Correction>Discreet CC Basics*.

    We will need to darken the rocks side and add shadows to it. There is not much color information here so the area quickly turns solid, but this will not be seen once the area is blurred.

3. Rename the layer **Background Color**.
4. Adjust the settings of the color basics until you get a result that blends the rock side to the rest of the picture. Try to avoid getting a solid color. You can gain a sense of depth by dropping down the gain and gamma of the shadows.

    Here are our settings used in the final composite:

    Master settings:

Saturate	55
Contrast	82
RGB gamma	0.90
R gamma	0.96
G gamma	0.95
B gamma	0.94
RGB gain	105
RGB offset	9
Red offset	3
Green offset	-1
Blue offset	5

Shadows settings:

Contrast	110
RGB gamma	0.40
RGB gain	56
R gain	103
G gain	89
B gain	67
RGB offset	-19

Midtones settings:

Saturate	20
Contrast	115
RGB gamma	0.98
R gamma	1.33
G gamma	0.78
B gamma	0.61
R gain	81
G gain	60
B gain	69
RGB offset	76
R offset	-12
G offset	5
B offset	-29

Highlights settings:

Saturate	92
Contrast	135
RGB gamma	0.52
R gamma	0.83
G gamma	0.40
B gamma	0.53
RGB gain	185
R gain	55
G gain	46
B gain	57
RGB offset	-12
R offset	13
G offset	-20
B offset	2

As you can see, a lot of parameters were adjusted.

The best way to achieve an acceptable look is to remove as much subjective thought from the process. Unless you are an experienced color corrector you can easily start accepting colors and gradations that you normally would not.

There are two ways of getting past this barrier. One is to frequently look away and then gaze back at the monitor. If you do not do this, green mountains quickly start looking great in your eyes, but when you have returned from lunch and take another look, you will be amazed how much damage you have done and will be forced to start over.

The second is to zoom in extremely close. Zoom in where the huge left background rock is cut in half by the sun. Here you can see the color you are aiming for and you can continuously see how close you are. By doing it this way, you start looking at colors and shading rather than rocks and skies.

After having setup the Discreet color basics operator, you should see a close to perfect blend between the overexposed rock side and the rest of the rocks.

You might see a soft dark or black line around the area you color corrected. This happens if you made your selection too large. To avoid this, shrink your mask manually.

If you shrink the selection too much, you will see a soft yellow halo around the rock side. Its a very fine balance, but with a little practice and some feathering, you should be able to create a close to perfect blend.

Removing the Selection

We will now apply a Remove Selection operator to stop the Background Selection operator from being transferred up the stack. If we did not do this, all of the subsequent operations would only affect the isolated area defined by our background isolation stroke.

1. Click the **Quarry Layer** in the Workspace.
2. From the menu, select *Operators/Selections/Remove Selection*.
3. Name the operator **Remove Background Selection**.
4. Save your workspace to the file *AlienComposite02.cws*.

COMBUSTION TRACKING TECHNIQUES

Before continuing, we will introduce an important technique that you will have to use throughout the composite. If you have not already discovered/considered/feared it, then you are probably about to.

If you change the time in your composite you will see that your selection does not follow the movement of the camera. So far all that needs to be fixed is the background isolation stroke, however as we complete the composite you will create many paint strokes that will need to be fixed.

You can make your job easier by grouping these strokes. Unfortunately, strokes cannot be parented, nor can you add elements to an already created group. We recommend that you complete as many of the paint and selection jobs as possible and then track them all together. You may choose to copy and paste track data, however it is not recommended. It all comes down to personal preference.

We also have another problem. What about parts of the footage that have not yet entered the viewport? Our current selection is not set up to cover these areas. Fortunately for us, this can be easily fixed by simply editing the selection. After having tracked the selection, go to a frame where more of the rock side is revealed and edit the mask to expand it until the entire area is covered. You may add more points, move points and change tangency. This keeps you from having to generate various selections until the entire area is covered and then tracking those selections one by one. This would be extremely tedious and you would have to make sure that there was no overlapping or gaps appear in the masked areas.

THE TRACKER

The Tracker records the movement of a reference feature across the screen and applies motion data to the active selection in the Workspace. This movement can then be applied to another element in your composite. In this way you can make them appear to be locked together.

The Tracker has its own panel in the main control area. If the Tracker panel is not visible, you can make it visible by selecting *Window/Palettes/Show Tracker* from the menu.

Tracker Panel

On the Tracker panel are a series of options. From here you can do two types of tracks:

- **Position** using 1 Tracker
- **Rotation** or **Scale** using 2 Trackers

Under some circumstances it can be useful to restrict the track to one axis only. In this project we will track in both the X and Y directions, so you will need to make sure that both are enabled.

You can also choose between *relative* and *absolute* tracking. By using the relative option we will be able to create an offset between the reference feature and the element we are tracking.

The smaller box defines the reference feature. The Tracker looks at contrast and tries to find the same feature in subsequent frames. The Tracker will also search within the larger boundary box. Both boxes can be resized, but keep in mind that the larger the box, the longer the processing time.

You can also choose to use a *roaming* or *fixed* reference. When set to roaming, the reference is updated every time the Tracker increments in time. A fixed reference always refers to the first frame the Tracker was initialized on.

The snap option is available in case the reference feature changes form radically during the track. You can have **Combustion** do this for you by enabling Auto Snap. We will not use this option in our session.

The Tracker has a tolerance slider for use in finding matches. If this value is set to 100%, the Tracker will always find a match. Lowering the tolerance may prevent the Tracker from generating keyframes on all frames.

When the Tracker is launched, keep an eye on the preview window in the bottom right corner. This image represents the reference feature, which should not move throughout the track. If the preview disappears this means the feature is lost, and no keyframes will be generated.

Using the Tracker

When the Tracker is launched, you can look in the timeline to see the confidence of the Tracker. This tells you how sure **Combustion** is of the current keyframe. Depending on the number of active Trackers, they will show up in the timeline with dimension, position, shift, confidence and tolerance properties. It is especially useful to follow the confidence value as the track progresses. Keep in mind that only selected Trackers will be analyzed.

The Tracker has a separate time control where you can specify in which direction the Tracker should track, and if a single frame or multiple frames should be tracked. These keys are found under the Analyze caption.

You can also change both magnification mode and strength. This is strictly for helping to position the reference and boundary box, without affecting the actual track.

The Reset Shift Only option will clear all tracking data, but leave the size of the reference and boundary box intact. The Reset All Shifts option will completely reinitialize the Tracker.

The import and export data buttons bind **Combustion**'s Tracker to other Discreet products such as Flint, Fire, Flame and Inferno. In this way, Tracker data can be moved from one package to another.

Tracker Workflow

A typical track workflow works as follows:

1. Select the object or layer that you wish to track.
2. Select the source.
3. Choose the type of track: Position, Rotation, or Scale.
4. Locate reference features that have the desired motion that you want to track
5. Adjust Trackers as needed. You can modify multiple Trackers at once by selecting them. Trackers can be adjusted numerically by using the timeline.
6. Make sure Trackers are selected.
7. Click one of the **Analyze** buttons to perform the track.
8. Click back space to delete keys if necessary.
9. Change tolerance and other parameters as necessary to get proper tracking.

You will be using this procedure in the tutorial that follows.

TUTORIAL 13.3 TRACKING THE SELECTIONS

Now we will use the selections made earlier for the tracking process.

Accessing the Tracker Panel

1. In the Workspace, double click the **Background Selection** operator to set the focus to this level.
2. Expand the **Background Selection** operator and single click on the **Background Isolation** stroke.
3. With the background isolation stroke selected, click the **Tracker** panel tab located under the timeline next to Selection Controls.

Setting up for Tracking

First we will track the position of the Tracker.

1. Make sure **Source** is set to **Background Clip**.
2. Since we are tracking based on position, click the **Position** button to enable it.

   After Scale is enabled, a set of boxes appears in the viewport. Positioned in the inner box is the reference feature. The Tracker will use the outer box as a search area for the pattern inside the reference box.

   The reference box's center point determines the shift, and the boundary box moves with it. The shift can be seen in the timeline while the Tracker is active.

Figure 13.21 *Source and Track type options*

3. Move both boxes at the same time to a point in the scene that contains a high contrast. In our scene, move them to the centermost tennis ball in the view.

   It doesn't matter what frame you are currently on as long as you are within the frame ranges we are using. You can always track backward if needed.

Figure 13.22 *Tracker*

4. Set the Axis to **X & Y**.
5. Set the Mode to **Relative**.
6. Verify the Reference type is set to **Fixed**.

   You can leave the Tolerance at 100%.

Performing the Tracking

1. Click **Select All Trackers**.

2. Under Analyze, click the **Play Forward** button.

   This will play the footage and read the tracking coordinates. From now on your cliff selection should follow the cliff as it moves through the scene.

Figure 13.23 *Tracked path*

Having the objects hidden prevents the Tracker from becoming confused by the paint strokes you are tracking to the footage, and saves processing time.

You can also track backward by clicking Play Backward under Analyze.

If you are not satisfied with the result, you can use the **<Backspace>** key to delete key frames. Make sure the Tracker is selected before attempting to delete the key frames. Otherwise, the entire layer will be deleted.

When the Tracker is analyzing the footage, the selection disappears. This is because the **Hide Objects** option is checked by default.

There is one point in the footage where the camera zooms in. During this period, you will have to track the scale first, then track the position in this section in order for it to track correctly.

To track the scale at this time, go to the beginning of the time period over which the zoom takes place. Click the **Scale** button on the **Tracker** panel, position the two Trackers over high-contrast points in the footage, then click **Select All Trackers** and click **Play Forward**. Then retrack the position for that period.

3. Continue working with the tracking feature until you are happy with the result.
4. When you are satisfied with the tracking results, commit to the data by clicking **Off**.

   This moves the keys from the temporary Tracker object to the paint stroke.
5. Save your workspace to the file ***AlienComposite03.cws***.

PAINT

Combustion includes a fully featured paint module called **Paint**. A paint object is created when you paint on your footage using any of the painting tools.

Strokes

When referring to paint operations, the term *stroke* is often used. (A stroke of paint, get it?) A paint stroke can be a solid brush stroke, a text object, a fill, a clone operation, a reveal or a gradient.

All Paint operations are vector-based and therefore resolution-independent, allowing you to animate strokes and without being concerned about the final output size or other issues.

All strokes are non-destructible, which means that they can be modified at any point in the production. You can delete, resize, reposition, change the strokes color or opacity, and even the stroke type may be changed after creation. Your footage is always left untouched and the stroke appears in your stack. Here you can select it, re-arrange it, copy/paste it, and even apply filters to it.

Using the Paint Operator

To create a paint object you must first select a layer, then choose *Operators/Paint* from the menu. This will change the view and control focus to a paint mode, and your Toolbar will contain the painting tools.

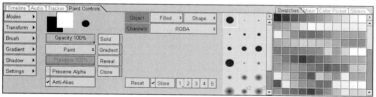

Figure 13.24 *Paint Controls*

In the Modes section, you can choose foreground and background color. Also, the drawing mode can be changed. We will use this option to desaturate part of the footage. In the Paint Controls you can also change the opacity of the strokes.

When you draw a stroke it will, by default, only exist on the current frame. Since we plan to track the strokes and hereby generate key frames on each frame, their duration will automatically be updated. However, since we will be putting strokes in groups, you should set the strokes to exist on all frames by changing this option on the panel.

On the Paint Controls panel you can also set the channels the stroke should affect: red, green, blue and/or alpha.

In the Transform section you can change, position, rotation, scale, shearing, and the pivots position.

In the brush section you can define the brush's shape, size, and falloff.

The Gradient section gives you access to color and opacity gradients. Furthermore, you can edit gradients by changing gradient type and gradient direction.

Besides painting solid colors, **Combustion** also offers access to a reveal and cloning tool. The reveal tool is used to reveal a layer underneath or the same layer offset in time. To see the reveal source, choose show reveal source from the windows menu. Here you can also set the level of transparency, through the reveal overlay options.

Using the cloning tool, you can copy from one part of the footage, offset in both space and time, and paste it to another. For example you may clone the same flower ten thousand times to create an entire field of flowers. You use the ALT key to select where the cloning will begin at, this is just like the rubber stamp tool in Photoshop.

In our scene we will be using all the tools described above.

Rotoscoping

In **Combustion**, *rotoscoping* refers to erasing or painting in elements in 2D. For example, when an actor is filmed doing wild flying stunts while wearing a harness supported by wires, the wires must be painted out at the post production stage with rotoscoping.

In the earlier years of filmmaking, rotoscoping was a time-consuming manual technique, with the artist working on each frame individually. The results depended largely on the talents and patience of the rotoscoping artist. The original Superman and Star Trek television series show some excellent examples of early rotoscoping.

We're sure you don't want to go through all that work (we sure don't!). With **Combustion**'s Tracker feature and its Paint operator, we can avoid the time consuming task of manual rotoscoping to paint out the tennis balls and make other adjustments to our footage.

Using Paint on our Project

The next operator we will apply will be a Paint operator with multiple paint strokes and selections. Each will be used with the Tracker to clean up problems with the background footage. The cleanup activities are:

- Tennis balls used as tracking objects must be painted out.
- Rocks that the alien will be sitting on should be lighter in color compared to those around him to bring them forward and highlight our star.
- The cliff face on the left needs to be colored gray.
- The sky needs to be painted in.
- Depth of field needs to be faked.

All these tasks will be no problem for **Combustion** and the Paint operator.

TUTORIAL 13.4 PAINTING THE FOOTAGE

Our first task is to get rid of the markers we don't want to see in the final scene, namely the tennis balls and the trees.

Preparing to Paint

1. Select the **Quarry Layer**. From the menu, choose **Operators/Paint**.

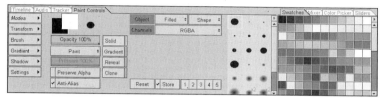

Figure 13.25 *Paint Controls*

The focus in the Workspace is changed to the new Paint operator.

2. Name this operator **Rocks, Tennis Balls, Sky**.

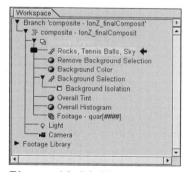

Figure 13.26 *The current Workspace*

3. Switch to the Toolbar and choose the **Freehand** paint tool. Make sure the mode is set to **Stroked**, not Filled. If it is set to Filled, click the Freehand paint tool to toggle to **Stroked** mode.

   There are two ways you can tell which mode you are in. If the paintbrush tip is solid, then it is currently set to Filled mode. You can also look at the Paint Controls panel.

4. Set the mode to **Clone**.

5. Click the **Lock** button to lock the frame on the current frame.

6. Locate an area on the image containing nothing but ground rocks. Hold down the **<Alt>** key on your keyboard and click your selected spot.

7. Click the **Brush** button in the Paint Controls panel, and set the brush's **Diameter** to 20.

8. Click the **Line type** button to set the profile to **Bezier**.

9. Edit the **Profile** graph so that the opacity profile has the stroke falling off softly. Watch the Brush preview window as you adjust the edit points. The circular edge should appear soft.

Painting Out the Tennis Balls

1. Paint out all the tennis balls in view. Each tennis ball only requires one stroke (one click of the mouse).
2. When you are done, click the **Workspace** tab and expand the **Paint** entry.
3. Select all of the strokes in the workspace.

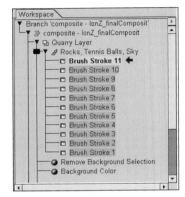

Figure 13.27 *Brush preview window*

TIP
You can also press <Ctrl-G> to group items together.

4. From the menu, choose **Object/Group** to group them together.
5. Name the group **Tennis Balls**.

   The new group needs to be tracked so that the tennis balls will remain hidden when the camera moves.

6. Use the techniques shown previously to track the group.

   You can hide the strokes by clicking the small yellow box next to the Tennis Balls group in the Workspace. This will allow you to see all of the tennis balls so that you can have a bright object to track. After you have completed the tracking process, click the small icon again so that the Tennis Balls group becomes visible.

   There are tennis balls that are not in view and need to also be removed via this painting method.

7. Go to a frame that shows the remaining tennis balls, paint them out, group those strokes to a new group, and track the new group.

Removing Color from Rocks

To add more detail to the scenery, we will remove color from the rocks on both the left and right.

1. Use the **Polygon Selection** tool to draw an outline around the target rocks, using the same method you used to select the background rocks earlier.
2. From the filters color corrections, choose **Gray**.
3. Set the **Amount** to 40.

In our final composite we used the second method. However, the amount of gray was dropped to 20 for the rocks on the right. The more color variance and contrast in the image, the more detail and depth will appear.

Painting the Canyon Wall

1. Verify that you are at the time 00:00:07:05, where are a good number of rocks in view.

2. Click the **Rocks, Tennis Balls, Sky** operator in the Workspace.
3. Click the **Toolbar** tab and select the **Polygon Selection** tool.
4. Draw a selection around the larger rock wall on the left of the scene closest to the camera.

Figure 13.28 *The selection*

5. From the menu, choose ***Effects/Color Correction/Gray***.
6. Set the **Amount** to 25.

This changes the selected wall area to more of a gray color.

7. In the Workspace, select both the new **Gray** and **Polygon Selection** entries.

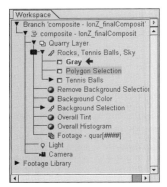

Figure 13.29 *The selection in the Workspace*

8. Group the selections with the name **Canyon Wall Gray**.

Painting the Foreground Rocks

1. In the Workspace, click on the **Rocks, Tennis Balls, Sky** entry.
2. Click on the **Toolbar** and select the **Stroked Polygon** tool.
3. Draw a selection around the foreground rocks (the large ones).

Figure 13.30 *The selection*

4. From the menu bar, select *Effects/Color Correction/Gray*.

5. Set the **Amount** to 22.
6. In the Workspace, select both the new **Gray** and **Polygon Selection** entry and group them together. Name the new group **Front Rocks Gray**.

Grouping the Gray Rocks Together

1. Select both **Front Rocks Gray** and **Canyon Wall Gray** from the Workspace.

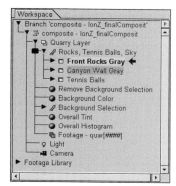

Figure 13.31 *The selection for grouping*

2. Group the two selected items together with the name **Gray Rocks**.
3. Save your workspace to the file *AlienComposite04.cws*.

Tracking and Selecting Gray Rocks

There are two problems with the two selections that we used for making the rocks gray. They are not tracked so they will not move with the footage, and the selection is not large enough to cover the areas of rocks that are offscreen at this particular frame.

Use the techniques shown earlier to track the group that the selections are in. Try tracking back toward frame 0 first.

After tracking the scene, go to various frames and modify the two selections to accommodate the new footage that has entered the viewport. Typically you will need only to add edit points and move them around.

Removing the Tree Line

Since this scene is supposed to be a Martian landscape, we cannot have trees growing in the atmosphere. To remove the trees we will use the paint selection tools and the fill bucket. We will cut out the sky roughly and then add a Gaussian blur to make the horizon appear hazy.

1. In the Workspace, click the **Rocks, Tennis Balls, Sky** entry.

2. Apply a selection around the sky. Be sure to include the tree and grass line as well.

> ☠ **TIP** ☠
>
> With the Polygon Selection entry selected, you can click on Modes in the Paint Controls and set the Feather parameter to 2 or 3. This will prevent a hard edge from developing between the sky and the rocks.

Figure 13.32 *Sky selection area*

3. Apply a selection to the sky, by drawing around its border with the lasso tool. Name this selection **Sky Selection**.

4. From the Toolbar, select the **Paint Bucket** tool.
5. Set the **Tolerance** to 100% at the bottom of the Toolbar.

    We do this so the selected area will be filled by one mouse click.

6. In the Paint Controls panel, select the foreground color and set it to:

    R 96%
    G 70%
    B 60%

7. From the menu, choose **Effects/Blur/Sharpen/Gaussian Blur** and set the **Strength** to 34. Name the Gaussian Blur entry **Horizon**.

    This will blend the sky with the far background, creating a haze around the edge of the rock side.

8. Group the Horizon, flood fill, and polygon selections into a single group named **Sky**.
9. Track the **Sky** entry.

Figure 13.33 *The finished sky*

Finalizing the Color Adjustment

At this point we need to do one more color adjustment, we need a bit more contrast to the image and we need to turn the red domination down a bit. You may not have noticed it, but the image actually appears too dim. It looks like its evening time. We can quickly fix this by applying a Gama/Pedestal/Gain and a Balance operator.

To lighten the image and to gain a little bit of contrast we will use a Gamma/pedestal/gain control.

1. Select the **Quarry Layer**.

2. From the **Operator** menu, select *Color Correction/Gamma/Pedestal/Gain* and set the following settings:

Gamma	1
Pedestal	-10
Gain	1.2

3. Select the **Quarry Layer**.

4. From the **Operator** menu select *Color Correction/Balance* and set the following settings:

Red	-12%
Green	-6%
Blue	-6%

Selecting for a Manual Depth of Field Effect

To give our composition more of a realistic look, we need to blur the background to a degree in some areas to simulate the real-life effect of a camera not being able to focus on every distance at once. Later we will use a depth of field operator to intensify this effect, but for now we can simulate it simply by blurring some of the background.

1. Click the **Quarry Layer** in the Workspace. From the **Operator** menu, choose *Selections/Draw Selection*.

   We will use our selection tools once more to select the areas that need to be blurred.

2. Click the Toolbar and select the **Polygon Selection** tool.
3. From the Selection Controls panel, set the Modes to **Replace**.
4. Draw a selection that covers about 1/4 of the bottom of the screen.

   Make sure you don't cut through any of the close rocks. Also, take care not to select too much, as the rover will be driving on the ground.

   These selections should be feathered to look more natural.

5. Set the **Feather** setting to about 25.
6. Name the selection **Front DOF Selection**.
7. From the Selection Controls panel, set the Modes to **Add**. This will add to the current selection.
8. Using the same procedure, select about ¼ of the screen from the bottom. Name the selection **Back DOF Selection**.

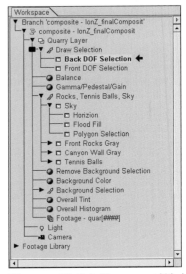

Figure 13.34 *The current Workspace*

9. Select and group these two new selections into one group called **DOF Selection Group**.

Figure 13.35 *The two selections*

Adding a Blur to the Selection

The blur selections are now set up. All that remains is to apply a Gaussian Blur operator.

1. In the Workspace, click the **Quarry Layer**.
2. From the Operators menu, select *Blur/Sharpen/Gaussian Blur*.
3. Set the **Radius** parameter to 1.5.

   Now we can remove the selection.

4. In the Workspace, click the **Quarry Layer**.
5. From the *Operators* menu, select *Selections/Remove Selection*.

Compositing the Alien into the Shot

Now that the background is fully prepared, we can add the CG footage of the alien with his rover using the tracking data created earlier.

1. In the Workspace, click on the **Composite - IonZ_finalComposite** entry.

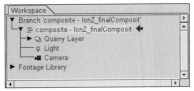

Figure 13.36 *Composite entry selected, Quarry layer collapsed to conserve space*

2. From the menu, select *File/Import Footage*.
3. On the Import Footage dialog, select the rendered RPF files that you rendered in *Chapter 12*, and click **OK**.
4. Locate the new layer that you just imported and rename it **CGI Layer**. This layer will be directly above the Quarry Layer.

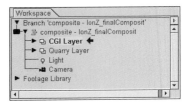

Figure 13.37 *The new CGI Layer*

5. If you do not see the alien over the background footage, double-click the **Composite - IonZ_finalComposite** entry in the Workspace.

Figure 13.38 *CG footage imported*

Color Correcting the Alien

The alien must now be color corrected to fit in with the scene. It is important to keep in mind that the alien is bluish. Be careful not to tint him too red. The atmosphere around him would make him appear more red, but there is a fine line between painting an element another color to make it match and adjusting color ranges to make it appear to be in the actual scenery.

The Discreet Color Corrector will give us the controls for this entire process.

1. In the Workspace, click the **CGI Layer.**

2. From the **Operators** menu, choose **Color Correction/discreet CC Color Wheel** and name the new operator **Alien Color Match**.

3. In the CC Color Wheel Controls, verify that **Master** is clicked, and set the **Strength** to 7 and **Hue Tint** to 26.

   This will give the overall layer a slightly yellow cast.

   The midtones will be drastically affected by our color correction. The alien is coming into the scene extremely blue. In reality he would appear to be more of an orange cast because of the lighting on the landscape.

4. Click **Midtones** and set the **Strength** to 20 and **Hue Tint** to 17.

   This will produce an orange-brown mix. The effect will be complete, with the soft edges of the layer blending into the others and the colors of the layer blending with the background.

Adding Further Depth of Field

The absence of depth of field and motion blur effects can make an animation look artificial. We added a little depth of field manually by blurring part of the background, but now we will use the depth, velocity and material ID information stored in the RPF file to intensify the realistic effect.

We will use the RPF Channels to add the following postproduction effects:

- Depth of field
- Motion blur
- Glow

1. Click the **CGI Layer** in the Workspace.

2. From the **Operators** menu, select *3D Post/3D Depth of Field*.

3. Set the following parameters:

Near Focused Plane	-4243
Far Focused Plane -	65535
Maximum Radius	1.60
Blur Type	**Box**

With these values we are focusing on the animated objects in the scene, and blurring the scene in front of and behind them. Because the 3D depth information was saved in our RPF files, Combustion knows how far these objects are from the camera and focuses accordingly.

Note that the near and far focused plane numbers are easy to determine. You can click on the button with the plus symbol next to these two parameters and then click the alien or the rover in the viewport. While holding the mouse button down, you can interactively adjust the focused planes.

Blurring the Alien

Even with the depth of field effect, the alien remains a little too sharp near his center compared to the rest of the environment. We will apply a Gaussian Blur to fix this problem.

1. In the Workspace, click on the **CGI Layer** and expand it.
2. From the **Operators** menu, select **Blur/Sharpen/Gaussian Blur**.

   This operation will soften the entire layer, making it match the landscape more closely. The blur will be calculated in a bounding box that is the same dimension as the layer. This means that the blur will not reach all the way to the edge of the layer with equal effect.

3. To correct for this, click the **Resize** Image option in the Gaussian Blur Options panel.
4. Set the **Radius** to 0.55.

Figure 13.39 *CG footage slightly blurred*

Adding Motion Blur

1. Click the **CGI Layer** in the Workspace.
2. From the **Operators** menu, select **3D Post/RPF Motion Blur**.
3. Set the **Amount** to 5.

Adding the Light Glows

Now you will add a glow to the rover lights. Recall that in Chapter 11, for the rover lights material, you set the Material Effects Channel to 1. You will now tell Combustion to render a soft glow on all materials with this Material Effects Channel, which Combustion refers to as a Material ID.

1. Click the **CGI Layer** in the Workspace.
2. From the **Operators** menu, select **3D Post/3D Glow**.
3. Set the Glow source to **Material ID**.
4. Set the **Radius** to 2.30.
5. Set the **Intensity Boost** to 10%.
6. Check **Extend into Mask**.
7. Set **Material ID** to 1.
8. Set the **Minimum Luminance** to 6%.
9. Set the **Falloff** to 62%.

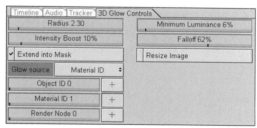

Figure 13.40 *3D Glow controls*

Creating a Sun

Finally we will create a sun that will cause a lens flare effect in the camera. In order for the sun and the lens flare to affect both layers, we must nest the two layers together. Nesting a layer, as mentioned in the introduction is like flattening layers in Photoshop; however, they are still accessible through the stack.

1. In the Workspace, select the **CGI Layer** and the **Quarry Layer**.
2. From the menu, select *Object/Nesting*.
3. In the Nesting Options dialog, set the Composite Name to **Alien Scene** and click **Selected Layers**. Click **OK**.

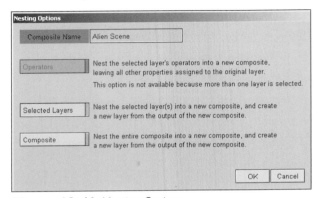

Figure 13.41 *Nesting Options*

This nests the layers, and the nested composite appears as the layer **Alien Scene**.

4. Select the **Alien Scene** layer from the Workspace.
5. From the *Operators* menu, choose *Stylize/Lens Flare*.
6. Set the **Strength** to 15%.
7. Click the **Flare Center** interactive placement button [+] and position the flare just over the cliff side in the back right corner.
8. Click the **Axis** interactive placement button [+] and position the axis so the flare catches the screen.
9. Verify that Elements is set to **Both** and **Flare Type** is 50-300 mm Zoom.

Figure 13.42 *The Sun in place*

Tracking the Sun

As you can see when you play back the composite, the lens flare needs to be tracked.

1. Enable the position crosshair, and in the **Tracker** panel, choose the **Quarry Layer** as your source.

2. Disable all the operators on the quarry layer and hide the alien layer.

3. Choose two tennis balls as reference features and start the Tracker with the default settings.

4. Once the track is complete, click off, turn all the operators back on, and unhide the alien layer.

5. Go to the transforms of the nested layers and set the scale value to 120%.

    The composite is now complete.

6. Save the workspace as *AlienComposite05.cws*.

Rendering the Final Images

You're finally at the finish line —- all that's left to do is render the sequence of files for delivery to the client.

Rendering the final images effectively flattens all the layers and saves them to 2D image files. There are a number of file formats we could render to, but we're going to render a sequence of TIFF files for delivery to the client on CD-ROM. This was the medial format we agreed upon at the start of the job (remember back then?). The client's editor will then import the TIFF files to their in-house editing system and put it together with the logo, sound effects and voiceover.

After all you've been through on this project, the steps required to render the final images are really simple.

1. From the menu, select *File/Render*.

   This will open the Combustion Render Queue dialog.

2. In the Combustion Render Queue, set the **Format** to **TIFF Sequence**.

3. Click the **Options** button next to the format setting, and inside the TIFF Options dialog, click **None** for compression. Click **Close**.

4. Set the **Quality** to **Best**.

5. Set the **Frame Size** to **Full**.

   This will render the frames at 720 x 486, the resolution used by most editing systems. Of course, you checked with the client at the beginning of the job to make sure this is the resolution they want.

6. Set **Fields** to **Lower First**.

   This is the usual setting when rendering to NTSC, but again, you should always check with the client. Sometimes, the only way to find out for sure is to send a few short test animations rendered with different field settings to their editors, and have them try them out.

7. Click the **Filename** button and set the appropriate file name and destination path. Make sure you have lots of disk space at the destination!

8. Set the **Start** and **End** fields appropriately. We started our animation at frame 44.

9. Click the **Process** button.

   Combustion begins processing the workspace and outputting single TIFF images for each frame.

   After the rendering is complete, review the rendered files. We used the RAM Player in **3ds max** to make sure everything rendered properly. If you find frames with problems, go back into Combustion, correct the problem and render again.

FINISHING THE JOB

You're just about there. Time to wrap it up with some tried and true finishing practices.

PICKING YOUR BATTLES

In checking your rendered images for problems, you will most likely find some errors. It is important at this point to distinguish between large errors that must be fixed and small errors that can be left alone.

If you're like us, you are never quite satisfied. For example, after looking at the animation for days (which you will often do when working on a job), you detect a tiny flickering shadow at the corner of the alien's eye, visible only in the long shot. The shadow bothers you so much that you fix it, but in the process you make the job three days late. If you had just taken the time to put the animation on a VHS tape and play it on a regular VCR, you would have found that video degradation had made the problem invisible.

At some point you have to stop fiddling with errors and call it "done". The animation will never be absolutely perfect, even if you worked on it for 100 years. Learn to live with this fact — it could mean the difference between success and failure for your career. It's great that you want the job to be perfect, but if you never complete a project, a reputation for perfection won't do you any good.

If you have trouble figuring out when animation is good enough for delivery, ask an experienced animator friend to help. You can also ship the job to the client early as part of the approval process and see what they say.

DELIVERING THE JOB

When you've decided that the job is done, put the files on the media specified at the beginning of the job and ship it to the client. Pat yourself on the back. You did great!

Now you can sit back and relax... for five minutes or so. There's one last important step.

BACKING UP THE JOB

Before you can call it a wrap, you must perform a comprehensive backup of all the files you used on the job. You never know when the client will call back for minor revisions or additions to the job for another commercial or project. For example, they might request a low-resolution AVI to post on their website.

When preparing to back up, it is very tempting to delete entire folders of background material, false starts and textures that didn't make the grade, but you might accidentally delete something essential if you do this. Better to back up too much than too little!

Label all your backups and put them in a safe place. Only then can you delete the job files from your system.

Okay, now you can kick back and take a break. Until your next job, that is!

OUR VERSION OF THE ANIMATION

If you want to see our rendered version with a few sound effects and titles, view the file *Ionz.avi* in the *Movies/Final* folder on the CD.

We invite you to come see us on the Internet at **www.The3DBuzz.com/Mastering** and send us your own version of the animation. There you will also find additional resources regarding this book.

It was a pleasure going through this project with you. May it help you with your own work, now and in the future.

Index

A

alien
    blocking animation 562-569
    concept sketch 10-11
    control script
        feet controls 461-464
        hand controls 450-455
        morph targets 446-449
        poses 457-460
    facial expressions 422
    materials 490-494
        body 492
        eyes 486
    merging into final scene 545
    object names 385-386
    scaling 546-547
    virtual studio 102-103
Alien - Setup and Deformation.max 431, 436
Alien - Setup and Textured.max 545
Alien - Setup with UI.max 463, 486
Alien - Setup.max 388, 399
Alien Model.max 160, 332
Alien UI.ms 446, 463
Alien Virtual Studio.max 102
aligning normals 88-89
Arc Rotate 30

B

background plates 526, 535, 544, 591
    location 12-13
baking animation 320-327
BaseColor.tif 495
Bevel 49-50
Bezier Corner vertex 163
Bezier vertex. See sub-objects, face
blocking animation 562-569
Bones 333-349
    clavicle 340-342
    creating linked 335-336
    shaping/scaling 337-340
border sub-object. See *sub-objects, border*
border, definition 222
Bottom viewport 145
BrandX.jpg 522

C

camera tracking
    features 468
        use in tracking 476-480
    in 3ds max
        alternative to 482
        creating CamPoints for 473
        error handling 479-480
    in Combustion 602-606
        Tracker panel 602-604
cameras, non-tracked 554-556, 570-571
Cap 222
CD files
    Alien - Setup and Deformation.max 431, 436
    Alien - Setup and Textured.max 545
    Alien - Setup with UI.max 463, 486
    Alien - Setup.max 388, 399
    Alien Model.max 332
    Alien Model.max. 160
    alien script pictures 447
    Alien UI.ms 446, 463
    Alien Virtual Studio.max *102*
    Alien_front.jpg *102*
    Alien_side.jpg *102*
    BaseColor.tif 495
    BrandX.jpg 522
    CD Tutorial 1.pdf 495
    CD Tutorial 2.pdf 515
    CD Tutorial 3.pdf 573
    CD Tutorial 4.pdf 573
    ColorMap.jpg 515
    low-res footage.zip 526, 535-537, 544, 591
    Minus.jpg 519
    Plus.jpg 517
    Remote Virtual Studio.max 36
    Remote_Side.jpg 30
    Remote_Top.jpg 24
    Rover - No Tracks.max 237
    Rover - Setup.max 327, 494
    Rover Callback.ms 301
    Rover UI.ms 317
    Rover_Front.jpg 166
    Rover_Side.jpg 166
    Rover_Top.jpg 166
    Set - Lights and Ground Plane.max 535
    Set - Lights Ground and Rocks.max 539, 543
    Set - Tracked Camera.max 526
CD Tutorial 1.pdf 495
CD Tutorial 2.pdf 515
CD Tutorial 3.pdf 573
CD Tutorial 4.pdf 573
Chamfer 70
character design *10*
clavicle, definition 340
client approvals 4-5
collapse to Editable Mesh 22
ColorMap.jpg 515
Combustion 583
    and RPF files 573
    blurring 617, 620
    color wheel 596
    compositing animation in 617-618
    depth of field effect
        with manual blur 616-617
        with RPF depth information 619-620

glow, adding 621
importing footage 591
interlacing, removing 592-594
motion blur, adding 621
rotoscoping 608
selecting footage areas 597-598
sun, adding 622-623
concept sketches 9-12
controllers
    definition 287
    in Track View 294-295
    Motion Capture 287-288, 289-299
    Reactor 367-373
Corner vertex 163
    in lofting 91-92
custom attributes 296-298
    for alien hand controls 364-365
    for rover position and rotation 296-298, 548-549
        wiring 298-300
Cut 55-56

D

deformers in skinning 392
design, characters and objects 10-12
Divide 109
dummy object 288

E

edge sub-object. See *sub-objects, edge*
edge, definition 33-34
Edged Faces mode 37
Editable Mesh 26
envelopes 390-391
    adjusting 395-397
    deformers 392

F

face sub-object. See *sub-objects, face*
face, definition 33-34
features
    definition 468
footage 526, 535, 544, 591
Flex 428-429

G

global illumination 526
Grid and Snap Settings dialog 45-46
grouping 97

H

histogram
    definition 593
    using to correct colors 594-595

I

IFL file 470-471
Ignore Backfacing 39, 55
IK chain 350
Ion-Z batteries client
    project description 2
    final animation 626

L

Lathe 84-86
lights
    direct 524
    dome lights 527-531
    global illumination 526
    omni 524-525
    simulating bounced light 531-532
    spot 523-524
Link constraint 552-554
lofting 89-97
    Scale Deformation 93-97
Look-At constraint 431
low-res footage.zip 526, 535, 544, 591

M

Make Planar 65
mapping coordinates 26-27
marquee selection 35
Material Editor 25
    Cellular map 493-494
    Checker map 502
    Glossiness parameter 487
    Gradient Ramp map 489
    map channel ID 514
    Material/Map Navigator 503-504
    Matte/Shadow 535, 536, 539
    Noise map 496
    Show Map in Viewport 25
    Speckle map 493-494
    Specular Level parameter 487
MAXScript
    baking animation 320-327
    button 261-262, 319-320, 445-446
    comments 252-258
    detecting time slider changes 301-303
    error checking 265-266
    floater 268-270
    global variables 264-266
    listing parameters 246-248
    pictures in UI 447-449
    position parameter 249-250
    Reset Position not working 320, 560
    rollout 267-268, 317-318, 444-445
    slider 263
    utility 317

merging final scene 543
   alien 545-546
   remote control 549-550
   rover 548
   set 543-544
Min/Max Toggle 167
Minus.jpg 519
morph targets
   definition 420
   for alien 422-426
   limiting vertices for 427
   loading in Morpher modifier 427
   with alien UI 434-435, 439-443, 446-449
Morpher 427-428
Motion Capture 289-290
   testing 295-296
Multi/Sub-Object material
   creating by assigning materials to polygon selections 497-499
   definition 499-500
   picking from object 499-500

N

network rendering. See *CD Tutorial 4*
normals
   definition 85
   flipped on Surface 185
   flipping on Lathe 85
NURMS subdivision *104*
   shaping an object with 186-188

O

Outline 112-113

P

parameter wiring 298-300
PathDeform 277-279
pivot point, moving 190
plane 22
Plus.jpg 517
polygon sub-object. See *sub-objects, polygon*
polygon, definition 33-34
previewing animation 571-572
proxy 550

Q

quarry 12-13

R

Reactor controller 367-373
reference images 20
Region Zoom 49

remote control
   concept sketch 11-12
   grouping 97
   merging into final scene 549-550
   parts 35
   proxy 550
   slicing, mirroring and attaching 76-78
   virtual studio 22-33
Remote Virtual Studio.max 36
Remote_Side.jpg 30
Remote_Top.jpg 24
rendering
   final images 624-626
   network. See *CD Tutorial 4*
   to RPF files 573, 581-582, 589-590
Reset Position not working 320, 560
rotoscoping 608
rover
   concept sketch 11-12
   control by dummy object 288
   custom attributes 296-300, 548-549
   fuel tank 231-232
   materials 494-514
      body 495
      engine 505
      fuel tank 511
      light housing 509
      lights 509
      reactor 500
      track wheels 507
      tracks 507
      wheels 513
      windshield 498
      windshield rim 497
   merging into final scene 548
   parts 165
   springs 229-230
   struts 226-228
   throttle 288-289
   treads 276-286
Rover - No Tracks.max 237
Rover - Setup.max 327
Rover Callback.ms 301
Rover Setup.max 494
Rover UI.ms 317
Rover_Front.jpg 166
Rover_Side.jpg 166
Rover_Top.jpg 166
RPF files 573-574
   for depth of field in Combustion 619-620
   for glows in Combustion 621
   for motion blur in Combustion 621
   rendering in Video Post 577-578

S

Scale Deformation 93-97
scripting, See *MAXScript*
See-Through mode 37
selection
    polygons shaded red *43*
    with a marquee 35
semi-transparent display 37
set
    ground plane 526-527
    lights 527-535
    rocks 537-539
Set - Lights and Ground Plane.max 535
Set - Lights Ground and Rocks.max 539, 543
Set - Tracked Camera.max 526
shaded view, displaying 25
ShapeMerge 71-73
Show Map in Viewport 25
Skin 400
skinning
    applying the Skin modifier 400
    basics 393-419
    deformers 392
    envelopes 390-391
        adjusting 395-397, 398
    testing with key poses 403
    vertex weights 391
        setting 398-399, 407-410
        viewing 397
Slice 76
slicing, mirroring and attaching
    alien 121-122
    remote control 76-78
Smooth + Highlights 25
Smooth vertex 163
smoothing groups 79-81
snapping
    3D Snap 46
    to vertices 45-46
soft selection 151
sub-objects 33-35
    accessing 34-35
    border
        capping 222
        creating a shape from 274-276
    definitions 33-34
    edge 33-34
        chamfering 70
        cutting 55-56
        dividing 109-110
        moving 57-60
        tessellating 147
    face 33-34
    polygon 33-34
        beveling 49-50
        drawing from scratch 191-193
        extruding 43-44
        making planar 65
        outlining 112-113
        shading when selected 43
    selecting 35
    vertex 33-34
        changing type 164
        collapsing 117-118
        removing isolated 56
        setting type while drawing 164
        welding 48-49
sun, adding in Combustion 622-623

T

Tessellate 147
tracking. See *camera tracking*
treads
    for tracks 282-286
    for wheel 198-201, 206-207

U

UVW Map modifier 26
    Bitmap Fit 27

V

vertex sub-object. See *sub-objects, vertex*
Vertex Ticks 168
vertex weights in skinning 391
    setting 398-399, 407-410
    viewing 397
vertex, definition 33-34
Video Post 574-581
    Add Image Output Event 577-578
    Add Scene Event 575-576
    Render Options 576
    rendering 581-582
    RPF files 577-578
viewport
    Bottom 145
    Edged Faces mode 37
    maximizing 167, 334
    rotating 30
    Smooth + Highlights 25
    turning off grids 33
    zooming 49

W

welding vertices 48-49
wiring parameters 298-300
    with Reactor controller 367-373
workflow
    administrative 8-9

X

XRef 542-543, 549